Lec   n
Co   ter Science

Edited b   nd J. Hartmanis

401

H. Djidjev (Ed.)

Optimal Algorithms

International Symposium
Varna, Bulgaria, May 29–June 2, 1989
Proceedings

Springer-Verlag

Berlin Heidelberg New York London Paris Tokyo Hong Kong

Editor

Hristo Djidjev
Center of Informatics and Computer Technology
Bulgarian Academy of Sciences
Acad. G. Bonchev str., bl. 25-A, Sofia 1113, Bulgaria

CR Subject Classification (1987): F.1.2–3, F.2.2, G.2.2, E.1

ISBN 3-540-51859-2 Springer-Verlag Berlin Heidelberg New York
ISBN 0-387-51859-2 Springer-Verlag New York Berlin Heidelberg

Printed in Germany

Printing and binding: Druckhaus Beltz, Hemsbach/Bergstr.
2145/3140-543210 – Printed on acid-free paper

FOREWORD

The papers included in this volume are a subset of the papers presented at the Second International Symposium on Optimal Algorithms held in Varna, Bulgaria, May 29-June 2, 1989. The symposium was organized and funded by the Center of Informatics and Computer Technology of the Bulgarian Academy of Sciences. There were two major sections: Algorithms and Optimal Recovery. This volume includes only papers presented in the section Algorithms.

Several people were invited to give special lectures at this symposium and all of them were invited to contribute papers for the proceedings. These papers are identified as invited papers in the table of contents. For the remaining papers, 51 were submitted for consideration, and of these, 37 were selected for formal presentation at the symposium. From these 37 contributed papers, after their presentation at the symposium, 13 were selected for inclusion in these proceedings. The papers in this volume have not been fully refereed, and those which contain original research results should be expected to appear later in a more complete form in regular refereed journals.

I would like to thank to the following members of the advisory committee for their help in making the final selection: F. Dehne, I. Ipsen, V. Ramachandran, J.-R. Sack, P. Spirakis, P. Young, as well to J. Reif for his cooperation. Finally I would like to thank to the members of the local organization committee, especially to V. Tzelkova, B. Stoyanova, and L. Aleksandrov for their efforts during the organization of the symposium and the preparation of these proceedings.

Sofia, June 1989

Hristo N. Djidjev

TABLE OF CONTENTS

RANDOMIZATION IN PARALLEL ALGORITHMS AND ITS IMPACT ON COMPUTATIONAL GEOMETRY†

John H. Reif and Sandeep Sen
Computer Science Department
Duke University,
Durham, N.C. 27706,
U.S.A.

Abstract

Randomization offers elegant solutions to some problems in parallel computing. In addition to improved efficiency it often leads to simpler and practical algorithms. In this paper we discuss some of the characteristics of randomized algorithms and also give applications in computational geometry where use of randomization gives us significant advantage over the best known deterministic parallel algorithms.

Motivation

Designing parallel algorithms for various fundamental problems in computational geometry has received much attention in the last few years. After some early work by Anita Chow in her thesis, Aggarwal et al.[13] developed some general techniques for designing efficient parallel algorithms for a number of fundamental problems. These included convex hulls in two and three dimensions, voronoi diagram for planar point sites, triangulation and planar point-location among others. Although most of these problems have a sequential time-complexity of $\Theta(nlogn)$, the authors presented parallel algorithms which uses a linear number of processors and runs in $O(\log^k n)$ time (k being typically 2,3 or 4) in a PRAM model. Consequently, the problem of designing optimal (in the processor-time product sense) algorithms were left open. Since then a number of the open problems in the original list have been settled due to the work by Atallah, Cole and Goodrich[14] who were able to apply Cole's elegant techniques for parallel mergesort to a number of these problems.

We present techniques for obtaining optimal parallel algorithms for problems in computational geometry using randomization. As applications of our methods, we derive efficient parallel algorithms for planar-point location, convex-hull and trapezoidal decomposition. These algorithms run in time T = O(logn) using O(n) processors for problem size n and terminate in the claimed time bound with probability $1 - n^{-c}$ for any integer c. These bounds are worst-case and do not depend on any input

† Supported in part by Air Force Contract AFSOR-87-0386, ONR contract N00014-87-K-0310, NSF grant CCR-8696134, DARPA/ARO contract DAAL03-88-K-0185, DARPA/ISTO contract N00014-88-K-0458.

distribution. The main contribution of our work is a new random sampling technique called **Polling** which can be used for doing divide-and-conquer efficiently on various problems in computational geometry. Our techniques lead to algorithms that are considerably simpler than their methods and appear to have wider applications. For example we have derived an optimal O(logn) time n processors algorithm for constructing the convex hull of points in three dimensions (Reif and Sen[18]). Presently the best known deterministic algorithm for this problem takes $O(\log^2 n \log^* n)$ time using n processors (Dadoun and Kirkpatrick[17]).

Basics

Randomization was formally introduced by Rabin[6] and independently by Solovay & Strassen[8] as a tool for improving the efficiency of certain algorithms. In a nutshell, a randomized algorithm uses coin-flips to make decisions at different steps of the algorithm. Therefore a randomized algorithm is actually a family of algorithms where each member of this family corresponds to a fixed sequence of outcomes of the coin-flip. Two of the most commonly used forms of randomization in literature are the *Las Vegas* algorithms and *Monte Carlo* algorithms. The former kind ensures that the output of the algorithm is always correct - however only a fraction (usually greater than 1/2) of the family of algorithms halt within a certain time bound (as well as with respect to some other resources like space). In contrast, the Monte Carlo procedures always halt in a pre-determined time period; however the final output is correct with a certain probability (typically > 1/2). This lends itself very naturally to decision algorithms (Rabin's primality testing being a good example). For the purpose of this discussion we shall limit ourselves to the Las Vegas algorithms which have been more popular with the algorithm designers. For a general algorithm which produces more than just 'yes-no' output, the precise meaning of an incorrect output becomes subjective; for example we may need to know how close are we to the correct output in order to decide if the output is acceptable. Although, this is one of the reasons for bias towards Las Vegas algorithms, the use of either kind of algorithms depends on the particular application.

Complexity measures of randomized algorithms

Before we discuss the applications of these algorithms in parallel computing, it is important to review some of the performance measures used by these algorithms. This will enable us to compare the relative merits of different randomized algorithms. To begin, we must emphasize the distinctions between a randomized algorithm and probabilistic algorithm. By probabilistic algorithms, we imply those algorithms whose performance depend on the input distribution. For such algorithms, we are often interested in the average resources used over all inputs (assuming a fixed probability distribution of the input). A randomized algorithm does not necessarily depend on the input distribution. A randomized algorithm uses a certain amount of resources for the worst-case input with probability $1-\varepsilon$ $(0 < \varepsilon < 1)$, i.e. the bound holds for any input (which is a stronger bound than the average bounds). This can be very well illustrated with the example of Hoare's *Quicksort* algorithm. In its original form, it is a probabilistic algorithm which performs very well on certain inputs and deteriorates sharply on some other inputs. By assuming that all inputs are equally likely (known as random-input assumption), the algorithm performs very well on the average. By introducing randomization in the algorithm itself, it has been shown to perform very well on **all** inputs with high probability. This is certainly a more desirable property since a *malicious oracle* who could control the performance of the original algorithm by giving it worst case inputs, can no longer affect it. Of course, the onus of a successful run of the algorithm is now shifted to the outcome of the coin-flips. This depends on certain randomness properties of the random-number generator, which is a topic in itself. Also note that this discussion does not preclude designing randomized algorithms which are dependent on the input distribution but these algorithms are no different from their deterministic counterparts.

Until now we have characterized the randomized algorithms with a success probability of 1 - ε without specifying the possible forms of ε. It can be a fixed constant or a function (which takes values between (0,1)). It must be clear that ε should be minimized (compare this with deterministic algorithms where ε is 0). Intuitively we can expect a trade-off between ε and the amount of resource used. In other words, the failure probability ε must decrease with increasing amount of resources. Let us consider a concrete example. Suppose T_A(n) is the *expected* running time of the randomized algorithm A for input size n. What can we say about ε? If we don't have any bounds other than the expectation we can only use Markov's inequality. From Markov's inequality, the probability that running time exceeds kT_A(n) is less than 1/k. For example, if k=2, ε = 1/2. Compare this with an algorithm B for the same problem whose running time exceeds $k\alpha T_B(n)$ with probability less than $1/n^{\alpha}$ and suppose that for any given α, k is a constant independent of n. This implies that the probability of failure diminishes rapidly as n increases and vanishes asymptotically. We have characterized the failure probability ε as a decreasing function of the problem size, n and resources used by the algorithm. The reader will recognize that the faster ε decreases with these parameters the better is the algorithm. This makes algorithm B superior to algorithm A if $T_A(n)$ and $T_B(n)$ represent the same function. The basic idea is that depending on the application, the user chooses a certain value of ε and accordingly chooses k (given the value of n) with the objective of minimizing k. There is no reason to be pedantic about the kind of function ε should be except that a failure probability of the second form (that of algorithm B) has been very widely used in literature and such algorithms have been termed as having *high probability* of success. This kind of failure probability function is quite robust with respect to a polynomial number of procedures i.e. the union of a polynomial number of events, each with high probability of success, succeeds with high probability. It may be a non-trivial task to transform an algorithm like A to an algorithm like B (which succeeds with high probability). The reader must also appreciate that randomized algorithms like B which have such high probability of success should be competitive with deterministic algorithms for the same problem. According to Adleman & Manders[1], a randomized algorithm with success probability more than 1 - 2^{-k} (for some large fixed k) has a lower probability of failure than the hardware itself.

Parallel computation and randomization

Randomization has proven to be an extremely effective in parallel algorithm design. One of the earliest treatment of this topic can be found in Reif[12]. For a more recent and extensive survey the reader is encouraged to read the first two chapters of Rajasekaran[7]. A commonly accepted measure of efficiency of parallel algorithms is the *processor-time* product (in short $P \cdot T$). Although the primary objective of parallel algorithms is to minimize time complexity (the number of parallel time steps), in practice one also has to be careful about the number of processors needed to achieve this speed-up. The *efficiency* of a parallel algorithm is a measure of how expensive is the speed-up compared to the sequential algorithm. Clearly $P \cdot T$ product cannot be better than the sequential time complexity of the algorithm. Ideally one would like the speed-up to be linear with the number of processors used; however this is far from true in most cases. This also gives an abstract measure of how 'hard' it is to parallelize a particular problem. We say that a parallel algorithm is *efficient* if $P \cdot T \leq O(Seq(n) \cdot \log^k n)$ for some constant k where n is the input size. The class NC is defined to be the class of problems which admit poly-logarithmic time parallel algorithm using a polynomial number of processors. Note that while these algorithms admit fast parallel algorithms they may not be necessarily *efficient*.

Use of Random Sampling

Randomized sampling techniques have been used extensively in cases of divide-and-conquer algorithms (most parallel algorithms would fall under this category). The idea is to divide up the problem 'almost evenly' into smaller sub-problems using a randomly chosen subset of the input. This random subset is called splitters and because of the process of random selection, various probabilistic arguments can be used to bound the size of the sub-problems. For example, in parallel sorting, we can choose $\sqrt{n}$ keys randomly and partition the input into $\sqrt{n}$ subsets using the partitions induced by the random keys. Using simple probabilistic arguments, it is not difficult to bound the size of the subproblems to approximately $O(\sqrt{n} \, logn)$ with high probability. The main algorithm is then used recursively on each of the partitions.

Random sampling in computational geometry was first introduced by Clarkson and since then he has published a series of results leading to improvements and simplification of a number of sequential algorithms in computational geometry. However his time bounds are *expected* in contrast to our *high-likelihood* bounds (these have success probabilities 1 - n^{-c} for any integer c) which aside from being weaker are of little use for obtaining parallel alorithms. The reason being that for sequential algorithms, he was able to use the linearity property of expectation (i.e. the expectation of the sum is the sum of expectations) and so it was enough to bound the expected running time of each individual step. For a parallel algorithm, one is looking for the maximum of the expectation from the expectation of a family of random variables and there exists no known method of obtaining this value. Among other techniques, we introduce a new random-sampling technique called 'Polling' which enables us to obtain *high probability* bounds for our algorithms and complements Clarkson's work to a large extent.

More specifically, it also allows us to use random sampling recursively without blowing up the problem size.

Resampling and Polling

The informal idea behind 'Polling' is the following: For a given problem of size n, we choose randomly a small (typically n^{ε} for $\varepsilon < 1/2$) subset of the given input and use it to divide the original problem. It is not difficult to show that the size of each sub-problem is no larger than $n^{1-\varepsilon} logn$ but the total size of the subproblems may be considerably larger than n. For example, if the input is line segments on the plane, a large number of the line segments may be broken up into smaller pieces during the divide step. This phenomenon is not witnessed in a problem like sorting where the total size of the subproblems is always exactly equal to the input size. Increase in the problem size at every recursive call would result in grossly inefficient algorithms. Clarkson[15] had shown that the total sum of the subproblems has *expected* value O(n). This implies that with probability atmost 1/2, this value would exceed $k_{total} n$ for some constant k_{total}. Consequently, if we choose independently O(logn) random subsets, at least one of them will be 'good' (i.e. sum of subproblems does not exceed k_{total} n) with high probability. If we resample O(logn) times, we may end up doing non-optimal number of operations. Instead we test the 'goodness' of a sample on only $\frac{n}{\log^r n}$ of the input. It can be proved that this gives estimates within a constant factor with very high probability (Reif and Sen[18]). Thus we are able to choose a random sample such that the size of the subproblems does not exceed more than a constant times the input size. We call this technique 'Polling' (as if we are polling a fraction of the input to test the 'goodness' of a sample).

Notice that even with Polling the sum of sub-problems may grow by a constant factor which could lead to a polylogarithmic factor increase over O(loglogn) levels of recursion. With some additional filtering methods (which can be efficiently applied since the size is still O(n)), we are able to bound the problem size by $k_{\max} n$ at any level where $k_{\max}$ is a constant. For a parallel algorithm, we are able to get a recursion which is roughly of the form
$T(n) = T(n^{1-\varepsilon}) + O(logn)$. This has a solution T(n) = O(logn) and the processor bound is simultaneously O(n).

The overall algorithm can be summarized as following:

(i) Choose independently O(logn) random subsets each of size O(n^{ε}).

(ii) Use 'Polling' to identify a 'good' sample.

(iii) Partition the problem by the random sample.

(iv) Use 'Filtering' to control the sum of the sub-problems. (The implementation of this step is problem dependent. For example it is different in the case of trapezoidal decomposition and convex hulls).

(v) If the largest sub-problem size is larger than a certain size apply algorithm recursively.

Randomization as a resource

Some recent efforts directed towards 'derandomization' of randomized algorithms have received attention. While such research has a lot of theoretical ramifications (in understanding relationship between the classes NC and Random NC), these methods, almost without exception lead to loss in efficiency of the algorithms. This may not be be very fruitful from a practical viewpoint. There are formal methods for proving lower bounds for randomized algorithms which gives a strong basis for 'optimality' in randomized algorithms. Moreover, it has been observed quite often that for a given problem, the lower-bounds for deterministic algorithms turn out to be identical to the lower bounds for randomized algorithms (within constant multiplicative factor).

Perhaps a more fruitful area of investigation could be directed towards reduction of the number of random bits used in an algorithm (without affecting the asymptotic bounds) since perfect 'randomness' has been recognized as an expensive resource. This has been demonstrated by some recent work due to Raghavan & Karloff[10] and the authors feel that further research in this direction could be very rewarding from a theoretical perspective as well as from a practical viewpoint. For all our algorithms we are able to bound the number of purely random bits to $O(\log^2 n)$.

Geometry on Interconnection networks

Note that the underlying model for all these algorithms is the parallel analogue of the sequential RAM (Random Access Memory) model, PRAM. This model has become the standard model (within minor variations) for theoretical work on design and analysis of parallel algorithms. The state of art of parallel geometric algorithms for feasible models such as butterfly or hypercubes is lagging far behind the PRAM models; the only known optimal O(logn) time n processor algorithm exists for 2-D convex hulls.

One of our primary motivation for research in randomized parallel algorithms is the absence of optimal deterministic sorting algorithm on the commonly used interconnection networks. The only known optimal sorting network is the *AKS* network which sorts in O(logn) depth but has horrendous constants. Moreover, the algorithm is not suited for the widely used architectures like butterfly or hypercube, which are more suitable for general-purpose computations. In contrast, Flashsort[11], an O(logn) depth randomized sorting algorithm has much lower constants and runs on the standard architectures. The other optimal parallel mergesort algorithm of Cole, which has low constants and is very elegant is of little use in practice since it is tailor-made for PRAM model. This model is very popular among algorithm designers from a theoretical perspective because of its simplicity. However, for such algorithms to be of any use, there must be a general mechanism for mapping algorithms from PRAM on to interconnection networks. These involve routing on interconnection networks and the best known routing algorithms (O(logn) time) involve use of randomization in some form. Moreover, such emulation usually lead to loss of efficiency by O(logn) multiplicative factor. Consequently, Cole's algorithm deteriorates to an $O(\log^2 n)$ algorithm on the interconnection networks. Unlike the cascading divide-and-conquer paradigm used in [14], we have reasons to believe that some of our methods can be extended to the fixed connection networks without loss of efficiency.

Conclusion

In summary, we note that randomized algorithms offer a very pragmatic alternative (to deterministic algorithms) in the area of parallel algorithms. Apart from being simpler than their deterministic counter-parts they usually have smaller constants. In addition, they provide a bridge between the algorithm-designer who designs algorithms for abstract models like PRAM and its realistic implementation on feasible architectures.

Bibliography

[1] L. Adleman and K. Manders, 'Reducibility, Randomness and Untractability,' Proc. 9th ACM STOC, 1977, pp. 151-163.

[2] A. Aggarwal and R. Anderson,'A Random NC Algorithm for Depth First Search,' Proc of the 19th ACM STOC, 1987, pp. 325-334.

[3] Aggarwal et al., 'Parallel Computational Geometry,' Proc. of the 26th Annual Symp on F.O.C.S., 1985, pp. 468-477. Also appears in ALGORITHMICA, Vol. 3, No. 3, 1988, pp. 293-327.

[4] Atallah, Cole and Goodrich, 'Cascading Divide-and-conquer: A technique for designing parallel algorithms, Proc. of the 28th Annual Symp. on F.O.C.S., 1987, pp. 151-160.

[5] Clarkson, 'Applications of random sampling in Computational Geometry II,' Proc. of the 4th Annual Symp. on Computational Geometry, June 1988, pp. 1-11.

[6] N. Dadoun and D. Kirkpatrick, 'Parallel Processing for efficient subdivision search,' Proc. of the 3rd Annual Symp. on Computational Geometry, pp. 205-214, 1987.

[7] H. Gazit, 'An optimal randomized parallel algorithm for finding connected components in a graph,' Proc of the IEEE FOCS, 1986, pp. 492-501.

[8[C.A.R. Hoare, 'Quicksort,' Computer Journal, 5(1), 1962, pp.10-15.

[9] H. Karloff and P. Raghavan, 'Randomized algorithms and Pseudorandom number generation,' Proc. of the 20th Annual STOC, 1988.

[10] G. Miller and J.H. Reif,'Parallel Tree contraction and its applications,' Proc of the IEEE FOCS, 1985, pp. 478-489.

[11] M.O. Rabin, 'Probabilistic Algorithms,' in: J.F. Traub, ed., Algorithms and Complexity, Academic Press, 1976, pp. 21-36.

[12] S. Rajasekaran, 'Randomized Parallel Computation,' Ph.D. Thesis, Aiken Computing Lab, Harvard University, 1988.

[13] J.H. Reif, 'On synchronous parallel computations with independent probabilistic choice,' SIAM J. Comput., Vol. 13, No. 1, Feb 1984, pp. 46-56.

[14] Reif and Sen, 'Optimal randomized parallel algorithms for computational geometry,' Proc. of the 16th Intl. Conf. on Parallel Processing, Aug 1987. Revised version available as Tech Rept CS-88-01, Computer Science Dept, Duke University.

[15] J. Reif and S. Sen, 'Polling: A new random sampling technique for Computational Geometry,' Proc. of the 21st STOC, 1989.

[16] J.H. Reif and L. Valiant, 'A Logarithmic Time Sort for Linear Size networks,' J. of ACM, Vol. 34, No.1, Jan '87, pp. 60-76.

[17] R. Solovay and V. Strassen,'A fast Monte-Carlo test for primality,' SIAM Journal of Computing, 1977, pp. 84-85.

[18] L.G. Valiant, 'A scheme for fast parallel communication,' SIAM Journal of Computing, vol. 11, no. 2, 1982, pp. 350-361.

THERE ARE PLANAR GRAPHS ALMOST AS GOOD AS THE COMPLETE GRAPHS AND AS SHORT AS MINIMUM SPANNING TREES

Christos Levcopoulos
Andrzej Lingas
Department of Computer and Information Science
Linköping University, 581 83 Linköping, Sweden, and
Department of Computer Science and Numerical Analysis
Lund University, Box 118, 221 00 Lund, Sweden

Abstract: Let S be a set of n points in the plane. For an arbitrary positive rational r, we construct a planar straight-line graph on S that approximates the complete Euclidean graph on S within the factor $(1+\frac{1}{r})\frac{2\pi}{3\cos(\frac{\pi}{6})}$, and it has length bounded by $2r+1$ times the length of a minimum Euclidean spanning tree on S. Given the Delaunay triangulation of S, the graph can be constructed in linear time.

1. Introduction

Consider a set S of n points in the plane. We would like to design a planar network between the points in S that approximates the complete Euclidean graph on S in the following sense: there exists a constant c such that for any pair of points in S there is a path in the network that connects the points and is of length bounded by c times the straight-line distance between them. Planar networks approximating the complete Euclidean graph can be applied in the design of route networks and transmission networks. They have also potential applications in the design of algorithms and heuristics that use shortest or almost shortest distances in the plane.

Chew showed that the Delaunay triangulation of S in the L_1 metric approximates the complete graph on S within a factor bounded from above by $\sqrt{10}$ [Ch]. Then, Dobkin *et al.* [DFS] showed that the Delaunay triangulation of S in the L_2 metric gives an approximation of the complete graph on S within a factor bounded from above by $\frac{(1+\sqrt{5})}{2}\pi$. Keil and Gutwin have recently decreased this upper bound to $\frac{2\pi}{3\cos(\frac{\pi}{6})}\approx 2.42$ [KG]. Keil considered also another family of networks on S (drawn in the plane with possible crossings) with the number of edges linear in the size of S, and showed that they can very closely approximate the complete Euclidean graph [Ke].

In the design of route or transmission networks, both the goodness of approximating the complete Euclidean graph and the cost of the resulting network are important. In the simplest case, the cost of a planar network is proportional to the total length of its edges. Clearly, if the network is connected, in particular, if it approximates the complete Euclidean graph on S, it has length not smaller than that of a minimum Euclidean spanning tree of S. Therefore, it is of interest to ask whether there exists a planar straight-line graph on S that (1) approximates the complete graph on S and (2) has length within a constant factor from the length of a minimum Euclidean spanning tree of S.

It is not difficult to construct examples of sets S where the Delaunay triangulation of S has length of order n times the length of a minimum Euclidean spanning tree of S (see [K]). Thus, the Delaunay triangulation of S does not satisfy the second requirement above. In this paper, we construct a planar straight-line graph that satisfies both requirements. We do it for an arbitrary, positive approximation parameter r by pruning the Delaunay triangulation of S. The graph approximates the complete Euclidean graph within the factor $(1+\frac{1}{r})\frac{2\pi}{3\cos(\frac{\pi}{6})}$ and has length not greater than $2r+1$ times the length of a minimum Euclidean spanning tree of S. Given the Delaunay triangulation of S, the graph can be constructed in linear time.

2. Preliminaries

We shall use standard set and graph theoretic notation and definitions (for instance, see [AHU]). As for computational geometry notation and definitions, we rely on [PS]. Among others, we assume the following conventions:

1) A planar straight-line graph (PSLG for short) is a pair (V, E) such that V is a set of points in the plane and E is a set of non-intersecting, open straight-line segments whose endpoints are in V. The points in V are called vertices of G, whereas the segments in E are called edges of G.
2) For a straight-line segment s, $\mid s \mid$ denotes the length of s. For a PSLG G, $\mid G \mid$ denotes the total length of edges of G.
3) Let v, w be two vertices of a connected PSLG G. The length of a shortest path in G connecting v with w is denoted by $d_G(v,w)$.
2) Let $G_1 = (V_1, E_1)$, $G_2 = (V_2, E_2)$ be two PSLG such that V_2 is a subset of V_1. Let c be a positive real number. G_1 approximates G_2 with factor c if for any edge (v,w) of G_2, $dist_{G_1}(v,w)/dist_{G_2}(v,w) \leq c$.
3) The Delaunay triangulation of a finite set S of points in the plane is denoted by $DT(S)$. The convex hull of S is denoted by $CH(S)$.

Fact 2.1 [DFS]: Let S be a set of n points in the plane. $DT(S)$ approximates the complete Euclidean graph on S with factor $\frac{2\pi}{3\cos(\frac{\pi}{6})}$.

Fact 2.2 [PS]: Let S be a set of n points in the plane. There exists a minimum Euclidean spanning tree of S that is a subgraph of $DT(S)$.

3. Constructing the network

Our algorithm for constructing the planar network that both approximates the complete Euclidean complete graph and is almost as short as a minimum Euclidean spanning tree is as follows.

Algorithm 3.1
Input: a set S of n points in the plane, and a positive rational r.
Output: a subgraph G of $DT(S)$.

1. Compute $DT(S)$ and mark all edges on $CH(S)$;
2. **for** each edge (v,w) of $DT(S)$ **do**
   begin
   $next(v,w) \leftarrow$ the next edge to (v,w) in clockwise order around w;
   $cnext(v,w) \leftarrow$ the next edge to (v,w) in counter-clockwise order around w;
   $weight(v,w) \leftarrow \mid (v,w) \mid$

end;

3. $G \leftarrow$ a minimum spanning tree of $DT(S)$;
4. $P \leftarrow$ the degenerate polygon obtained from G by doubling the edges of G;
5. $Q \leftarrow$ a stack of edges (v,w) of P such that (v,w) is not in $CH(S)$, and $next(v,w)$ is the edge of P that follows (v,w) on the perimeter of P in clockwise order;
6. **while** Q is non-empty **do**

begin

pop an edge (v,w) from Q;

$(w,u) \leftarrow next(v,w)$;

replace (v,w), (w,u) with (v,u) in P;

if $weight(v,w) + weight(w,u) > (1+\frac{1}{r}) \mid (v,u) \mid$ **then** augment G with (v,u)

else $weight(v,u) \leftarrow weight(v,w) + weight(w,u)$;

if $next(v,u)$ is in P and not in $CH(S)$ **then** push (v,u) on Q;

if $cnext(v,u)$ is in P and not in $CH(S)$ **then** push $cnext(v,u)$ on Q;

end

We prove the correctness of the above algorithm, characterize the graph G produced by the algorithm, and estimate its running time in the three following lemmas.

Lemma 3.1: Algorithm 3.1 produces a subgraph of $DT(S)$ which approximates the complete Euclidean graph on S with factor $(1+\frac{1}{r})\frac{2\pi}{3\cos(\frac{\pi}{6})}$.
Sketch: Algorithm 3.1 terminates since an edge of $DT(S)$ can occur at the top of the stack in at most two iterations of the while-statement. By induction on the number of iterations of the while-statement, we show that $weight(v,u)$ is never smaller than the length of a shortest path in G connecting v with u. On the other hand, by the first instruction in the block of the while-statement, $weight(v,u)$ is never greater than $(1+\frac{1}{r}) \mid (v,u) \mid$. It follows that for any edge (v,u) of $DT(S)$, there is a path in G of length $\leq (1+\frac{1}{r}) \mid (v,u) \mid$ connecting v with u. This combined with Fact 2.1 yields the lemma. ∎

Lemma 3.2: The subgraph G of $DT(S)$ produced by Algorithm 3.1 has length not greater than $2r+1$ times the length of a minimum Euclidean spanning tree of G.
Sketch: The idea of the proof is to give to the edges of the starting minimum spanning tree credits to pay for the lengths and credits of the edges subsequently added to G such that the total sum of edges in the current graph G and current credits is bounded by $2r+1$ times the length of a minimum spanning tree of $DT(S)$. Note that any minimum spanning tree of $DT(S)$ is a minimum Euclidean spanning tree of S by Fact 2.2. To define the propagation of credits, we augment Algorithm 3.1 as follows.

a) between Step 4 and Step 5:

for each edge e in $DT(S)$ **do**

if e is in P **then** $credit(e) \leftarrow \mid e \mid r$ **else** $credit(e) \leftarrow 0$;

b) in the **then** branch of the first **if** instruction in the while-block.

$credit(v,u) \leftarrow credit(v,w) + credit(w,u)$

c) in the **else** branch of the first **if** instruction in the while-block.

$credit(v,u) \leftarrow credit(v,w) + credit(w,u) - \mid (v,u) \mid$

d) after the first **if** instruction in the while-block.

$credit(v,w) \leftarrow 0$; $credit(w,u) \leftarrow 0$;

At the beginning, each of the two copies of an edge e of the spanning tree which appears in the polygon P obtains the credit of $r \mid e \mid$. Thus, the total sum of the lengths of edges in G and the current credits is $2r+1$ times the length of a minimum spanning tree of $DT(S)$, initially. We can prove by induction on the number of iterations of the while-statement, and by the algorithm of credit propagation, that after each such iteration, again the total sum of lengths of all edges of the current G plus the sum of all current credits does not increase.

Now, it is sufficient to prove that the sum of the current credits is never negative. For this purpose, we show that for any edge (v,u) in P, $credit(v,u) \geq r \cdot weight(v,u)$ which implies $credit(v,u) \geq 0$. The proof is again by induction on the number of iterations of the while-statement. In the inductive step, there are two steps to consider.

Case 1: G is augmented with (v,u) (**then** branch). Then, by the inductive hypothesis and the definition of the new weight of (v,u), we have:

$$credit(v,u) = credit(v,w) + credit(w,u) - \mid (v,u) \mid$$

$$\geq r \cdot weight(v,w) + r \cdot weight(w,u) - \mid (v,u) \mid$$

$$\geq r(1+\frac{1}{r}) \mid (v,u) \mid - \mid (v,u) \mid = r \mid (v,u) \mid = r \cdot weight(v,u)$$

Case 2: G is not augmented with (v,u) (**else** branch). By the inductive hypothesis and the definition of the new weight of (v,u), we have:

$$credit(v,u) = credit(v,w) + credit(w,u) \geq r \cdot weight(u,w) + r \cdot weight(w,u) = r \cdot weight(w,u)$$

∎

Lemma 3.3: Given $DT(S)$, Algorithm 3.1 can be implemented in linear time.

Proof: Given a DCEL representation (see [PS]) of $DT(S)$, Steps 1,2,4,5 can be done in linear time. Step 3 can be also done in linear time by [CT]. All the instructions in the block of the while-statement can be implemented in constant time. An edge of $DT(S)$ can occur at the top of the stack only during two iterations of the while-statement. Therefore, the number of iterations of the while-statement is $O(n)$. Thus, the while-statement takes linear time. ∎

Combining the three above lemmas, we obtain the main result of the paper.

Theorem 3.1: Let S be a set of n points in the plane, and let r be a positive rational. There exists a PSLG that approximates the complete Euclidean graph on S with factor $(1+\frac{1}{r})\frac{2\pi}{3\cos(\frac{\pi}{6})}$ and has length bounded by $2r+1$ times the length of a minimum Euclidean spanning tree of S.

4. Final Remark

The method presented in Section 3 is quite general. It can be applied to any planar graphs H, F embedded in the plane, where H is a connected subgraph of F and F is triangulated outside H, in order to construct a subgraph G of H that approximates H and has length $O(\mid H \mid)$.

Acknowledgements: We would like to express our appreciation to Andrzej Proskurowski for useful comments.

References

[AHU] A.V. Aho, J.E. Hopcroft and J.D. Ullman, *The Design and Analysis of Computer Algorithms* (Addison-Wesley, Reading, Massachusetts, 1974).

[C] L. Paul Chew, *There is a Planar Graph Almost as Good as the Complete Graph*, Proc. of the 2nd Ann. ACM Symp. on Computational Geometry, Yorktown Heights, 1986.

[CT] D. Cheriton, R.E. Tarjan, *Finding Minimum Spanning Trees*, SIAM J. Comput., 5 (1976), pp. 724-742.

[DFS] D.P. Dobkin, S.J. Friedman, and K.J. Supowit, *Delaunay Graphs are Almost as Good as Complete Graphs*, Proc. of the 28th Ann. IEEE Symposium on Foundations of Computer Science, Los Angeles, 1987.

[Ke] M. Keil, *Approximating the Complete Euclidean Graph*, Proc. of the 1st Scandinavian Workshop on Algorithm Theory, Halmstad, Sweden, 1988.

[KG] M. Keil and C. Gutwin, *The Delaunay Triangulation Closely Approximates the Complete Euclidean Graph*, to appear in Proc. of the 1st Canadian Workshop on Algorithms and Data Structures, Ottawa, August 1989.

[Ki] D.G. Kirkpatrick, *A Note on Delaunay and Optimal Triangulations*, Information Processing Letters, Vol. 10, No. 3, 1980.

[PS] F.P. Preparata and M.I. Shamos, *Computational Geometry, An Introduction*, Texts and Monographs in Computer Science, Springer Verlag, New York.

COMPUTING DIGITIZED VORONOI DIAGRAMS ON A SYSTOLIC SCREEN AND APPLICATIONS TO CLUSTERING *

FRANK DEHNE

Center for Parallel and Distributed Computing
School of Computer Science, Carleton University, Ottawa, Canada K1S 5B6

Abstract. A *systolic screen* of size M is a $\sqrt{M} \times \sqrt{M}$ mesh-of-processors where each processing element P_{ij} represents the pixel (i,j) of a *digitized plane* Π of $\sqrt{M} \times \sqrt{M}$ pixels. In this paper we study the computation of the Voronoi diagram of a set of n planar objects represented by disjoint images contained in Π. We present $O(\sqrt{M})$ time algorithms to compute the Voronoi diagram for a large class of object types (e.g., points, line segments, circles, ellipses, and polygons of constant size) and distance functions (e.g., all L_p metrices).
Since the Voronoi diagram is used in many geometric applications, the above result has numerous consequences for the design of efficient image processing algorithms on a systolic screen. We obtain, e.g., an $O(\sqrt{M})$ time systolic screen algorithm for "optical clustering"; i.e., identifying those groups of objects in a digitized picture that are "close" in the sense of human perception.

1 INTRODUCTION

Consider a *digitized plane* Π of size M, i.e. a rectangular array of M lattice points, or *pixels*, with integer coordinates $(i,j) \in \{1,..., \sqrt{M}\}^2$, and a set $I_1, ..., I_n$ of n disjoint *images* in Π where an image (or digitized picture) I_i is defined as an arbitrary subset $I_i \subseteq \Pi$.

In this paper we study efficient parallel algorithms for processing such images. We consider the *mesh-of-processors* architecture; i.e., a set of m processors P_{ij} $(i,j \in \{1,..., \sqrt{M}\})$ arranged on a $\sqrt{M} \times \sqrt{M}$ grid where each processor is connected to its four direct neighbors, if exist. This architecture is particularly useful for image processing, since n disjoint images $I_1, ..., I_n$ in Π can be naturally represented on a mesh-of-processors of size M: Every processor P_{ij} has a *color-register* C-Reg(i,j) with value

$$\text{C-Reg}(i,j) = \begin{cases} k & \text{if } (i,j) \in I_k \ (1 \le k \le n) \\ 0 & \text{otherwise} \end{cases} .$$

---

* Research partially supported by the Natural Sciences and Engineering Research Council of Canada under Grant A9173.

For the remainder we will refer to a mesh-of-processors that represents a set of images as described above as a *systolic screen* (see Figure 1).

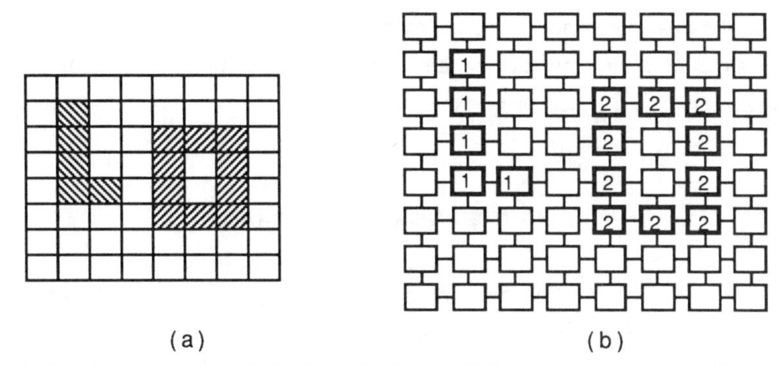

Figure 1: (a) Two Images in Π. (b) Systolic Screen Representation of these Two Images.

Systolic screen architectures have already been extensively used to manipulate images. A well known existing system is the MPP designed by NASA for analysing LANDSAT satellite data [Re84]. The MPP consists of 16,384 processing units organized in a 128x128 matrix where each processing unit, which has a local memory between 1K and 16K bits, represents a subsquare of pixels.

While most of the early applications of systolic screens considered "low level" image processing operations such as contour extraction or connected component labeling, recent research has also focussed on computing "high level" geometric operations on images. Miller and Stout [SM84], [MS85] have proposed $O(\sqrt{M})$ time algorithms for computing, e.g., the distance between two images, the convex hull, diameter, and smallest enclosing circle of an image. Dehne, Sack, and Santoro [DSS87] and Dehne, Hasenklover, Sack, and Sanotoro [DHSS87] have introduced $O(\sqrt{M})$ time algorithms for computing all nested rectilinear convex hulls of an image and for solving visibility problems on a systolic screen, respectively.

In this paper, we continue the study of algorithm design on a systolic screen, and consider the problem of computing the digitized Voronoi Diagram. We present an $O(\sqrt{M})$ time solution for computing the (digitized) Voronoi diagram of a set of n disjoint objects for a large class of object types (e.g., points, line segments, circles, ellipses, and polygons of constant size). The algorithm can compute the (digitized) Voronoi diagram for a number of distance functions which include, e.g., all L_p metrices. [1]

Since the Voronoi diagram is used in many geometric applications, the above result has numerous consequences for the design of efficient image processing algorithms on a systolic screen. In this paper we will present an $O(\sqrt{M})$ time systolic screen algorithm for "optical clustering"; i.e., identifying those groups of objects in a digitized picture that are "close" in the sense of human perception.

2 DIGITIZED VORONOI DIAGRAMS

Consider a set $S=\{s_1, \ldots, s_n\}$ of n geometric objects in $\mathbf{R}^2$ (e.g., points, line segments, polygons, cicles, ellipses) and let $d : \mathbf{R}^2 \times \mathbf{R}^2 \rightarrow \mathbf{R}^+$ be a distance function.
The well known *Voronoi diagram* V(S) (see, e.g., [SH75]) partitions $\mathbf{R}^2$ into n *Voronoi regions*

$V(s_i) := \{x \in \mathbf{R}^2 \mid d(x,s_i) \leq d(x,s_j) \text{ for all } j \neq i\}$.

Every Voronoi region $V(s_i)$ consists of two disjoint parts, the *interior*

$IV(s_i) := \{x \in \mathbf{R}^2 \mid d(x,s_i) < d(x,s_j) \text{ for all } j \neq i\}$,

and the *border*

$$BV(s_i) := V(s_i) - IV(s_i).$$

$BV(S) := \bigcup_{1 \leq i \leq n} BV(s_i)$, the union of all borders, is usually referred to as the set of *Voronoi points* of V(S).

[1] The problem of computing the digitized Voronoi diagram for point sets (for Euclidean and L_1 metric) on a mesh-of-trees architecture has recently been studied by Schwarzkopf [S88].

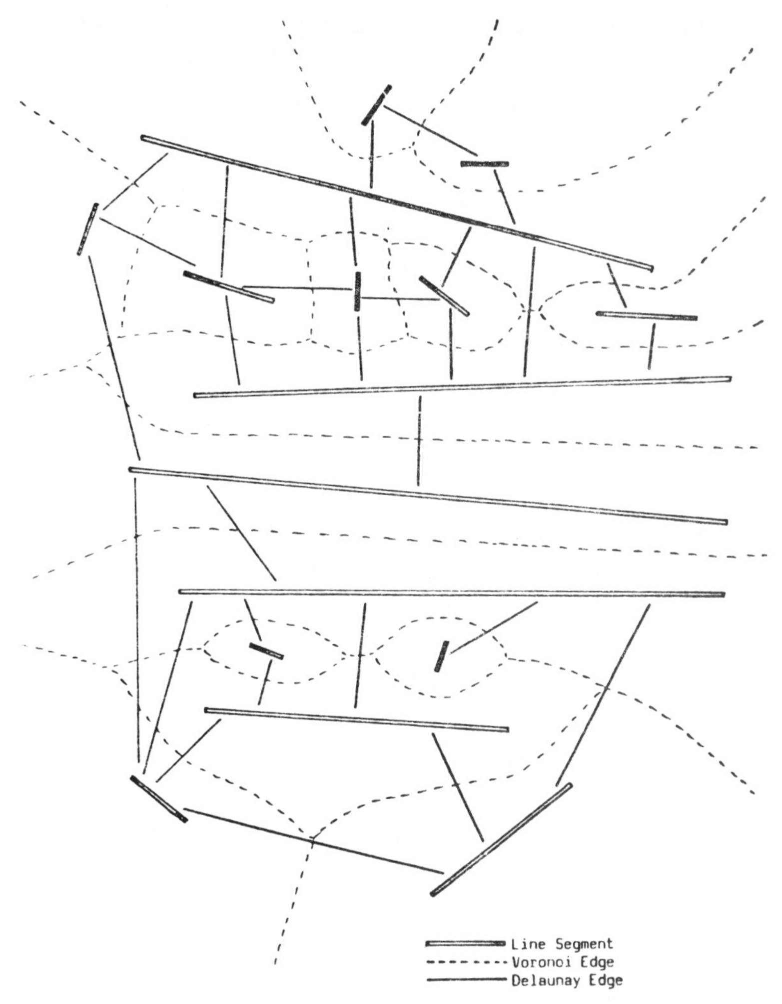

Figure 2: Voronoi Diagram for a Set of Line Segments.

In the remainder of this section, we will present how to translate the Voronoi diagram definiton to the digitized environment (see also [S88]).

We first introduce some definitions (cf., e.g., [Ro79] and [Ki82]):

- The *direct neighbors* of a pixel $(x,y) \in \Pi$ are the eight pixels $(x\pm 1, y)$, $(x, y\pm 1)$, $(x+1,y\pm 1)$, and $(x-1,y\pm 1)$. The *border* I^o of an image I is the set of all pixel of I which have a direct neighbor in Π-I. The *interior* of I, I - I^o, is denoted by $I^{\bullet}$.

- A *path* from $p \in \Pi$ to $q \in \Pi$ is a sequence of points $p=p_0,...,p_r=q$ such that p_i is a neighbor of p_{i-1}, $1 \le i \le r$. An image I is *connected* if for every $p,q \in I$ there exists a path from p to q consisting entirely of pixels of I. An image I which is connected is referred to as an *object*.
- With each pixel $p=(i,j) \in \Pi$ we associate its *cell* $<p> := [i-0.5, i+0.5] \times [j-0.5, j+0.5] \subseteq \mathbf{R}^2$ and with each image $I \subseteq \Pi$ its *region*

  $$<I> := \bigcup_{p \in I} <p>.$$

- Conversely, we define for a set $R \subseteq \mathbf{R}^2$ its *image* $Im(R) := \{ p \in \Pi \mid <p> \cap R \neq \emptyset \}$.

The *digitized Voronoi diagram* $V_d(S)$ can now be defined as follows:

Consider a set $S=\{s_1,...,s_n\}$ of n geometric objects $s_i \in <\Pi>$ such that $<s_i> \cap <s_j> = \emptyset$ for $i \neq j$; that is, consider a set S of n objects from "real geometry" such that their image representations in Π do not intersect. Let $d : \mathbf{R}^2 \times \mathbf{R}^2 \rightarrow \mathbf{R}^+$ be a distance function.

As described above, the standard Voronoi diagram V(S) induces a Voronoi region $V(s_i)$ for each object which consists of an interior $IV(s_i)$ and a border $BV(s_i)$.

The *digitized Voronoi diagram* $V_d(S)$ again consists of n *digitized Voronoi regions* $V_d(s_i)$, one for each object s_i. Each digitized Voronoi region consists of an *interior* $IV_d(s_i)$ and a *border* $BV_d(s_i)$ defined as follows:

- $BV_d(s_i) := Im(BV(s_i))$
- $IV_d(s_i) := Im(V(s_i)) - BV_d(s_i)$.

That is, the border of a digitized Voronoi region is the image of the border of the respective standard Voronoi region; the interior of a digitized Voronoi region consists of the remaining pixels in the image of the respective standard Voronoi region.

Consequently, the set $BV_d(S)$ of all *Voronoi pixels* of the digitized Voronoi diagram is defined as follows:

$$BV_d(S) := \bigcup_{1 \le i \le n} B_d(s_i) ;$$

i.e., the Voronoi pixels of $V_d(S)$ are obtained by computing the image of the Voronoi points of V(S).

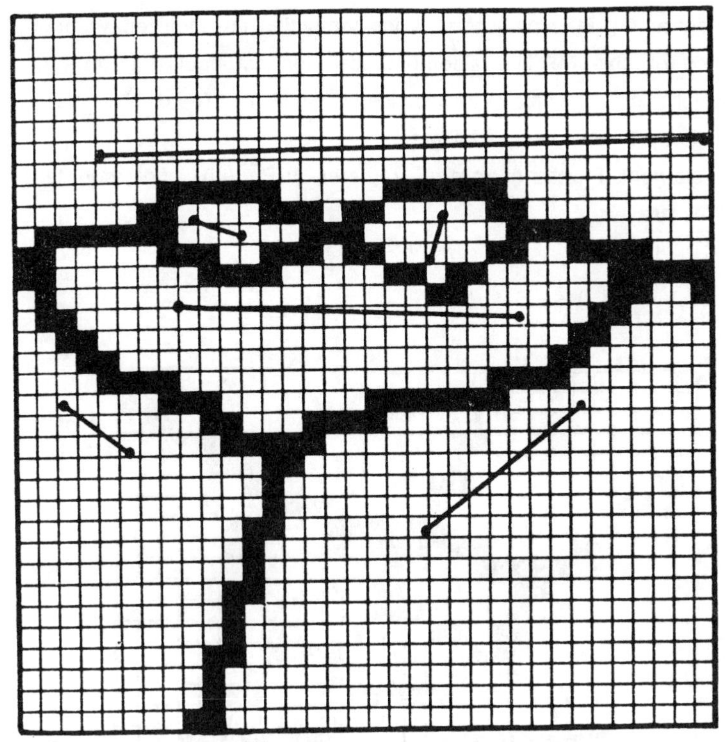

Figure 3: Digitized Voronoi Diagram for the Set of Line Segments of Figure 2. (The Black Pixels Represent the Voronoi Pixels.)

Note, that Voronoi points which do not intersect $<\Pi>$ are not represented in the digitized Voronoi diagram V(S) and that all Voronoi points which are contained in a cell $<p>$, $p \in \Pi$, are represented by one Voronoi pixel only.

In the following Sections 3 and 4 we will describe how to compute digitized Voronoi diagrams on a systolic screen. To simplify exposition, we will first consider the basic case of a set of points and Euclidean metric, and will then generalize our result to more general sets of objects and distance functions.

3 Computing Digitized Voronoi Diagrams for Point Sets and Euclidean Metric

Let $S=\{s_1,...,s_n\}$ be a set of n points in $<\Pi>$, and consider the Euclidean metric. (We assume that $Im(s_i) \cap Im(s_j) = \emptyset$ for $i \neq j$.)

We will now present an $O(\sqrt{M})$ time algorithm for computing the digitized Voronoi diagram $V_d(S)$ on a systolic screen of size M. The algorithm assumes as input that $Im(s_1),\ldots,Im(s_n)$ are represented on a systolic screen of size M as described in Section 1. The digitized Voronoi diagram of S will be reported by the systolic screen as follows:

Every processing element P_{ij} has a *Voronoi register*, V-Reg(i,j), and upon termination of the algorithm their values are

$$\text{V-Reg}(i,j) = \begin{cases} k & \text{if } (i,j) \in IV_d(s_k) \ (1 \le k \le n) \\ * & \text{if } (i,j) \in BV_d(S) \end{cases}$$

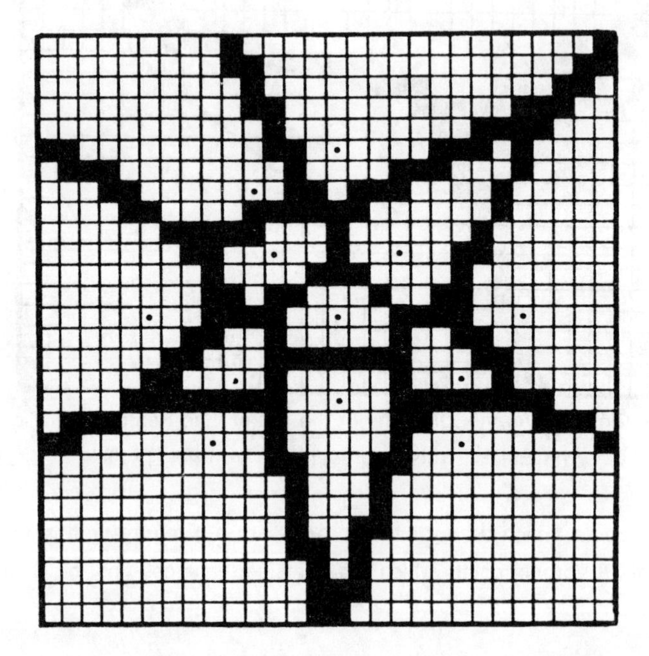

Figure 4: Digitized Voronoi Diagram for a Set of Points.

Before describing the algorithm, we need to introduce the following notations:

- Given a point $s \in \mathbf{R}^2$ and radius $r \in \mathbf{R}$, then $B(s,r) := \{x \in \mathbf{R}^2 / \ d(s,x) \le r\}$ denotes the *ball* with center s and radius r.
- The *processor distance* between two processors P_{ij} and $P_{i'j'}$ is their manhatten distance $|i-i'| + |j-j'|$.

Algorithm DIG-VOR:

(1) All Pij initialize their V-Register: V-Reg(i,j) $\leftarrow$ C-Reg(i,j)

(2) For t := 1 to $\sqrt{M}$ do

(a) All P_{ij} with V-Reg(i,j) = k > 0 send a message "k" to all $P_{i'j'}$ within processor distance $\lambda \in O(1)$ with V-Reg(i',j')=0 and <(i',j')> $\cap$ B(s_k,t) $\neq \emptyset$.

(b) All P_{ij} with V-Reg(i,j) = 0 which receive only messages "k" set V-Reg(i,j)$\leftarrow$ k.

(c) All P_{ij} with V-Reg(i,j) = 0 which receive at least two different messages "k_1" and "k_2" set V-Reg(i,j) $\leftarrow$ *.

(d) All P_{ij} with V-Reg(i,j) = k_1 which receive a message "k_2" such that B(s_{k_1},t) $\cap$ <(i,j)> $\neq \emptyset$ and B(s_{k_2},t) $\cap$ <(i,j)> $\neq \emptyset$ set V-Reg(i,j)$\leftarrow$*.

Theorem 1. *Algorithm DIG-VOR computes, on a systolic screen of size M, the digitized Voronoi diagram of a set S of n points in <Π>, for Euclidean metric, in time* $O(\sqrt{M})$.

Proof. The minimum distance of a pixel (i',j')$\in$ B(s_k,t+1) from some (i,j)$\in$ B(s_k,t) is at most $\lambda \in O(1)$. Thus, in oder to send a message "k" from all pixels in B(s_k,t) to the pixels in B(s_k,t+1)-B(s_k,t), it suffices that each P_{ij} with <(i,j)>$\cap$B(s_k,t)$\neq\emptyset$ sends a message "k" to all processors within distance λ. Hence, at time t, a processor P_{ij} with <(i,j)>$\cap$B(s_k,t)$\neq\emptyset$ has either already received some message or does now receive a message "k".

Consider a processor P_{ij} with all points in <(i,j)> closer to s_k than to any other $s_{k'}$. There exists some minimum t$\in$ {1, ..., $\sqrt{M}$} such that <(i,j)> $\subseteq$ B(s_k,t) but <(i,j)>$\cap$B($s_{k'}$,t')=$\emptyset$ for all k'$\neq$k. Hence, before time t, processors P_{ij} has not received any message yet and, at time t, gets only "k" messages. Thus, at time t, P_{ij} sets its Voronoi register V-Reg(i,j) to k (Step 2b).

On the other hand, consider a processor P_{ij} with points x$\in$ <(i,j)> that have the same distance to two objects s_k and $s_{k'}$. For such a P_{ij}, there exists some time t$\in$ {1, ..., $\sqrt{M}$} such that at that time it receives a message "k_1" and either at the same time or later receives a message "k_2". In both cases, P_{ij} sets its Voronoi register V-Reg(i,j) to * (Step 2c and 2d, respectively).

Thus, the correctness of algorithm DIG-VOR follows.

Since the execution of Step 1 and Parts a, b, c and d of Step 2 take time O(1), each, the running time of algorithm DIG-VOR is $O(\sqrt{M})$. ♦

4 Computing Digitized Voronoi Diagrams for Sets of Objects and Convex Distance Functions

After having solved the basic case of point sets and Euclidean metric, we will now generalize our result to other classes of objects and convex distance functions. It turns out that algorithm DIG-VOR does not need many modifications to handle more general cases, too.

Theorem 2. The digitized Voronoi diagram of a set $S=\{w_1,...,w_n\}$ of n objects $w_i \subseteq <\Pi>$ for any convex distance function can be computed on a systolic screen of size M in time $O(\sqrt{M})$ provided that the following conditions hold:

(i) For any two objects $w,w' \in S$, $Im(w) \cap Im(w') = \emptyset$.

(ii) For any object $w \in S$ there exists an O(1) space description such the from this deecription it can be decided for every $p \in \Pi$ and $t \in \{1,...,\sqrt{M}\}$ in O(1) time whether $<p> \cap B(w,t) = \emptyset$.

(iii) There exists a constant $\lambda \in O(1)$ such that for every $w \in S$, $t \in \{1,...,\sqrt{M}\}$, and $p \in \Pi$ with $<p> \cap B(w,t) \neq \emptyset$:

$$\min \{ d_1(p,p') | \ p' \in \Pi, <p'> \cap B(w,t-1) \neq \emptyset \} < \lambda,$$

where d_1 refers to the L_1-metric (processor distance).

Proof: Algorithm DIG-VOR needs only two minor modifications to handle the generalized case: B(s,r) needs to be generalized to the given type of objects and the given distance function, and the value of λ needs to be adjusted to the particular case. While Condition i ensures that, again, the images of two objects do not intersect, we need however two more conditions to show that algorithm DIG-VOR performs corretcly and terminates after $O(\sqrt{M})$ steps.

In Steps 2a and 2d of the algorithm, intersection tests between a ball B(s,r) and a rectangle <p> are performed. While such a test can clearly be executed in O(1) time for point objects and Euclidean metric, this may no longer be the case for arbitrary objects and distance functions. In fact, the processor performing this test does not only need the number of the oject but also the necessary information about the object to compute the intersection test. Therefore, an O(1) space decription of this information must be available, and the test must be executable in O(1) time; i.e., Condition ii must hold.

For Step 2a of algorithm DIG-VOR, the processor distance λ within that each P_{ij} has to scan all neighbors and send a message k to all $P_{i'j'}$ with V-Reg(i',j')=0 and $<(i',j')> \cap B(s_k,t) \neq \emptyset$ has to be modified according to the type of objects and the given metric. Condition iii ensures that λ is still O(1), which may not be the case in general.

However, with the above conditions, the correctness of this modified algorithm DIG-VOR follows in the same way as in the proof of Theorem 1, and its asymptotic running time does not change. Thus, Theorem 2 follows. ♦

The number of classes and object types for which the conditions in Theorem 2 apply is faily large. It contains all "simple" geometric objects that have an O(1) description and most of the standard distance function; in particular, all L_p metrices.

Corollary 3. *On a systolic screen of size M, the digitized Voronoi diagram of a set of points, line segments, circles, ellipses, and polygons of constant size can be computed, for any L_p-metric, in time $O(\sqrt{M})$ provided that their images do not intersect.*

5 OPTICAL CLUSTERING ON A SYSTOLIC SCREEN

In [De86] we have presented a sequential technique for "optical clustering", i.e., identifying those groups of objects in a digitized picture that are "close" in the sense of human perception. Given a set of n line segments in the Euclidean plane and a separation parameter $r \in \mathbf{R}$, two line segments s and s' are called *r-connected* if and only if there exists a ball with radius less than or equal to r intersecting s and s'. The optical clustering with respect to separation parameter r is then defined as the partitioning of the set of lines segment into equivalence classes with respect to the relation *r-connected* . In [De86] it is shown that the transitive closure of the relation *r-connected* is equivalent to the transitive closure of the relation *Delaunay connected with respect to r* obtained as follows:

> Compute the Voronoi diagram of the set of line segments. Two line segments s and s' are Delaunay connected if the Voronoi polygons of s and s' share a point that has distance of at most r from either line segment.

Therefore, Theorem 2 provides a way of computing, on a systolic screen, the optical clustering with respect to separation parameter r for any set of objects and distance function that have the properties listed in Theorem 2. We compute the digitized Voronoi diagram as described in Sections 3 and 4 with the following minor modification:

> For every message originating at an object and travelling through the systolic screen, as described in algorithm DIG-VOR, its current distance from the object (i.e., minimum distance, with respect to the given distance funtion, of the current processor to the object) is constantly updated. When a message changes the register V-Reg(i,j) of a processor P_{ij}, the message's current distance from its object is also stored in an additional register D-Reg(i,j). Every processor P_{ij} with C-Reg(i,j)$\neq 0$ sets D-Reg(i,j)=0.

Consider for a given separation parameter $r \in \mathbf{R}$ the image I(r) consisting of all those pixels (i,j) with D-Reg(i,j)$\leq$r, then the optical clustering of the object set with respect to separation parameter r corresponds exactly to the set of connected components of the image I(r) (see [NS80] for a definition of connected components of an image). In [NS80] it is shown that on a systolic screen of size M, the connected components of an image can be computed in time $O(\sqrt{M})$; hence, we obtain

Corollary 4. *For any set $S=\{w_1,...,w_n\}$ of n objects $w_i \subseteq <\Pi>$ and any convex distance function that have the properties listed in Theorem 2, the optical clustering with respect to separation parameter r can be computed on a systolic screen of size M in time $O(\sqrt{M})$.*

REFERENCES

[AH86] M.J.Atallah, S.E.Hambrusch, "Solving tree problems on a mesh-connected processor array", Information and Control, Vol.69, Nos.1-3, 1986, pp.168-186.

[AK84] M.J.Atallah, S.R.Kosaraju, "Graph problems on a mesh-connected processor array", J. of the ACM, Vol.31:3, 1984, pp.649-667.

[De86] F. Dehne, "Optical clustering", The Visual Computer 2:1, 1986, pp. 39-43.

[DHSS87] F. Dehne, A. Hassenklover, J.-R. Sack, and N. Santoro, "Parallel visibility on a mesh-connected parallel computer", in Proc. International Conference on Parallel Processing and Applications, L'Aquila (Italy), 1987, North Holland 1988, pp. 203-210.

[DSS87] F. Dehne, J.-R. Sack and N. Santoro, "Computing on a systolic screen: hulls, contours and applications", in Proc. Conference on Parallel Architectures and Languages Europe, Eindhoven (The Netherland), 1987, Vol. 1, Lecture Notes in Computer Science 258, Springer Verlag, pp. 121-133.

[Ki82] C.E. Kim, "Digital disks", Report CS-82-104, Computer Science Dept., Washington State University, Dec. 1982.

[Mi84] P.L.Mills, "The systolic pixel: A visible surface algorithm for VLSI", Computer Graphics Forum 3, 1984, pp.47-60.

[Mo70] G.U. Montanari, "On limit properties of digitization schemes", J. ACM 17, 1970, pp 348-360.

[MS85] R.Miller and Q.F.Stout, " Geometric algorithms for digitised pictures on a mesh-connected computer", IEEE Trans. on PAMI 7:2, 1985, pp.216-228.

[NS80] D.Nassimi and S.Sahni, "Finding connected components and connected ones on a mesh-connected parallel computer", SIAM J. Computing 9:4, 1980, pp.744-757.

[Re84] A.P.Reeves, "Survey parallel computer architectures for image processing", Computer Vision, Graphics, and Image Processing 25, 1984, pp.68-88.

[Ro79] A. Rosenfeld, "Digital topology", Amer. Math. Monthly 86, 1979, pp 621-630.

[S88] O.Schwarzkopf, "Parallel computation of discrete Voronoi diagrams", Tech. Rep., Fachbereich Mathematik, Freie Universität Berlin (W.-Germany), 1988.

[SH75] M.I.Shamos, D.Hoey, "Closest Point Problems", Proc. 7th Ann. IEEE Symp. on Found. of Comp. Sci., 1975.

[SM84] Q.F.Stout, R.Miller, "Mesh-connected computer algorithms for determining geometric properties of figures", in Proc. 7th Int. Conf. on Pattern Recognition, Montreal, 1984, pp.475-477.

[Un58] S.H.Unger, "A computer oriented towards spatial problems", Proc. IRE 46, 1958, pp.1744-1750.

PRAM algorithms for identifying polygon similarity

Costas S. Iliopoulos † § and W.F. Smyth ‡ §

ABSTRACT

The computation of the least lexicographic rotation of a string leads to the identification of polygon similarity. An $O(logn)$ time CRCW PRAM algorithm for computing the least lexicographic rotation of a circular string (of length n) over a fixed alphabet is presented here. The logarithmic running time is achieved by using $O(n/logn)$ processors and its space complexity is linear. A second algorithm for unbounded alphabets requires $O(lognloglogn)$ units of time, uses $O(n/logn)$ processors.

1.Introduction

The computational model used here is CRCW PRAM (Concurrent Read- Concurrent Parallel RAM). The processors are unit-cost RAM's that can access a common memory. Some processors can access the same memory location: they can concurrently read and write; when two or more processors are attempting to write in the same memory location one succeeds in a non- deterministic fashion.

Let $Seq(n)$ be the fastest known worst-case running time sequential algorithm,where n is the length of the input for the problem at hand. Obviously, the best upper bound on the parallel time achievable using p processors without improving the sequential result is of the form $O(Seq(n)/p)$. A parallel algorithm that achieves this running time is said to have *optimal speed-up* or simply *optimal.* A primary goal in parallel computation is to design optimal algorithms that run as fast as possible.

Here we consider the question of determining a *minimum lexicographic rotation* of a *circular* string, i.e said to be the *canonical form* of the string. Formally the problem can be stated as follows: given a string $a_1, ..., a_n$ of length n and over an alphabet Σ, we wish to compute an index k in the range $1 \leq k \leq n$which satisfies the condition

$$a_k...a_n a_1...a_{k-1} \leq a_i...a_n a_1....a_{i-1} \qquad for\ all\ 1 \leq i \leq n$$

The string to the left side is the original string rotated to the left k positions.

Booth ([B]) gave a linear algorithm for computing the canonical form of a circular string by generalizing the Knuth-Morris-Pratt ([KMP]) linear time pattern matching algorithm.

---

† Royal Holloway College, University of London, Department of Computer Science, Egham TW20 0EX, England

§ This work of both authors was partially supported by the GR/E 75752 grant of the Science and Engineering Reserch Council of the UK. The first author was supported in part by the UK SERC GR/F 00898 grant and a Royal Society grant. The work of the second author was partially supported by grant No A8180 of the Natural Sciences and Engineering Council of Canada

‡ McMaster University, Department of Computer Science and Systems,Hamilton, Ontario, Canada L8S 4K1

Moreover Shiloah ([S]) gave another linear algorithm improving the constant factor of the running time of Booth's algorithm. Both algorithm seem to be inherently sequential. In Booths algorithm the computation of the "failure function" is is the main osticle of the parallelization, and Shiloach's series of novel series of comparisons are strictly sequential. Apostolico et al ([AIP]) gave a parallel algorithm for the llr problem that requires $O(logn)$ units of time and $O(n)$ processors to compute the canonical form of a string of length n over an alphabet of size $O(n)$. Given the linearity of the sequential complexity there is obviously room for improvement in the parallel complexity bounds.

Theorem 1.1† Given a string x over a fixed alphabet , one can compute the canonical form of x in $O(logn)$ units of time on a CRCW PRAM with $O(n/logn)$ processors requiring linear space. □

Theorem 1.2 Given a string x of length n over an alphabet of size $O(n)$, one can compute the canonical form of x in $O(lognloglogn)$ units of time on a CRCW PRAM with $O(n/logn)$ processors. □

The algorithm for both the fixed unbounded alphabet make use of (i) Preprocessing and (2) A *duelling* between points procedure; the technique of the main algorithm is shared by both problems.

Preprocessing is necessary for reducing the problem to one of size $n/logn$. In the constrained version of the problem, where the alphabet is fixed,we preprocess the string by dividing it into blocks of size $logn$ and computing a local msp. In the case of fixed alphabet we create a set of all prefixes and suffixes of the substring defined by two consecutive local msps each packed in a single word. In the case of unbounded alphabets we present two versions one that makes use of the suffix trees and one that makes use of lexicographic sorting.

String comparisons. We need a procedure that compares two substring of x over the local msps that requires constant time. The comparison is done my means of the prefixes and suffixes supplied from the preprocessing. Here we also use a stronger version of the CRCW PRAM model which has all of its processors labeled with an integer and when two or more processors attempt to write in the same memory location the one with the smallest label succeeds; this is done in conjunction with the constant time simulation procedure of the two version of the two models given in [AIP]

Lemma 1.3 Let $\alpha_i, ..., \alpha_j$ and $\alpha_j, ..., \alpha_{j+k}$ with $k = j - i + 1$ be substrings of a circular string x. If $w \geq z$ lexicographicly , then α_j is not a starting point of a mlr. □

2.Preprocessing

Given the string x we split it into blocks x_i of size $k = \lceil logn \rceil$,

$$x = x_1, .., x_m, \quad m = \lceil n/k \rceil$$

Furthermore we compute *local minimum starting points*, i.e., a_r is said to be the lmsp of $x_i = a_1, ..., a_t$ if and only if r is the smallest index such that

$$a_r, a_{r+1}, ..., a_l \leq a_j, ..., a_{j+l-r} \quad for\ any\ j\ with\ 2t \geq l \geq 2r - j$$

† This theorem is also cited in [AI], where a different construction can be found

Now let $s_1, ..., s_m$ be the lmsp of the blocks $x_1, ..., x_m$ respectively. The s_i's split the string x into m substrings whose prefixes and suffixes into encoding into single words; here $prefix_l(z)$ ($suffix_l(z)$) denotes the prefix (suffix) of the string x of length l. The detailed preprocessing procedure is as follows::

Algorithm 2.1
begin
  forall $1 \leq i \leq n/k$ **pardo**
    processor p_i computes the lmsp s_i of x_i;
    comment This computation can be done by means of a linear sequential algorithm.
    processor p_i is assigned to the string z_i and w_i;
    comment The string w_i is defined to be the substring of x that starts at position s_i and terminates at position s_{i+2}.The string z_i is defined to be the substring of x that starts at position s_{i-2} and terminates at position s_i.
    processor p_i "packs" $prefix_l(w_i), suffix_l(z_i)$;
    comment The number of prefixes and suffixes is $O(logn)$. The processor can sequentially pack them into single words of length at most $O(logn)$ in $O(logn)$ units of time.
  odpar
end □

Theorem 2.2 Procedure packs all prefixes and suffixes of the blocks of the string x in $O(logn)$ time using $O(n/logn)$ processors and linear space. □

3.The duel of two local msp's

Let s_i and s_j be local msps and wlog let$j \geq i$. We have to find a certificate that give us evidence that one of the two local msp's is not a global one; there is the case that both can be global msp in which we keep the leftmost one. The duel between s_i and s_j can be performed by comparing the strings

$$w = (s_i, s_{i+1}, ..., s_j) \text{ and } z = (s_j, s_{j+1}, ..., s_{j+k}) \text{ with } k = j - i + 1$$

Let $C = \{c_r, i \leq r \leq j\}$, where c_r denotes the position of s_r in the string w (here $c_i = 1$). Similarly let $D = \{d_q, j \leq q \leq j + k\}$ where d_q denotes the position of s_q in the string z (here also $d_j = 1$). Moreover let $A = \{\alpha_1, ..., \alpha_{2k}\}$, denote the ordered sequence of the elements of the set $C \cup D$. Then one can easily show the following lemma:

Lemma 3.1 Every three consecutive elements of the A sequence are not all members of the C set nor the D set. □

Let B be the sequence A by removing the second c of every two consecutive c's and the second d of every two consecutive d's. Now consider splitting the strings w and z into the substrings defined by the positions of the B sequence. One can observe that all substrings of w and z are members of the prefixes and suffixes sets computed in the preprocessing stage. A detailed account of the duel between two strings is given below:

Procedure 3.2
begin

1. Processor p_m is assigned to the point $s_m, i \leq m \leq j+k$;
2. Processor $p_r, i \leq r \leq j$ *writes* c_r *into a doubly-linked list* C ;
3. Processor $p_q, j \leq q \leq j+k$ *writes the position* d_q *into a doubly-linked list* D ;
4. Processor $p_r, i \leq r \leq j$ *computes its position in the* A *sequence (as defined above);*

   comment Processor p_r inserts c_r into the D list; it is not difficult to see that c_r has to be inserted in one of the positions between d_{r-2} and d_{r+2} and furthermore processor p_r checks whether c_{r-1} and c_{r+1} must be linked with c_r.
5. Let $A = \{\alpha_1, ..., \alpha_{2k}\}$ and a processor is attached to each entry of A;
6. **if** α_r and α_{r-1} are both in C or D **then**

   The processor attached to α_r *deletes* α_r *and marks itself idle* ;

   comment The newlist created is the list B defined above.
7. Processor p_m *compares the necessary strings as they are defined by the* B *sequence;*

   commentAll comparisons can be done in constant time, since all strings involved have been packed into single words at the preprocessing stage.
8. **if** *processor* p_m *finds an inequality* **then** hfill
   Processor p_m *is marked "winner"*
9. Compute the smallest $r(q)$ such that $p_r(p_q)$ is marked winner;

   comment This can be done in constant time, see [].
10. Let p_v be the processor attached to the smallest α;
11. Processor p_v returns $min\{w, z\}$.

   comment Processor p_r compares the strings that are inequal.

end □

Lemma 3.3 The computation of the smallest indexed processor at step 9 above can be done in constant time using O(k) processors. □

Lemma 3.4 The procedure above requires $O(1)$ units of time and $O(k)$ processors for executing the duel between w and z. □

4.Computing the msp over fixed alphabets

The main algorithm for computing the canonical structure of a circular string obeys the following framework. It employs a reducing procedure to produce a smaller instance of the problem (of size $\leq n/2$, say). The smaller problem is solved recursively until it brings below some threshold for the size of the problem. First preprocessing reduces the problem from size $O(n)$ to one of size $O(n/logn)$. This will allow $O(logn)$ units of time for solving the problem and achieving optimal speed-up.

From (1.1) and procedure 2.1 one can see that the input string can be expressed as a string of length $O(n/logn)$over the alphabet defined by the set $\{prefix_l(w), suffix_l(z)$, with w and z as in procedure 2.1.$\}$. Thus the problem has been reduced to a $O(n/logn)$ one.

The reduction procedure used here is that of *prefix sum* algorithm. All symbols of the string x are paired with a processor and they are attached as leaves to a full binary tree(wlog we assume that the number of symbols is a power of 2). Initially every symbol is considered to be a candidate for the starting point of the llr. Every pair of siblings duels for the occupation of the parent node. The duel is executed by means of the procedure 3.1. The winner moves to the parent node and prepares for a new duel with its new sibling; the loser teams up with the winner for the new duel. The index conqueror of the root is the required starting point of the least lexicographic rotation of the input string.

Initially $O(n/logn)$ processors are attached to the leaves. Each duel at the i-th level involves strings of length 2^i and there are as many processors available to perform it; this can be done in constant time by means of procedure 3.1 (see lemma 3.2). Since the height of the tree is at most $logn$, the overall time required is $O(logn)$,and no more that $O(n/logn)$ processors are needed at any time;this together with lemma 2.2 lead us to the proof of Theorem 1.1.

5.Computing the msp over general alphabets

The preprocessing procedure of section 2 applied on a string over a general alphabet will not necessarily lead to a space reduction , since the packing into single words will take $O(log^2 n)$ bits of space.

One can see from the core algorithm of section 4 that it suffices to have procedures for

(i) Comparing the strings at step 9, actually we only need evidence that the two strings involved are different.

(ii) At step 10 we need to know the order of two strings of length $O(logn)$.
Both (i) and (ii) require constant time in the case of constraint alphabets.Here we employ a data structure for all prefixes and suffixes -as they defined in section 2- that allows the issue of certificates of difference in constant time. The date a structure itself requires linear space but its construction requires $O(n^2/logn)$ space; its takes $O(logn)$ units of time and $O(n/logn)$ processors to be constructed. But by searching for the lowest common ancestor of two strings we can order the strings (as in (ii) above) in $O(loglogn)$ units of time and thus increasing the total time complexity by this factor.

The procedure below is used into constructing a *merged prefix tree* which is a data structure defined as follows::

(i) The merged prefix tree is a rooted tree.

(ii) The label of a path from the root to a leaf is the string $v_i, 1 \leq i \leq n/logn$.

(iii) Two equal strings are labels of the same path.

(iv) Every prefix of a v_i is the label of a path from the root to a non-leaf.

Procedure 5.1
begin
  Processor p_i, $1 \leq i \leq n/logn$ is assigned to the string v_i;
  for $j = 1$ **to** $logn$ **do**

Let BB be an $1 \times n$ matrix and let a_{i_j} denote the j-th symbol of v_i;
Processor p_i writes its index i in $BB(a_{i_j})$;
Processor p_i stores $BB(a_{i_j})$ in its own memory;
od
All processors are attached to the root (marked new);
for $i = 1$ **to** *logn* **do**
forall processors attached to node marked new **pardo**
An $1 \times n/logn$ bulletin board NBB is attached to each node marked new
Processor p_i writes is index i in $NBB(BB(a_{i_j}))$;
Processor p_i with $i \neq NBB(BB(a_{i_j}))$ is marked "looser";
Processor $p_{NBB(BB(a_{i_j}))}$ is marked as "winner";
The "winner" creates a new node as child of its associated node marked new ;
The "winner" marks the new node as "new";
The "winner" labels the new edge witha_{i_j};
The "winner" gets attached to the new node;
The "losers" get attached to the new node created by the "winner" processor $p_{NBB(BB(a_{i_j}))}$;
odpar
od
end □

Lemma 5.2 The procedure above can construct the merged prefix tree in $O(logn)$ units of time, $O(n/logn)$ processors and uses $O(n^2/log^2n)$ space. □

The procedure below is used for the construction of*merged suffix tree*, a data structure defined as follows:

(i) It is a rooted tree.

(ii) There are two types of leaves the *real leaves* and *pseudo leaves*. The pseudo-leaves include a list of suffixes and a pointer pointing to an edge of the root.

(iii) A label of path to a real leaf is a $suffix(v_i)$

(iv) Two equal suffixes are labels of the same path.

Procedure 5.3
Procedure MST
begin
for $i = 1$ **to** $n/logn$ **pardo**
Processor p_i is assigned to the string v_i;
Processor p_i computes the Suffix tree of v_i;
od
forall $i = 1$ **to** $n/logn$ **pardo**
for $j = 1$ **to** *logn* **do**
Let $BB_\alpha, \alpha \in \Sigma$ be an $1 \times n/logn$ bulletin board.
Processor p_i writes its index in BB_{a_j} Where a_j is the first letter of v_i
od
odpar
Compact $BB_\alpha, \alpha \in \Sigma$into CBB_α
comment This can be done in $O(logn)$ units of time.

```
forall i = 1 to n/logn pardo
   for j = 1 to logn do
      processor p_i is attached to suffix_j(v_i);
      if |BB_α| = 1 then
         procesor p_i marks the list JOB(j);
      else
         The winner at CBB_{a_j} created a new node as a child of the root of MST;
         The winner labels the new node with a_j;
         The winner associates the new node with CBB_{a_j};
         processor p_i creates a "suffixlink" bettwen the new node and suffix_{j-1}(v_i);
   od
odpar
forall i = 1 to n/logn pardo
   for j = 1 to logn do
      if JOB(J) is marked  then
         Procesor p_i attatches the subtree labelled with suffix_j(v_i) of the
         suffix tree of v_i to the root of MST;
   od
odpar
end □
```

Lemma 5.4 The procedure above computes the merged suffix tree in $O(logn)$ units of time, uses $O(n/logn)$ processors and $O(n^2)$ space. □
The Procedure COMPARE below compares every prefix versus a suffix. This is done by means of the data structures Merged Prifix and suffix tree already known.

Procedure 5.5

```
begin
   forall i = 1 to n/logn pardo
      CUR_i = {α : α is the first symbol of a label of a rooted edge of MST}
      mark(i) = ∅
      for j = 1 to logn do
         procesor p_i is attached to prefix_j(v_i)
         if prefix_j(v_i) ∉ CUR_i then exit
         Let cur_i ∈ CUR_i be the symbol of the label that macthes prefix_j(v_i)
         ifcur_i has a suffix-link then
            The symbol pointed by the matching suffixlink becomes cur_i
            if mark(i) = ∅ then
               mark(i) = CBB(suffixlink)
         if $ ∈ CUR_i then
            Let m be the index of the suffix stored at the leaf;
            processor p_i marks prefix(i, j − 1) = suffix({m} ∩ mark(i), j − 1
         if cur_i is last symbol of label (above a node) then
            CUR_i = {α : αis the first symbol of a label of an egde atached to the node}
         else
            CUR_i = {cur_i}
      od
```

odpar

end □

Corollary 5.6 One can compare a "suffix" and a "prefix" in $O(loglogn)$ units of time, using $O(n/logn)$ processors and $O(n^2)$. □

The space requirements can be reduced to $O(n^{1+\epsilon})$ using the techniques presented in [AILSV].

6.References

[AT] Akl, A. G., Toussaint, G.T., *An improved algorithmic check for polygon similarity*, Inf. Proc. Lett. , 7-3 (1978) 127-128

[AI] Apostolico A., Iliopoulos, C., Unpublished manuscript.

[AILSV] Apostolico, A., Iliopoulos, C., Landau, G.M., Schieber, B., Vishkin, U., *Parallel construction of a suffix tree with applications*,Acta Algorithmica (1988)3: 347-365

[AIP] Apostolico, A., Iliopoulos, C.S., Paige, R.,*An $O(nlogn)$ cost parallel algorithm for the one function partitioning problem*, Proc. of Int. Conf. on Parallel architectures, ed.Jung and Mehlhorn, Academie-Verlag, (1987) pp70-76

[B] Booth, K.S., *Lexicographically least circular substrings*,Inf. Proc. Letters, 10, 4, (1980) pp240-242

[KMP] Knuth, D., Morris, J., Pratt, V., *Fast pattern maching in strings*, SIAM J. Computing, 6 (2) (1977) pp 323-350

[S] Shiloach, Y.,*Fast canonization of circular strings* , J. Algorithms 2 (1981) 107-121

A FRAMEWORK FOR PARALLEL GRAPH ALGORITHM DESIGN

Vijaya Ramachandran

Department of Computer Sciences
University of Texas
Austin, TX 78712, USA

ABSTRACT

We describe a graph search technique called open ear decomposition and its applicability in the design of algorithms for several connectivity problems on undirected graphs. All of these algorithms have optimal or efficient parallel and sequential implementations.

1. Introduction

The design of efficient algorithms for graph problems is a widely studied area in the context of sequential computation and optimal sequential algorithms have been designed for several graph problems. A vast majority of these algorithms are based on the well-known depth first search technique (see, e.g., [Tarjan, Even]).

In this paper we describe a search technique for undirected graphs called *open ear decomposition* for which we have developed a good parallel algorithm [Miller & Ramachandran 1986] (see also [Maon, Schieber & Vishkin]). We then describe our work in applying open ear decomposition to obtain efficient algorithms for testing k-vertex and k-edge connectivity of graphs for k=2, 3, and 4 and for deriving the biconnected and triconnected components of a graph. All of these algorithms have optimal or efficient parallel and sequential implementations. In this paper we will mainly discuss the parallel implementations.

2. Model of Parallel Computation

The model of parallel computation that we will be using is the *Parallel Random Access Machine* or *PRAM,* which consists of several independent sequential processors, each with its own private memory, communicating with one another through a global memory. In one unit of time, each processor can read one global or local memory, execute a single RAM operation, and write into one global or local memory location. We will use the ARBITRARY CRCW PRAM in which concurrent reads and concurrent writes are permitted at a memory location and any one processor participating in a concurrent write may succeed.

Let S be a problem that, on an input of size n, can be solved on a PRAM by a parallel algorithm in parallel time t(n) with p(n) processors. The quantity $w(n)=t(n)\cdot p(n)$ represents the *work* done by the parallel algorithm. Any PRAM algorithm that performs work w(n) can be converted into a sequential algorithm running in time w(n) by having a single processor simulate each parallel step of the PRAM in p(n) time units. More generally, a PRAM algorithm that runs in parallel time t(n) with p(n) processors also represents a PRAM algorithm performing O(w(n)) work for any processor count $P<p(n)$.

Define $\text{polylog}(n)=\bigcup_{k>0} O(\log^k n)$. Let S be a problem for which currently the best sequential algorithm runs in time T(n). A PRAM algorithm A for S, running in parallel time t(n) with p(n) processors is *optimal* if the work $w(n)=p(n)\cdot t(n)$ is O(T(n). Algorithm A is *efficient* if

a) t(n)=polylog(n); and

b) the work $w(n)=p(n)\cdot t(n)$ is $T(n)\cdot\text{polylog}(n)$.

An efficient parallel algorithm is one that achieves a high degree of parallelism and comes to within a polylog factor of optimal speed-up. A major goal in the design of parallel algorithms is to find optimal or efficient algorithms with t(n) as small as possible. For more on the PRAM model and PRAM algorithms, see Karp & Ramachandran.

Several problems including the n-element prefix sums and list ranking problems and the computation of various tree functions on an n-node tree have optimal O(log n) time algorithms. For the problem of computing the connected components and a spanning forest of an undirected graph Jung has announced an optimal O(log n) time algorithm, improving the 'almost-optimal' result of Cole & Vishkin. The algorithms we will survey in this paper will use these optimal algorithms as subroutines.

3. Open Ear Decomposition

An *ear decomposition* $D=[P_0,P_1,\cdots,P_{r-1}]$ of an undirected graph is a partition of its edge set into an ordered collection of simple paths called *ears* such that the first ear P_0 is a simple cycle and each P_i, $i>0$ is a simple path (possibly closed), each of whose endpoints belongs to some P_j, $j<i$, and none of whose internal vertices belongs to any P_j, $j<i$.

An *open ear decomposition* is an ear decomposition for which none of the P_i, $i>0$ is a cycle.

A *separating edge* of a graph G is an edge e such that G - {e} has at least two connected components. G is *2-edge connected* if it contains no separating edge.

A *separating vertex* or *cutpoint* of G is a vertex v such that G - {v} has at least two connected components. G is *biconnected* if it contains no cutpoint.

The following two classical results relate ear decomposition to 2-edge connectivity and biconnectivity.

Lemma 2.1 [Whitney] A graph is 2-edge connected if and only if it has an ear decomposition.[]

Lemma 2.2 [Whitney] A graph is biconnected if and only if it has an open ear decomposition.[]

We now describe a simple algorithm for finding an ear decomposition in a 2-edge connected undirected graph G [Lovasz; Maon, Schieber & Vishkin; Miller & Ramachandran 1986].

1. Find a spanning tree T for G and root it at a vertex r.

2. Assign labels to the nontree edges of G in nondecreasing order of the least common ancestors of the endpoints of the nontree edges. (These are the ear numbers of the nontree edges.)

3. Label each tree edge e by the label of the nontree edge with the smallest label whose fundamental cycle contains e. (These are the ear numbers of the tree edges.)

A proof by induction on ear number shows that this algorithm generates a sequence of paths satisfying the properties of an ear decomposition. Each of these steps can be performed optimally in O(log n) time using optimal O(log n) time algorithms for finding a spanning tree [Jung], least common ancestors [Schieber & Vishkin], and tree functions [Tarjan & Vishkin; Cole & Vishkin], provided ear labels are required to be merely an increasing sequence and not a sequence of consecutive integers.

The above algorithm can be refined to give an open ear decomposition for a biconnected graph by modifying step 2. Here we further order the nontree edges with the same least common ancestor by using local connectivity information at that least common ancestor [Maon, Schieber & Vishkin; Miller & Ramachandran 1986]. This in turn gives an optimal O(log n) time parallel algorithm for finding an open ear decomposition in an undirected biconnected graph.

In case the input graph is not biconnected the above algorithm will induce an open ear decomposition for each biconnected component of the graph.

4. Open Ear Decomposition and Triconnectivity

A *separating pair* in a graph is a pair of vertices whose removal results in a graph with more than one connected component. A graph is *triconnected* if it has no separating pair.

The following claim, shown in Miller & Ramachandran 1987, shows that an open ear decomposition serves to localize each separating pair on a single ear.

Claim 3.1 [Miller & Ramachandran 1987] Let D be an open ear decomposition of a graph G and let {x,y} be a separating pair of G. Then there exists an ear P in D such that x and y lie on P and the portion of P between x and y is separated from the rest of P in G - {x,y}.[]

Let G be a biconnected graph with an open ear decomposition D. A *bridge* of an ear P (see e.g., [Even]) is a maximal collection of edges in G - P such that there exists a path in G - P between any pair of edges in the collection. In this section we will only be interested in those edges in a bridge of G - P that are incident on P. Miller & Ramachandran 1987 show that by finding the bridges of P and ascertaining their 'interlacement' pattern we can identify the separating pairs of G that lie on P. (Informally, two bridges of an ear P *interlace* if they cannot be placed on the same side of P in any planar embedding of G.) We also show that it suffices to compute certain approximations to the bridges of P (called the *nonanchor bridges* and the *anchoring star)* that are easy to compute. This leads to the following high-level description of the triconnectivity algorithm.

1. Find an open ear decomposition D of the input graph G.
2. Find the nonanchor bridges and the anchoring star of each ear.
3. Determine the interlacement pattern of the bridges of each ear as computed in step 2 and hence determine the set of separating pairs on each ear.
4. If no separating pair is identified in step 3 then report that G is triconnected.

Miller & Ramachandran 1987 use a divide and conquer technique to implement steps 2 and 3, resulting in an $O(\log^2 n)$ time algorithm with a linear number of processors. Ramachandran & Vishkin reduce the time bound to O(log n) by pipelining the computation of Miller & Ramachandran 1987 in step 2 and by reducing step 3 to the problem of determining connectivity in an auxiliary linear-sized graph.

Fussell, Ramachandran & Thurimella use a technique called *local replacement* to eliminate the need for a divide and conquer approach for step 2. The local replacement technique removes the interaction between ears at their endpoints and allows the computation of step 2 to be performed simultaneously on all ears. This reduces the complexity of the entire triconnectivity algorithm to the the complexity of determining connectivity. By the result claimed by Jung this gives an optimal O(log n) time parallel algorithm for determining graph triconnectivity.

These results have been extended to the problem of finding the triconnected components of a graph. In the following we briefly describe this work.

Finding Triconnected Components

We first review some material from [Tutte; Hopcroft & Tarjan] relating to the definition of triconnected components. While this definition may appear contrived at first, in reality it decomposes a biconnected graph into substructures that preserve the triconnected structure of G. In particular, questions relating to graph planarity and isomorphism between a pair of graphs can be mapped onto related questions regarding the tree of triconnected components.

This material deals with multigraphs. An edge e in a multigraph is denoted by (a,b,i) to indicate that it is the ith edge between a and b; the third entry in the triplet may be omitted for one of the edges between a and b.

We extend the definition of separating pairs to multigraphs. A pair of vertices a,b in a multigraph G is a separating pair if and only if there are two nontrivial bridges, or at least three bridges, one of which is nontrivial, of {a,b} in G. If G has no separating pairs then G is triconnected. The pair a,b is a *nontrivial* separating pair if there are two nontrivial bridges of {a,b} in G.

Let {a,b} be a separating pair for a biconnected multigraph G, and let B be a bridge of {a,b} in G which is biconnected and has $|E(B)|\geq 2$. We can apply a *Tutte split* s(a,b,i) to G by forming G_1 and G_2 from G, where G_1 is $B \cup \{(a,b,i)\}$ and G_2 is the induced subgraph on $(V-V(B)) \cup \{a,b\}$, together with the edge (a,b,i). The graphs G_1 and G_2 are called *split graphs* of G with respect to {a,b}. The *Tutte components* of G are obtained by successively applying a Tutte split to split graphs until no Tutte split is possible. Every Tutte component is one of three types: i) a triconnected simple graph; ii) a simple cycle; or iii) a pair of vertices with at least three edges between them. The Tutte components of a biconnected multigraph G are the unique *triconnected components* of G.

Miller & Ramachandran 1987 show that an open ear decomposition gives a simple and natural method to perform Tutte splits. Each Tutte split s(a,b,i) is made at an ear P that contains the separating pair {a,b}. An edge (a,b,i) is added to the component containing the portion of P that lies between a and b and another edge (a,b,i) is added to the component containing the rest of P.

It is possible that there are several separating pairs containing a given vertex a and in such a case there will be several triconnected components containing a copy of vertex a. Miller & Ramachandran 1987 use a divide and conquer technique to generate the triconnected components in parallel without causing conflicts between different processors attempting to access the same vertex to generate different splits associated with that vertex. This leads to an $O(\log^2 n)$ time algorithm with a linear number of processors to obtain the triconnected components of a graph. Fussell, Ramachandran & Thurimella use the local replacement technique to avoid the use of divide and conquer. A vertex is converted into a tree of copies of itself after the application of local replacement; a given copy can appear as a 'real' separating pair in at most one ear. Thus all of the Tutte splits can be performed simultaneously. This leads to an algorithm for finding triconnected components with the same complexity as that of finding connected components, which by the result of Jung can be performed optimally in O(log n) time.

5. Four Connectivity

A vertex triple {x,y,z} is a separating triplet of a graph G if G - {x,y,z} contains at least two connected components. If G contains no separating triplet then G is said to be *four connected.*

Kanevsky & Ramachandran establish the following claim that relates each separating triplet in a triconnected graph G with an open ear decomposition D to a specific ear in D.

Claim 4.1 [Kanevsky & Ramachandran] Let G be a triconnected graph with an open ear decomposition D and let {x,y,z} be a separating triplet of G. Then there exists an ear P in D such that two of the vertices, say x and y, in {x,y,z} lie on P, and the portion of P between x and y is separated from the rest of P in G - {x,y,z}. Further, if z does not lie on P then z is a cutpoint in a bridge of P.[]

Given a triconnected graph G together with an open ear decomposition D, Kanevsky & Ramachandran represent any separating triplet of G by the notation ([x,y],z) where x and y represent an unordered pair of vertices that lie on a single ear and {x,y,z} is a separating triplet of G. We further classify separating triplets to be either Type 1 or Type 2. In a Type 1 separating triplet ([x,y],z), the vertex z belongs to the ear containing x and y; in a Type 2 separating triplet vertex z is a cutpoint in a bridge of that ear. Using this characterization Kanevsky & Ramachandran develop an algorithm for testing four connectivity and finding all separating triplets in a triconnected graph. This algorithm runs in $O(\log^2 n)$ parallel time with n^2 processors on an n node graph. A sequential version of the algorithm runs in $O(n^2)$ time which improves the asymptotic bounds obtained by previous algorithms for this problem on dense graphs by a linear factor.

6. Conclusion

Open ear decomposition has proved to be a powerful tool in the design of sequential and parallel algorithms for vertex connectivity of graphs. The same techniques lead to similar algorithms for edge connectivity by using ordinary ear decomposition in place of open ear decomposition.

Open ear decomposition has been used in the development of other graph algorithms, including s-t numbering and planarity [Maon, Schieber & Vishkin; Klein & Reif]. Recently we have developed a new planarity algorithm based on open ear decomposition (under preparation). It is to be anticipated that other algorithms based on open ear decomposition will be developed, strengthening its claim to be the basis of a new framework for the design of efficient and optimal sequential and parallel algorithms for undirected graphs.

References

Cole, R. & Vishkin, U., "Approximate and exact parallel techniques with applications to list, tree and graph problems," *Proc. 27th Ann. IEEE Symp. on Foundations of Comp. Sci.,* 1986, pp. 478-491.

Even, S. *Graph Algorithms,* Computer Science Press, Potomac, MD, 1979.

Even, S. & Tarjan, R. "Computing an st-numbering," *Theoretical Computer Science* 2, 1976, pp. 339-344.

Fussell, D., Ramachandran, V. & Thurimella, R., "Finding triconnected components by local replacements," *Proc ICALP,* Italy, Springer-Verlag LNCS, July 1989, to appear.

Hopcroft, J.E. & Tarjan, R.E. "Dividing a graph into triconnected components," *SIAM J. Computing* 2, 1973, pp. 135-158.

Jung, H. "An optimal parallel algorithm for computing connected components in a graph," Preprint. Humboldt-University Berlin, German Democratic Republic, 1989.

Kanevsky, A. & Ramachandran, V., "Improved algorithms for graph four-connectivity," *Proc. 28th Annual IEEE Symp. on Foundations of Comp. Sci.,* 1987, pp. 252-259; *Jour. of Computer and System Science,* to appear.

Karp, R.M. & Ramachandran, V. "Parallel algorithms for shared-memory machines," *Handbook of Theoretical Computer Science,* North-Holland, 1989, to appear.

Klein, P. & J.H. Reif, "An efficient algorithm for planarity," *Proc. 27th Annual IEEE Symp. on Foundations of Computer Science,* 1986, pp. 465-477.

Lempel, A., Even, S. & Cederbaum, I. "An algorithm for planarity testing of graphs," Theory of Graphs: International Symposium, Rome, July 1966, Gordon and Breach, New York, NY, pp. 215-232.

Lovasz, L. "Computing ears and branchings in parallel," *26th Annual IEEE Symposium on Foundations of Computer Science,* 1985, pp. 464-467.

Maon, Y., Schieber, B. & Vishkin, U. "Parallel ear decomposition search (EDS) and st-numbering in graphs," *VLSI Algorithms and Architectures,* Lecture Notes in Computer Science 227, 1986, pp. 34-45.

Miller, G.L. & Ramachandran, V. "Efficient parallel ear decomposition with applications," unpublished manuscript, MSRI, Berkeley, CA, January 1986.

Miller, G.L. & Ramachandran, V. "A new graph triconnectivity algorithm and its parallelization," *Proc. 19th Annual ACM Symp. on Theory of Computing,* New York, NY, 1987, pp. 335-344.

Ramachandran, V. & Vishkin, U. "Efficient parallel triconnectivity in logarithmic time," *VLSI Algorithms and Architectures,* Springer-Verlag LNCS 319, 1988, pp. 33-42.

Schieber, B. & Vishkin, U. "On finding lowest common ancestors: simplification and parallelization," *VLSI Algorithms and Architectures,* Springer-Verlag LNCS 319, 1988, pp. 111-123.

Tarjan, R. E. , "Depth first search and linear graph algorithms," *SIAM J. Comput.* 1, 1972, pp. 146-160.

Tarjan, R.E. & Vishkin, U. "An efficient parallel biconnectivity algorithm," *SIAM J. Computing* 14, 1984, pp. 862-874.

Tutte, W. T., *Connectivity in Graphs,* University of Toronto Press, 1966.

Whitney, H. "Non-separable and planar graphs," *Trans. Amer. Math. Soc.* 34, 1930, pp. 339-362.

FAST SOLITON AUTOMATA

T.S.Papatheodorou and N.B.Tsantanis
Computer Technology Institute (CTI)
P.O. Box 1122, 261 10 *Patras, Greece*

ABSTRACT.

Solitons, are special moving waves with the remarcable characteristic that when colliding they come out of the collision without loosing their initial properties. Soliton Automata (SA) are a particular class of Cellular Automata which support solitons.

In this work we present a survey of recent developments in the area of SA and some new results regarding algorithmic issues of sequential and parallel versions of the SA. We show that the parallel implementation of the SA in the CRCW model of computation is optimal.

KEY WORDS: Solitons, Cellular Automata, Optimal Parallel Algorithms.

1. INTRODUCTION.

Cellular Automata (CA) is a modern and continuously growing interdisciplinary area of the contemporary science. The range of their applications extends to a great diversity of subjects, including Physics, Mathematics, Computer Science and Neural Sciences. Many physical processes have been found to be conveniently modeled by CA and many efforts are made for even more.

In this paper we study a specific class of CA called Parity Filter Automata (FA) which exhibit soliton like behavior. Solitons are special waves which preserve their shape and velocity after colliding with other localized waves. Solitons have many interesting applications in Biology and Physics.

In Section 2 we present a brief review of the work done so far in the area of Soliton Automata (SA). Next we consider Serial and Parallel algorithms regarding their evolution. Specifically, in Section 3 we give the complexity of the (sequential) parity filter rule, while in Section 4 we exploit the inherent parallelism of the so called Fast Rule Theorem. The result is an optimal algorithm in the CRCW model of parallel computation.

2. SURVEY.

Park et.al. [1] introduced a new rule and called it the *"Parity Filter"* rule because the operation of the automata it supports, resembles the operation of the infinite impulse response digital filters. Its main characteristic is that contrary to the typical notion of CA it takes into consideration not only the values of the previous state of the automaton (time t) but also the newly computed ones (time $t+1$). This fact introduces a sequence in which the state values are to be updated, namely from left to right.

Specifically, let a_i^t denote the value of site i at time t. The CA we study is a binary one, that is a_i^t can be only 0 or 1. We assume an infinite number of sites, allowed to extend from $-\infty$ to $+\infty$, but we also assume that the initial configuration is made up of a finite number of 1's. The new state is calculated by the rule:

$$a_i^{t+1} = F(a_{i-r}^{t+1}, \dots, a_{i-1}^{t+1}, a_i^t, a_{i+1}^t, \dots, a_{i+r}^t), \tag{1}$$

where r is a fixed integer ≥ 2 called the *radius*.

The sites $a_{i-r}^{t+1}, \dots, a_{i+r}^t$ in formula (1) comprise the neighborhood of the cell a_i^{t+1} being evaluated and can be graphically depicted as in Figure 1.

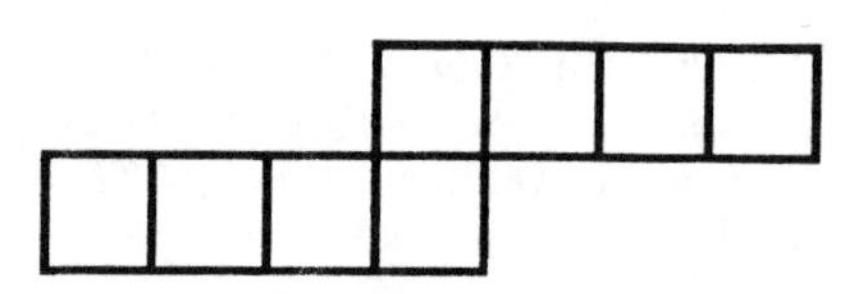

Fig.1: The Parity Filter Rule's neighbourhood

Thus we can think of a window that contains the sites participating in the rule and slides from left to right each time a new value at level $t+1$ is to be calculated by use of (1).

For each cell, the total number S of 1's in the sliding window is calculated i.e.

$$S = \sum_{j=-r}^{-1} a_{i+j}^{t+1} + \sum_{j=0}^{r} a_{i+j}^{t}. \tag{2}$$

The new state a_i^{t+1} is decided upon the parity of S, i.e

$$a_i^{t+1} = \begin{cases} 1, & \text{if } S \text{ even but not zero} \\ 0, & \text{otherwise.} \end{cases} \tag{3}$$

The significance of this rule stems from the fact that it supports moving periodic "particles" which move to the left in various velocities resulting to collisions. Park et.al. made statistics of the rule and found that a surprisingly large percentage of the collisions were solitonic, in the sence that colliding particles continue their motion maintaining their initial identities. In Figure 2 a typical case of solitonic collision is presented.

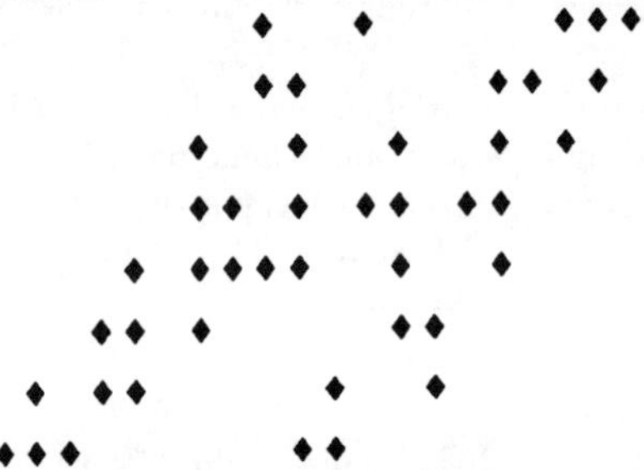

Fig.2 A solitonic collision

One serious drawback of the Parity Rule is that its use makes it almost impossible to analytically follow the evolution of the automaton, predict its behavior, or give criteria as to when two sites appearing next to each other should be considered to belong in separate or single entities, or to prove intuitively clear and experimentally tested facts such as the stability of the FA etc. By the term *stability* we refer to whether a finite number of 1's at one state can generate only a finite number of 1's in the next state. for instance the question of stability was left as an open problem in [1]. Nevertheless Papatheodorou et.al[2] introduced an equivalent explicit rule, which made it possible to analytically follow and predict the evolution of this type of SA, to easily detect cases of single entities or the presence of splitting of one entity into more particles, to easily prove results such as stability and to characterize cases of periodic particles and also predict solitonic collisions. This explicit equivalent of FA was named *"Fast Rule Theorem" (FRT)* in [2] and states that the parity rule used above is equivalent to first selecting a finite set of sites and then obtaining the next state by inverting the one bit in these sites, while leaving the remaining sites unchanged.

The finite set used in FRT is called the *box set* and can be determined by inspection as follows:

(0) Let σ be the site of the first 1 in a configuration.

(1) Place a box on site σ and on all sites following it in steps of $(r+1)$, up to and until a box is found that is followed by at least r consecutive zeroes.

(2) If there are more 1's to the right of these zeroes, let σ be the site of the first of them and go to (1).

As just stated, FRT computes the next state by use of the following:

(3) Copy all bits a_i^t into a_{i-r}^{t+1}, i.e. below and r positions to the left , if i is not a box site. Otherwise set $a_{i-r}^{t+1} = \overline{a}_i^t$, the complement of a_i^t.

When the condition in (2) is satisfied, two succesive box sites are found to be separated by at least r zeroes. In this case it turns out that we may compute the next state using the serial parity rule in two ways, either by considering two such sets of separated sites as one or by treating them separately i.e. by computing the next state for each one independently and ignoring the presence of the other. The next state will be the same for both calculations and we say that we have *splitting.* Note that FRT offers itself naturally for parallel treatment of SA.

In order to demonstrate the analytical power of FRT, we give one important and straighforward application of it i.e. an extremely simple proof of the stability of the FA. Specifically, we show that finite 1's in t result in finite 1's in $t+1$ as follows: Because of finite 1's in t the box set is finite (from its definition). The state at $t+1$ differs from the state at t only over the finite box set, hence it also contains a finite number of 1's. The proof is complete.

Another conribution of the FRT is that it gave insight in the understanding of the operation of the FA. The placement of boxes every $r+1$ cells indicates a reasonable grouping of the sites in groups of $r+1$ with a leading box. Papatheodorou et.al. [3] named these groups *basic strings* (b-strings) and showed that these strings and the interaction of their internal 1's played a dominant role in the evolution of the automaton. They stated their results in the *General Evolution Theorem.*

With the help of FRT they showed that b-strings disappear from the left and reappear to the right of the automaton, at specific instances which they called *"landmark times"*. They were then able to give a *"General Evolution Theorem"* for the analytic prediction of the future states. Furthermore, they completely characterized particles of period 1, i.e. such that $a^{t=1}$ is the same (shifted) string of 0's and 1's as a^t. Moreover, they predicted cases of solitonic collisions. The following simple example may provide the intuition for some of their results.

At $t=0$ let the particle be $(r=3)$

$$a^0 : \ \ldots 0\boxed{1}\ 0\ 1\ 1\boxed{0}\ 1\ 0\ 1\boxed{0}\ 0 \ldots$$

meaning that, in terms of b-strings, $a^0 = \ldots \bigcirc \ A \ B \ \bigcirc \ldots$, where

$$A = 1\,0\,1\,1,\ B = 0\,1\,0\,1,\ \bigcirc = 0\,0\,0\,0.$$

There are three 1's in the (leading) string A. According to the theory developed in [3], as it can also be observed directly by use of FRT, the basic strinf A will disappear from the left and reappear to the right at landmark time $\tau = 3$, i.e. $a^{\tau=3} = \ldots \bigcirc \ \bigcirc \ C \ A \ \bigcirc \ldots$, where the new triplet $\bigcirc \ C \ A$ is computed from the original $A \ B \ \bigcirc$ by bitwise XOR operation $\oplus$ with A, i.e. $\bigcirc = A \oplus A, C = A \oplus B, A = A \oplus \bigcirc$. Similarly consider a particle $\ldots \bigcirc \ A \bigcirc \ B \ \bigcirc \ldots$. At the first landmark time the particle becomes $\ldots \bigcirc \ \bigcirc \ A \ C \ A \ \bigcirc \ldots$, where $C = A \oplus B$. At the next landmark time it becomes $\ldots \bigcirc D \bigcirc A \bigcirc \ldots$, where $D = A \oplus C = A \oplus A \oplus B = B$. Hence the b-strings A and B collided and they reappeared with their relative positions interchanged. We thus observe a solitonic collision under certain conditions.

The study of the FA was enhanced by new results drawn by Fokas et.al. [4],[5]. The research was focused in the domain of simple (or small) particles, which exhibit a more easy to describe behavior.

A particle is called simple iff it consists of only one basic string. It is quite interesting that one is able to give a complete characterization of the interaction of any two simple particles. In particular:

(i) Every simple particle is periodic with period equal to the number of its 1's.

(ii) if l_A, l_B is the number of 1's in the two simple particles A, B then B is faster than A iff $l_B > l_A$.

(iii) if $l_B > l_A$ and A, B begin interacting with A in front of B, then after $2l_A$ time steps, A and B will emerge as two particles with B in front of A. This interaction is *solitonic*. Although the faster particle B is in front of A, the particles A and B may interact again; however eventually they will get separated with the faster particle moving in front of the slower particle. (The number of interactions can be given explicitly and depends on the relative positions of 1's in the two particles.)

(iv) If $l_B = l_A$, and the particles are close enough so that they can interact, there exist two cases. Either they will interact once and they get separate travelling independently of each other, or they will form a new periodic configuration by interacting for ever.

Results in the same direction were also presented by Goldberg [6], who related the behavior of FA with physical notions. Characteristically, the energy of a configuration was defined as the sum

$$\mathcal{E}^t = \sum_{i=-\infty}^{\infty} |a_i^t t - a_i^t - r - 1|.$$

It was proved that the energy obeys to the Second Law of Thermodynamics, i.e. the energy of a configuration is never increased. Interesting results were obtained from this, refering to the characterization of the behavior of the particles from the scope of energy. For example the energy of a particle is the same in all phases of its evolution and the particle's period is $\mathcal{E}^t$.

3. THE COMPLEXITY OF THE SEQUENTIAL PARITY FILTER RULE

In this section we study the complexity of the Parity Filter Rule, i.e. of the sliding window algorithm described by (2) and (3). Regarding the availability of processors, we consider separately the case of one and the case of as many as possible processors. An observation is made that significantly improves the complexity in both cases and, at the same time, demonstrates the sequential character of this rule.

We assume that the given particle consists of B basic strings i.e. of $O(B(r+1))$ cells.

We start our study by examining the complexity of the Parity Filter Rule. We shall consider separately the two possible cases occuring regarding the number of processing elements: The case of only one processor and the one of as many as possible processors. Finally we shall make a clever remark that significantly improves the complexity of both cases.

(i) One processor.

For the update of each cell the number of "live" ones in its neighborhood need to be counted. The size of the neighborhood is $2r+1$ cells so with only one processor available, $O(r)$ time is required for each cell. That means that a total of $B(r+1)$ cells (plus a constant extention on both sides of the particle) need to be updated. Therefore the total time needed for a single step in the evolution of the FA is $O(B(r+1)^2)$.

(ii) Many processors.

We assume now that an arbitrary number of processing elements is available. Because of the serial nature of the algorithm only one window can be evaluated each time and no window overlapping can occur. Consequently the maximum parallelization of the sequential algorithm can be obtained by at most $O(r)$ processors. In this case the evaluation of each neighborhood needs $O(\log(r))$ time and the total time for all cells is $O(B(r+1)\log(r+1))$.

(iii) Improved Variation.

The complexity found above can be improved by making a crucial observation: Whenever the update window advances from the previous cell to the next, the resulting new neighborhood differs from the old one in exactly four cells. Namely if i is the index of the new cell, then the new neighborhood excludes the previously included cells a_{i-r-1}^{t+1} and a_{i-1}^{t} and appends the cells a_{i-1}^{t+1} and a_{i+r}^{t}.

Thus, one needs to only evaluate the effect of the change in these four cells. Therefore the time required for each step becomes: $1 \cdot t_{count} + O(B(r+1)) \cdot O(1)$ where t_{count} is the time required for the evaluation of a neighborhood and is given (from above) by:

$$t_{count} = \begin{cases} O(r) & \text{if one processor is available,} \\ O(\log(r)) & \text{if } O(r) \text{ processors are available.} \end{cases}$$

The total time becomes $O(r + B(r+1))$ and $O(\log(r) + B(r+1))$ respectivelly. We observe that there is no improvement with the use of many processors in the dominant term Br, due to inherent sequential behaviour of the parity rule.

4. OPTIMAL PARALLEL ALGORITHM IN THE CRCW MODEL OF COMPUTATION

In this section we study the performance of the parallel version of Parity Filter Rule, that is the one resulting from an algorithm based on implementation of the FRT. We make use of the CRCW model of computation. We initially study the case in which the cell configuration consists of a single periodic particle, i.e. when no splitting occurs. Next, we extend this result into the general case in which splitting is allowed and arbitrary number of particles may be present. This implementation is highly parallel, as we consider that one processor is available for each cell. We omit the details of the description of each processor.locations in a CRCW manner.

(i) Study for a single particle.

The main issue arising from the treatment of a single particle is how to detect its boundaries, that is the first and the last box placed by the FRT. Knowing this, one can easily locate all the box cells simultaneously in $O(1)$ time. To solve it we consider two variables (memory locations) associated with each paricle. The one holds the index of the first box of the particle and the other holds the index of the last box. The determination of each index is made as follows:

Determination of the first box:

Each cell (processor) that is "alive" (that is its state is one) writes its index concurrently to the same memory location. The processor with the smallest index prevails. This value is assigned to a variable called *first* and marks the first box (and first ace) of the particle. The cost for this operation is $O(1)$.

Determination of the last box:

We first locate the index of the last ace of the particle. This is made in the same way as above and is kept in a variable called *end*. We now need to locate the index of the last box of the particle. Its distance from *first* is the nearest multiple of $r+1$ that is not less than *end*. This is called *last* and is given by:

$$last := first + (\ (end \text{ div } (r+1)) + 1\ ) \cdot (r+1).$$

The whole operation takes $O(1)$ time.

Update.

After determining *first* and *last* the update of the configuration is made in a simple manner as described by the FRT.

Each cell with index i concurrently reads $first$ and $last$ and checks whether $first \leq i \leq last$. This condition assures that the cell belongs to the particle. Then each cell of the particle has to determine whether it has a box or not. The corresponding condition is

$$i = start + k * (r + 1)$$

or equivalently

$$i \bmod (r + 1) = start \bmod (r + 1)$$

and is checked in $O(1)$ time.

The actual update is made by shifting and inverting (when necessary). Specifically, if a cell has a box then it inverts its state and sends it r cells to the left. If a cell does not have a box then it simply shifts its state to the left unchanged. Instead of actually shifting values to the left, each cell can alternatively reduce its index by r. The time required for this operation is again $O(1)$.

(ii) General case: multiple (two) particles.

We now consider the case in which splitting may occur (simple or multiple). For simpilcity let us assume that only simple splitting occurs, that is at most two particles are present at any time. A simple technique for splitting detection is the following:

The boxes are placed all the way from the first to the last ace, as if we had a simple particle. Next the cells of each basic string make a concurrent write and decide in $O(1)$ time whether they contain at least one ace or not. In the former case a 1 is the result, in the latter a 0. Because of the simple splitting hypothesis made, we expect either no 0's, if no splitting occurs, or one or more consecutive 0s denoting the place of null basic strings.

The index of the first and last basic strings having 0 can be determined in constant time by means of two Concurrent Writes (Let them be $First$ and $Last$ respectively). Now, the boxes for each cell prior to $First$ have been correctly marked. But the boxes after $Last$ need to be respecified. This can be made in constant time by the method described previously. Finally the two particles may be updated simultaneously in $O(1)$ time.

Summarizing we can describe the evolution of the FA for each step as consisting of two phases:

First phase: Determination of the number and location of particles.

Second phase: Update of each particle in constant time.

Note that the sequential algorithm of the preceding section requires $O(Br)$ time. The parallel algorithm presented here requires $O(1)$ time with $O(Br)$ processors and is, therefore, optimal.

6. DISCUSSION

In this paper we made a preliminary study of the implementation of the Parity Filter Automata using fully sequential, sequential with a small degree of parallelism and highly parallel architectures. In the last case we exploited the theoretical results that stem from the FRT.

The parallel implementation is optimal.

REFERENCES.

[1] J.Park, K.Steiglitz and W.Thurston, *"Soliton Like Behaviour in Automata"*, *Physica D 19 (1986).*

[2] T.S.Papatheodorou, M.J.Ablowitz and Y.G.Saridakis, *"A Rule for Fast Computation and Analysis of Soliton Automata"*, *Studies in Appl. Math., 79, pp. 173-184 (1988)* — initially *CTI Technical Report TR 87.05.01 (May 198).7*

[3] T.S.Papatheodorou and A.S.Fokas, *"Evolution Theory and Characterization of Periodic Particles and Soliton in cellular Automata"*, to appear in *Stud. of Appl. Math. (1989)* — initially *CTI Technical Report TR 87.05.02 (May 1987).*

[4] A.S.Fokas, E.P.Papadopoulou, Y.G.Saridakis and M.J.Ablowitz, *"Interaction of Simple Particles in Soliton Cellular Automata"*, to appear in *Stud. of Appl. Math. (1989).*

[5] A.S.Fokas, E.P.Papadopoulou, Y.Saridakis, *"Particles in Soliton Cellular Automata"*, Inst. *for Nonlin. Stud., Clarkson Un., INS#107 (November 1988)*

[6] C.H.Goldberg, *Parity Filter Automata,Manuscript Dec 1987*, Princeton Un.

An Upper Bound on the Order of Locally Testable Deterministic Finite Automata†

Sam Kim, Robert McNaughton, and Robert McCloskey

Computer Science Department
Rensselaer Polytechnic Institute
Troy, N.Y. 12180

Abstract: A locally testable language is a language with the property that for some nonnegative integer k, called the order of locality, whether or not a word w is in the language depends on (1) the prefix and suffix of w of length k, and (2) the set of intermediate substrings of w of length k + 1, without regard to the order in which these substrings occur. The local testability problem is, given a deterministic finite automaton, to decide whether it accepts a locally testable language or not. Recently, we introduced the first polynomial time algorithm for the local testability problem based on a simple characterization of locally testable deterministic automata. This paper investigates the upper bound on the order of locally testable automata. It shows that the order of a locally testable deterministic automaton is at most $n^4 + 1$, where n is the number of states of the automaton.

1. Introduction

The property of local testability is a mathematical idealization of a property that one encounters in pattern recognition. The situation arises when it is possible to verify that an image has a certain pattern by moving a small window of fixed size around the image and recording the set of window-size subimages that are seen without regard to any relationship between these subimages. The large image is classified on the basis of the set of

† Partial support for this research was provided by the Directorate of Computer and Information Science and Engineering of the National Science Foundation under Institutional Infrastructure Grant No. CDA-8805910.

such subimages. This procedure is called local testing.

Certainly, some patterns involve global constraints and therefore cannot be recognized by local testing. Nevertheless, for many patterns, local testing is sufficient. It is certainly worthwhile to study the concept mathematically. Our idealization of this concept begins by studying one-dimensional patterns, whereupon the images become character strings, and the patterns become languages.

The concept that one arrives at this way was first introduced by one of the authors and Papert [9] in 1971 and, since then, has been extensively investigated [2,5,8,9,11]. A precise definition is as follows: A locally testable language is a language with the property that for some nonnegative integer k, called the order of locality, whether or not a word w is in the language depends on (1) the prefix and suffix of w of length k, and (2) the set of intermediate substrings of w of length k + 1, without regard to the order in which these substrings occur. It is easy to show that all locally testable languages are regular. Among the several problems concerning this concept that have been investigated, the problem of deciding whether a given deterministic automaton accepts a locally testable language and computing the order of locality, if it does, has not yielded to a practical solution. Recently, in [6], we introduced a simple characterization of locally testable deterministic automata, and showed an efficient polynomial time algorithm for the local testability problem. In [2], it is shown that the order of locality of a locally testable language of a deterministic automaton is bounded above by the size of the semigroup of the automaton, which is potentially exponential. Based on the characterization of locally testable automata that we introduced in [6], this paper shows that the order of locality of a locally testable language is at most $n^4 + 1$, where n is the number of states of the automaton.

This paper has four sections including this section. Section 2 introduces definitions and notation for the basic properties of a state transition graph. Section 3 shows an upper bound on the order of locally testable languages. Finally, Section 4 gives concluding remarks.

2. Notation, Terminology, and Definitions

We assume that a problem instance is given in terms of the state transition graph of a reduced deterministic finite automaton, $M = (Q, \Sigma, \delta, q_0, F)$. Our notation follows [3]. In a state transition graph, a double circle denotes an accepting state. For simplicity, we will omit the non-accepting sink state, which takes a transition to itself on all input symbols, from the transition graph. We shall use the terminology **connected component** (sometimes just component) to refer to any subgraph (of a transition graph) whose underlying undirected graph is connected.

Definition 1 (substring vector) [2,9]. Let Σ be a finite alphabet and $x \in \Sigma^*$. For an integer $k \geq 0$ and $x \in \Sigma^*$, define

$f_k(x)$: prefix of x of length k,
$t_k(x)$: suffix of x of length k, and
$I_{k+1}(x)$: the set of substrings of x of length k + 1,
i.e. $\{v \mid u,v,w \in \Sigma^*, x = uvw, |v| = k + 1\}$.

The substring vector of x of order k + 1 is a tuple $SV_{k+1}(x) = (f_k(x), I_{k+1}(x), t_k(x))$.

Definition 2 (locally testable languages) [2,9]. Let $L \subseteq \Sigma^*$.

(a) L is 0-testable iff it is Σ^* or $\emptyset$.

(b) For an integer $k \geq 0$, L is (k+1)-testable iff, for all $x, y \in \Sigma^*$, $SV_{k+1}(x) = SV_{k+1}(y)$ implies that either both x,y in L or neither is in L.

(c) L is locally testable if and only if it is k-testable for some $k \geq 0$.

In [9], locally testable languages are defined using the set of intermediate substrings of length $k \geq 0$. In this paper, we follow the definition in [2] for the length of intermediate substrings for convenience of dealing with certain boundary conditions. Those two definitions are equivalent [2].

Definition 3 (locally testable automaton). A finite automaton is locally testable iff it accepts a locally testable language. If an automaton accepts a language which is k-testable, we say that the automaton is k-testable. Its order of locality is the smallest k such that it is k-testable. The local testability problem of deterministic finite automata is that of deciding, given an arbitrary deterministic automaton, whether the language accepted by the automaton is locally testable or not. Figure 1 shows examples. Automaton M_1 is 3-testable, but not 2-testable. For example, consider two words x = abbabab and y = abbab. They have the same substring vector of order 2, i.e., $f_1(x) = f_1(y) = a$, $t_1(x) = t_1(y) = b$, and $I_2(x) = I_2(y) = \{ab, bb, ba\}$. The word y is accepted, while x is rejected. In the figure, automaton M_2 is not k-testable, for any $k \geq 0$.

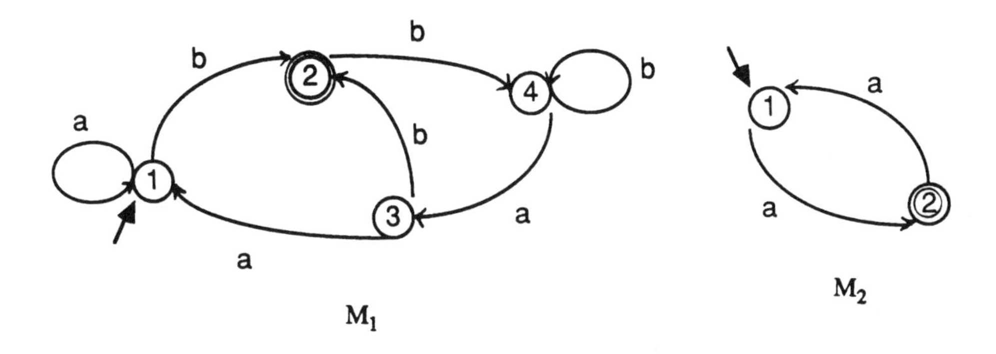

Figure 1.

Definition 4 (transition span, input span). Let $M = (Q, \Sigma, \delta, q_0, F)$ be an automaton, and let m be a connected component of the state transition graph of M. The **transition span** of m with respect to a state $p \in Q$ is the set $TS(m(p)) = \{x \mid x \in \Sigma^*$, for every prefix w of x, $\delta(p, w)$ is in m $\}$. Notice that if p is not in m, TS(m(p)) is empty. A state q has **input span** $x \in \Sigma^*$ iff $\delta(p, x) = q$, for some $p \in Q$.

Definition 5 (TS-equivalence). Let m be a connected component of the state transition graph of an automaton. States p and q are TS-equivalent in m iff TS(m(p)) = TS(m(q)). Figure 2 shows examples. In Figure 2-(a), for every pair of states $i,j \in \{1,2,3,4\}$, $TS(m_1(i)) = TS(m_1(j))$. In Figure 2-(b), $TS(m_2(1)) = TS(m_2(3))$, while $TS(m_2(1)) \neq TS(m_2(2))$. Notice that the transition $\delta(1, c)$ is not in m_2, while $\delta(2, c) = 3$ in m_2.

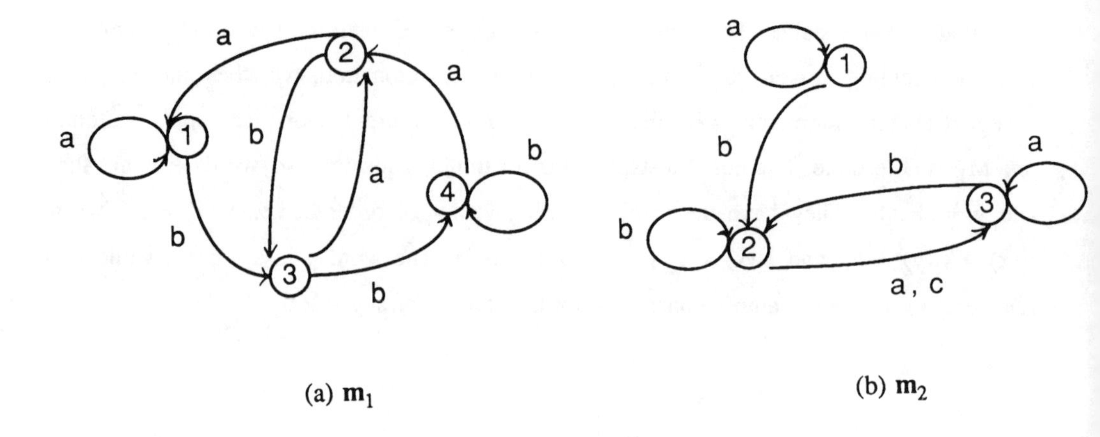

(a) $\mathbf{m}_1$

(b) $\mathbf{m}_2$

Figure 2.

Definition 6 (reaching component). Let m_i and m_j be two components of a state transition graph. The **reaching component** from m_i to m_j, written m_{ij}, is the subgraph which comprises m_i and m_j together with all directed paths which start in m_i and end in m_j. In this paper we are only interested in a reaching component m_{ij} of m_i and m_j which are both strongly connected components (SCC's). When component m_i is a single state q, then m_{qj} denotes the component which is reachable from q to m_j including m_j.

The following lemma is readily obtained from Definitions 5 and 6. We leave the proof for the reader.

Lemma 1. Let m be a component of a deterministic state transition graph.

(a) If states p and q are TS-equivalent in m, then so are r and s, where $r = \delta(p, w)$ and $s = \delta(q, w)$, for any $w \in \Sigma^*$.

(b) Let p, q, r, and s be states such that there is a path that goes through p, q, r, and s in that order. If p, r and s belong to some SCC's m_0, m_1, and m_2, respectively, and p and r are TS-equivalent in m_{01}, and q and s are equivalent in m_{q2}, the reaching component from q to m_2, then there is a state t in m_2 which is TS-equivalent to p in m_{02}.

(c) Let p and q be states which belong to SCC's m_0 and m_1, respectively. If p is an ancestor of q and TS-equivalent to q in m_{01}, then, for any state r in m_{01}, there exists a state t in m_1 which is TS-equivalent to r in m_{r1}.

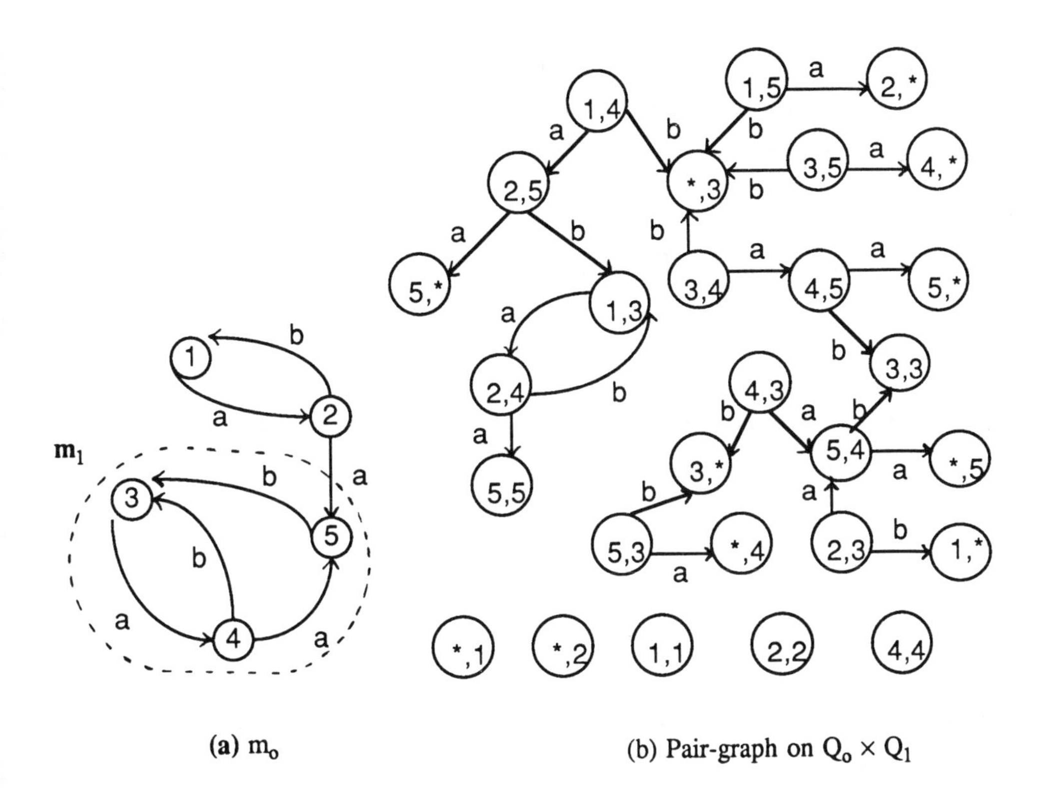

(a) m_o

(b) Pair-graph on $Q_o \times Q_1$

Figure 3.

Definition 7 (pair-graph) [10]. Let $M = (Q, \Sigma, \delta, q_0, F)$ be a deterministic automaton. Let $Q_1, Q_2 \subseteq Q$, which are not necessarily disjoint, and $*$ be a symbol not in Q. The pair-graph on $Q_1 \times Q_2$ is the edge-labeled graph G(V, E), where $V = (Q_1 \cup \{*\}) \times (Q_2 \cup \{*\})$, and E is defined as follows: Define δ_i: $Q_i \times \Sigma \rightarrow Q_i \cup \{*\}$, i = 1, 2, such that, for all $p \in Q_i$ and $a \in \Sigma$,

$$\delta_i(p, a) = \begin{cases} r & \text{if } \delta(p, a) = r \in Q_i \\ * & \text{otherwise.} \end{cases}$$

Let $p \in Q_1$ and $q \in Q_2$ be two distinct states. Then E has a directed edge from a node (p,q) to a node (r,s) with a label $a \in \Sigma$ iff $\delta_1(p, a) = r$ and $\delta_2(q, a) = s$. Notice that, for any $t \in Q_1 \cup Q_2$, the pair-graph does not have an outgoing edge from the nodes (t, *), (*, t), (t, t), and (*, *). Figure 3-(b) shows the pair-graph on $Q_o \times Q_1$, where $Q_o = \{1,2,3,4,5\}$ and $Q_1 = \{3,4,5\}$, the sets of states on m_o and m_1, respectively, in Figure 3-(a). In the figure, node (*,*) and all its incoming edges are not shown. A pair-graph is not necessarily connected.

Definition 8 (s-local, pairwise s-local). (a) Let m_1 be a component of a state transition graph whose set of states is Q_1. Then m_1 is s-local (for strongly local) iff the pair-graph on $Q_1 \times Q_1$ is acyclic.

(b) Let m_1 and m_2 be two disjoint components of a deterministic state transition graph which have sets of states Q_1 and Q_2, respectively. The components m_1 and m_2 are pairwise s-local iff the pair-graph on $Q_1 \times Q_2$ is acyclic.

From the definition it is easy to prove the following lemma. We leave the proof for the reader.

Lemma 2. Let m_i and m_j be SCC's of a state transition graph.

(a) SCC m_i is s-local if and only if m_i has no distinct states p and q for which there exists a word $x \in \Sigma^+$ such that $\delta(p, x) = p$ and $\delta(q, x) = q$.

(b) The components m_i and m_j are pairwise s-local if and only if there is no pair of states p in m_i, q in m_j such that $\delta(p, x) = p$ and $\delta(q, x) = q$, for some $x \in \Sigma^+$.

(c) Let p and q be distinct states which belong to an s-local SCC m, and let Q be the set of states in m. For any word x of length at least $|Q|^2$, if both $\delta(p, x)$ and $\delta(q, x)$ are in m, then $\delta(p, x) = \delta(q, x)$.

In [6], an automaton is defined as having finite input memory span if and only if no two states p and q of the automaton have transitions $\delta(p, x) = p$ and $\delta(q, x) = q$, for any nonnull input string x. Our definition of s-locality is a generalized version of the finite input memory span of [7].

Definition 9 (TS-local). The state transition graph of a deterministic finite automaton is TS-local iff it has the following property: For all SCC's m_i and m_j of the transition graph, if m_i is an ancestor of m_j, then either

(a) m_i and m_j are pairwise s-local, or otherwise,

(b) there exists a pair of states p and q, respectively, in m_i and m_j such that p and q are TS-equivalent in m_{ij}, the reaching component from m_i to m_j.

Let m be a component of a state transition graph which includes an SCC m_j. If conditions (a) and (b) above are satisfied for every ancestor SCC m_i in m with respect to m_j, then we say the component m is TS-local w.r.t. m_j.

Lemma 3. Let $M = (Q, \Sigma, \delta, q_0, F)$ be a finite automaton all of whose SCC's are s-local, and let $n = |Q|$, $p \in Q$ and $x \in \Sigma^+$. If $\delta(p, x^n) = q$, then for all $n' \geq n$, $\delta(p, x^{n'}) = q$. In particular, this implies that $\delta(q, x) = q$.

Proof. Let $n = |Q|$. We have $\delta(p, x^{n-j}) = \delta(p, x^n) = q$, for some $j > 0$, by a simple counting argument on $|Q|$. Thus $\delta(q, x^j) = q$. Suppose $\delta(q, x) = r$. Then $\delta(r, x^j) = r$. Since q and r are in the same s-local SCC, by Lemma 2-(a) it must be that $r = q$. The result follows. □

The following lemma provides a simple way of finding a pair of states p and q which belong to SCC's m_i and m_j, respectively, and are TS-equivalent in m_{ij}, if there exists such a pair.

Lemma 4. Let m_1 and m_2 be s-local SCC's such that m_1 is an ancestor of m_2. Let states p in m_1 and q in m_2 be TS-equivalent in m_{12}, and let r in m_1 and s in m_2 have a word x such that $\delta(r, x) = r$ and $\delta(s, x) = s$. Then r and s are TS-equivalent in m_{12}.

Proof. Find $w \in \Sigma^*$ such that $\delta(p, w) = r$. Then we have $\delta(p, wx^n) = r$ for all $n \geq 0$. Let $\delta(q, w) = q'$, $n = |Q|$, and $\delta(q', x^n) = s'$, i.e., $\delta(q, wx^n) = s'$. Since p and q are TS-equivalent in m_{12} and $\delta(p, wx^n) = r$ is in m_1, it follows that $\delta(q, wx^n) = s'$ is in m_{12} (hence, in m_2) and, from Lemma 1-(a), is TS-equivalent to r in m_{12}. By Lemma 3, $\delta(q', x^n) = s'$ implies $\delta(s', x) = s'$. By Lemma 2-(a), $s = s'$. □

3. An Upper Bound on the Order of a Locally Testable Deterministic Automaton

In [6], we show the following characterization theorem for locally testable automata, and introduce the first polynomial time algorithm for the local testability problem. In this paper we show a different proof, in particular, for the sufficiency of the characterization. As a corollary of this proof, we get an upper bound on the order of a locally testable deterministic automaton.

Theorem 1 (characterization). A reduced deterministic finite automaton is locally testable iff it satisfies the following conditions:

(a) All SCC's of the state transition graph are s-local.

(b) The transition graph is TS-local.

Proof (of necessity). We omit the proof which appears in [6].

For the proof of sufficiency of (a) and (b) of Theorem 1, we first introduce the following well known theorem.

Theorem 2.[3] Let $x \in \Sigma^*$ and $y, z \in \Sigma^+$ such that $xy = zx$. Then there exist $u \in \Sigma^*$ and $i \geq 0$ such that $y^i = uxy = uzx$.

Lemma 5. Let $M = (Q, \Sigma, \delta, q_0, F)$ be an automaton which satisfies conditions (a) and (b) of Theorem 1, and let $k = |Q|^4$. For $y,z \in \Sigma^+$ and $x \in \Sigma^k$ such that $xy = zx$, let $\delta(p, x) = q$ and $\delta(p, xy) = r$, for some $p \in Q$. Then

(a) r belongs to an SCC m_1 and is TS-equivalent to q in m_{q1}, i.e., the reaching component from q to m_1,

(b) $\delta(r, y) = r$, and

(c) r has input span x.

Proof. Consider the path on the transition graph that corresponds to $\delta(p, xy)$. Let this path be $P = p_0p_1.....p_n$, where $n = |xy|$, $p_0 = p$, $p_k = q$, and $p_n = r$. Notice that $\delta(p_0, x) = p_k$. If $q = r$, then obviously the lemma is true. We prove the lemma for the case $q \neq r$.

Consider paths $p_0p_1....p_k$, which corresponds to the transition $\delta(p_0, x) = p_k = q$, and $p_{n-k}p_{n-k+1}.....p_n$, which corresponds to $\delta(p_{n-k}, x) = p_n = r$. Notice that $\delta(p_0, zx) = \delta(p_{n-k}, x) = p_n$. Construct the pair-graph on $\{p_0, p_1, ..., p_k\} \times \{p_{n-k}, p_{n-k+1}, ..., p_n\}$. The pair-graph has a path $R = (p_0, p_{n-k})(p_1, p_{n-k+1}).... (p_k, p_n)$ which has span x. Notice that, for all i, $0 \le i \le k$, $p_i \ne p_{n-k+i}$. Otherwise, p_k and p_n, which are q and r, respectively, cannot be different. Since the pair-graph has at most $|Q|^2$ nodes and the path R has span x of length $|Q|^4$, in R there exists a closed path of length $r \ge |Q|^2$. Let this closed path be $(p_i, p_{n-k+i})(p_{i+1}, p_{n-k+i+1}) .... (p_{i+r}, p_{n-k+i+r})$, for some i, $0 \le i \le k\text{-}r$, where $p_i = p_{i+r}$ and $p_{n-k+i} = p_{n-k+i+r}$. Let u be the span corresponding to this closed path. Clearly, $\delta(p_i, u) = p_{i+r} = p_i$ and $\delta(p_{n-k+i}, u) = p_{n-k+i+r} = p_{n-k+i}$, which implies that on P there are two different closed paths P_0 and P_1 which have the same span u. Since all SCC's of M are s-local, P_0 and P_1 are closed paths on different SCC's. Let m_0 and m_1 be the two SCC's which include P_0 and P_1, respectively. Clearly, m_0 is an ancestor of m_1, but they are not pairwise s-local. Hence, by Lemma 4, p_i and p_{n-k+i} are TS-equivalent in m_{01}. Let w be the suffix of x of length k-i. Then $\delta(p_i, w) = p_k = q$ and $\delta(p_{n-k+i}, w) = p_n = r$. By Lemma 1-(a), r is in m_1 and TS-equivalent with q in m_{q1}.

Now, we prove part (b) of the lemma, i.e., $\delta(r, y) = r$. Notice that, since $|u| \ge |Q|^2$, we have $|w| \ge |Q|^2$. Since q and r are TS-equivalent in m_{01}, both $\delta(q, y) = r$ and $\delta(r, y) = t$, for some $t \in Q$, should also be TS-equivalent in m_{01}. Since r is in m_1, t should also be in m_1. Again, since $\delta(r, y) = t$ is in m_1, $\delta(t, y)$ should also be in m_1. It follows that for all $i \ge 1$, $\delta(r, y^i)$ is in m_1. Let $i_o = |Q|$. Then $\delta(r, y^{i_o}) = s$, for some state s in m_1, which implies, by Lemma 3, that $\delta(s, y) = s$. Now, we prove that $s = r$ to show $\delta(r, y) = r$. Recall that $\delta(p_{n-k+i}, w) = p_n = r$, where w is a suffix of x of length at least $|Q|^2$, and p_{n-k+i} is a state in m_1. By Theorem 2, we know that $y^{n_o} = vx$, for some $v \in \Sigma^*$ and $n_o \ge 0$. Since w is a suffix of x, $\delta(s, y^{n_o}) = \delta(s, vx) = s$ which implies that $\delta(t', w) = s$, for some state t' in m_1. Since m_1 is s-local, by Lemma 2-(c), $\delta(p_{n-k+i}, w) = r = \delta(t', w) = s$. Hence, we have $\delta(r, y) = r$. Part (c) follows imediately from Theorem 2 and part (b). □

Lemma 6. Let M be a finite automaton which satisfies properties (a) and (b) from Theorem 1, and let $(p_0, q_0), (p_1, q_1), (p_2, q_2), ..., (p_m, q_m)$, $m \ge 2$, be a sequence of state pairs from M such that

(1) p_0 is in an SCC m_0,

(2) for each n, $0 \le n < m$, there are paths from p_n to p_{n+1}, from p_{n+1} to q_n, and from q_n to q_{n+1},

(3) for each n, $0 \le n \le m$, q_n is in an SCC m_n, and

(4) for each n, $0 \le n \le m$, p_n is TS-equivalent to q_n in $m_{p_n n}$, the reaching component from p_n to m_n.

Then there exists a state t in m_m such that p_1 is TS-equivalent to t in $m_{p_1 m}$, the reaching component from p_1 to m_m.

Proof. It will be proven by induction on n that, for each n, $1 \le n \le m$, there is a state t_n in m_n such that p_1 is TS-equivalent to t_n (in $m_{p_1 n}$).

For the basis (n = 1), choose as p, q, r, and s in Lemma 1-(b) the states p_0, p_1, q_0 and q_1, respectively, and apply that lemma. We see that there exists a state t_1 which is TS-equivalent to p_0 in the reaching component from p_0 to m_1. the required properties.

For an induction hypothesis, assume that, for some n, $1 \le n < m$, there exists a state t_n having the specified properties. Now choose as p, q, r and s in Lemma 1-(b) the states p_0, p_{n+1}, t_n, and q_{n+1}, respectively, and apply that lemma. We see that there exists a state t_{n+1} which is TS-equivalent to p_0 in the reaching component from p_0 to SCC m_{n+1}.

Now choose as p, q, and r in Lemma 1-(c) the states p_0, t_m, and p_1, respectively, and apply that lemma. Then we see that there exists a state t in SCC m_m that is TS-equivalent to p_1 in the reaching component from p_1 to m_m. □

Now, we introduce a subclass of state transition graphs which will be conveniently used together with Lemmas 5 and 6 for the sufficiency proof.

Definition 12 (substring graph). For $x \in \Sigma^+$ and an integer $k \ge 1$, let $F_k(x)$ be the set of substrings of x of length k and prefixes of x of length less than k, i.e.,

$$F_k(x) = \{y \mid u,v,y \in \Sigma^*, |y| = k, x = uyv\} \cup \{w \mid w \in \Sigma^*, |w| < k, w \text{ is a prefix of } x\}.$$

Notice that ε, the null string, is a member of $F_k(x)$. Construct a state transition graph $G_k(x) = (V, \Sigma, \delta', q'_0, \{q'_f\})$, where the set of state $V = \{[w] \mid w \in F_k(x)\}$, the start state $q'_0 = [\varepsilon]$, the only accepting state is $q'_f = [t_k(x)]$, and the transition function δ′ is defined as follows: For all $[y],[z] \in V$ and $a \in \Sigma$, $\delta'([y], a) = [z]$, iff $t_{k-1}(y)a = z$ and ya is in $F_{k+1}(x)$.

We call $G_k(x)$ the **substring graph** of order k for word x. For example, Figure 4 shows $G_2(x)$, with x = abababbaacbbcbc. Notice that the start state [ε] of a substring graph is a transient state, i.e., a state which does not belong to an SCC.

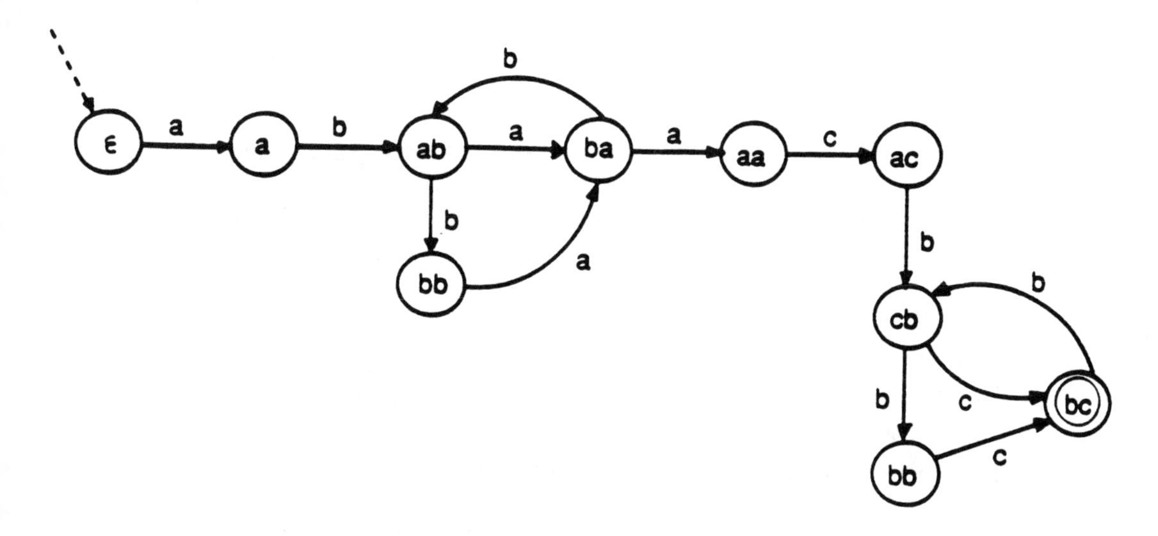

Figure 4. $G_2(x)$, x = abababbaacbbcbc

By $L(G_k(x))$ we denote the set of strings which are accepted by $G_k(x)$ after taking each transition (i.e., edge) at least once. In other words, $L(G_k(x))$ is the set of words whose substring graph is isomorphic to $G_k(x)$, i.e., $L(G_k(x)) = \{y \mid y \in \Sigma^+, G_k(y) = G_k(x)\}$. Note that $x \in L(G_k(x))$. Now, we have the following proposition which shows some interesting properties of substring graphs.

Lemma 7. (a) In a substring graph of order k, no two states have the same input span of length k, and each state has at most one such input span.

(b) In a substring graph, no transient state has more than one successor.

(c) All SCC's, if any, are linearly linked, i.e., there is a linear ordering of the SCC's m_1, m_2, ...,m_r, $r \geq 1$, such that m_i is reachable from the start state via m_{i-1} only, and there is only one path which links m_{i-1} to m_i.

(d) For all $x,y \in \Sigma^+$, $f_k(x)=f_k(y)$, $I_{k+1}(x)=I_{k+1}(y)$, and $t_k(x)=t_k(y)$ iff $G_k(x) = G_k(y)$.

Proof. (a) This is obvious from the definition of substring graphs.

(b) From the definition, $G_k(x)$ should have a unique path from the start state to the final state which has span x. This path should cover all the edges of the graph and end at the final state. Hence, if a transient state has more than one successor, then no single path exists that covers all the edges of the graph.

(c) This part of the lemma is also true by the same reason described in the proof for (b).

(d) For any pair of words $x,y \in \Sigma^*$, suppose that $f_k(x)=f_k(y)$, $I_{k+1}(x)=I_{k+1}(y)$, and $t_k(x)=t_k(y)$. Obviously, $G_k(x)$ and $G_k(y)$ should have the same set of nodes. Suppose that $G_k(x)$ has an edge ([u],a,[v]) for which there is no matching edge in $G_k(y)$. If $|u| < k$, then $f_k(x) \neq f_k(y)$. If $|u| = k$, then $I_{k+1}(y)$ does not have ua, while $I_{k+1}(x)$ has. Clearly, if the final states of $G_k(x)$ and $G_k(y)$ are different, $t_k(x) \neq t_k(y)$. It follows that $G_k(x) = G_k(y)$.

The converse of part (d) of the lemma is obvious. □

Now we use the substring graphs to prove the following lemma which shows that properties (a) and (b) of Theorem 1 are sufficient for an automaton to be locally testable.

Lemma 8. Let $M = (Q, \Sigma, \delta, q_0, F)$ be a deterministic automaton. If M has properties (a) and (b) of Theorem 1, then it is locally testable.

Proof. Let $k = |Q|^4$, and $x, y \in \Sigma^*$ such that $G_k(x) = G_k(y)$. We show that $\delta(q_0, x) = \delta(q_0, y)$. Let $G_k(x) = (V, \Sigma, \delta', q'_0, \{q_f\})$, and suppose $G_k(x)$ has r SCC's, m'_1, ..., m'_r, for some $r > 0$. (If $r = 0$, then $L(G_k(x)) = \{x\}$. The lemma is trivially true.) Recall that $G_k(x) = G_k(y)$. Let q'_i, $1 \leq i \leq r$, be the state on m'_i which has an incoming edge from a state outside of m'_i. Notice that, for each SCC m'_i, $1 \leq i \leq r$, q'_i is unique since the SCC's are linearly linked. Let $x_i, y_i \in \Sigma^+$ be the shortest prefixes of x and y, respectively, such that $\delta'(q'_0, x_i) = q'_i$ and $\delta'(q'_0, y_i) = q'_i$, $1 \le i \le r$, in $G_k(x)$. Let $\delta(q_0, x_i) = q_i$, $1 \le i \le r$. For any i, $1 \le i \le r$, let w_i and w'_i be such that x_iw and y_iw', respectively, the longest prefixes of x and y such that $\delta'(q'_0, x_iw_i)$ and $\delta'(q'_0, y_iw'_i)$ are in m'_i. We prove by induction that $\delta(q_0, x_rw_r) = \delta(q_0, y_rw'_r)$ in M.

Let x_0 and y_0 respectively, be the prefixes of x and y which correspond to the spans of the initial transient path on the substring graph. Since $x_0 = y_0$, $\delta(q_0, x_0) = \delta(q_0, y_0) = q_1$ in M.

Suppose that, for any $i < r$, $\delta(q_0, x_i) = \delta(q_0, y_i)$. Let $z_i, z'_i \in \Sigma^+$ be the shortest substrings of x and y, respectively, such that $\delta'(q'_0, x_iw_iz_i) = q'_{i+1}$ and $\delta'(q'_0, y_iw'_iz'_i) = q'_{i+1}$. By Lemma 7-(c), $z_i = z'_i$, because z_i corresponds to the span of the transient path from m'_i to m'_{i+1}. If $\delta(q_0, x_iw_i) = \delta(q_0, y_iw'_i)$, which is proven to be true in the Appendix, then in M we have $\delta(q_0, x_iw_iz_i) = \delta(q_0, x_{i+1}) = \delta(q_0, y_iw'_iz'_i) = \delta(q_0, y_{i+1}) = q_{i+1}$. This proves that $\delta(q_0, x_rw_r) = \delta(q_0, y_rw'_r)$.

Now, we show that $\delta(q_0, x) = \delta(q_0, y)$. If the accepting state q'_f of the substring graph is in m_r, then $x_rw_r = x$ and $y_rw'_r = y$. We have $\delta(q_0, x) = \delta(q_0, y)$. Otherwise, let z_r and z'_r be the suffices of x and y, respectively, such that $x_rw_rz_r = x$ and $y_rw'_rz'_r = y$. Then, by Lemma 7-(b), $z_r = z'_r$. Since $\delta(q_0, x_rw_r) = \delta(q_0, y_rw'_r)$, we have $\delta(q_0, x_rw_rz_r) = \delta(q_0, x) = \delta(q_0, y_rw'_rz'_r) = \delta(q_0, y)$. □

Recall that the proof of Lemma 8 is based on the substring graph of order $k = |Q|^4$. Hence, we have the following:

Corollary 1. Let $M = (Q, \Sigma, \delta, q_0, F)$ be a reduced deterministic automaton. If M is locally testable, then its order of locality is at most $|Q|^4 + 1$.

Proof. From the proof of Lemma 8, we see that if M is a locally testable automaton, then for any pair of words $x, y \in \Sigma^*$ such that $G_k(x) = G_k(y)$, $k = |Q|^4$, we have $\delta(q_0, x) = \delta(q_0, y)$. Lemma 7-(d) says that $G_k(x) = G_k(y)$ iff x and y have the same substring vector of order $k + 1$. We conclude M is $(k + 1)$-testable.

4. Conclusion

We have presented an upper bound of $n^4 + 1$ on the order of the locally testable language of a deterministic automaton having n states. We have the following interesting questions concerning locally testable automata:

(a) Can we improve the upper bound? It seems to us there is a large gap. Is there a practical algorithm which, given a locally testable deterministic automaton, finds k such that the automaton is properly k-testable (i.e., k-testable but not (k-1)-testable)?

(b) What is the upper bound on the size of the semigroup [2,9] of a locally testable automaton? Is there a polynomial time algorithm for computing the semigroup of a locally testable automaton?

By extending the ideas presented in this paper, we believe that (a) can be answered in the affirmative. For question (b), we have reason to suspect that the size of the semigroup of a locally testable automaton may be polynomial in the size of the state set.

References

(1) Aho, A., Hopcroft, J., and Ullman, J., The Design and Analysis of Computer Algorithms, Addison-Wesley (1974).

(2) Brzozowski, J. and Simon, I., Characterizations of locally testable events, *Discrete Mathematics* 4 (1973), pp. 243-271.

(3) Harrison, M., Introduction to Formal Language Theory, Addison Wesley (1978).

(4) Hopcroft, J., and Ullman, J., Introduction to Automata Theory, Languages, and Computation, Addison Weslely (1979).

(5) Hunt, H. and Rosenkrantz, D., Computational parallels between the regular and context-free languages, *SIAM J. COMPUT.*, 7 (1978), pp. 99-114.

(6) Kim, S., McNaughton, R., and McCloskey, R., A polynomial time algorithm for the local testability problem of deterministic finite automata, *Workshop on Algorithms and Data Structures*, (1989).

(7) Martin, R., Studies in Feedback-Shift-Register Synthesis of Sequential Machines, M.I.T. Press, (1969).

(8) McNaughton, R., Algebraic decision procedures for local testability, *Mathematical Systems Theory*, Vol.8 (1974), pp. 60-76.

(9) McNaughton, R. and Papert, S., Counter-free Automata, M.I.T. Press, (1971)

(10) Menon, P., and Friedman, A., Fault detection in iterative logic arrays, *IEEE Trans. on Computers,* C-20 (1971), pp. 524-535.

(11) Zalcstein, Y., Locally testable languages, *Journal of Computer and System Sciences*, 6 (1972), pp. 151-167.

APPENDIX

Let $M = (Q, \Sigma, \delta, q_0, F)$ be a deterministic automaton which has the properties (a) and (b) of Theorem 1. Let $G_k(x) = (V, \Sigma, \delta', q'_0, \{q'_f\})$ be the substring graph of order k of a word $x \in \Sigma^*$, where $k = |Q|^4$. Suppose that $G_k(x)$ has r SCC's $m'_1, m'_2, ...m'_r$, for some $r > 0$. Choose an arbitrary i, $1 \leq i \leq r$. Let q'_i be the state on m'_i that has an incoming edge from outside of m'_i. For any $y \in \Sigma^*$ such that $G_k(y) = G_k(x)$, let x_i and y_i be the shortest prefixes of x and y such that $\delta'(q'_0, x_i) = \delta'(q'_0, y_i) = q'_i$. Let w and w' be such that x_iw and y_iw', respectively, are the longest prefixes of x and y such that $\delta'(q'_0, x_iw)$ and $\delta'(q'_0, y_iw')$ are in m'_i. Then we prove the following lemma.

Lemma A. In automaton M, if $\delta(q_0, x_i) = \delta(q_0, y_i) = q_i$, then $\delta(q_0, x_iw) = \delta(q_0, y_iw')$.

Proof. Let $P' = s'_0\, s'_1 \ldots\, s'_{|w|}$ and $P = s_0\, s_1 \ldots\, s_{|w|}$ be the paths which correspond to span w, respectively, in m'_i and M, where $s'_0 = q'_i$ and $s_0 = q_i$. We show that $\delta(s_0, w) = \delta(s_0, w') = s_{|w|}$. On path P' find a sequence of state pairs $(s'_{i_1}, s'_{j_1}), (s'_{i_2}, s'_{j_2}), \ldots (s'_{i_m}, s'_{j_m})$, for some $m \leq |w|$, $1 \leq i_n \leq |w|$ such that,

(1) $i_1 = 0$, and $j_m = |w|$,

(2) $i_{n-1} < i_n < j_{n-1} < j_n$, $1 < n \leq m$, and

(3) $s'_{i_n} = s'_{j_n}$, $1 \leq n \leq m$.

Recall that m'_i is an SCC and the path P' covers all the edges on m'_i. For s'_{i_1}, which is s'_0, find a state s'_{j_1} of the largest index j_1 such that $s'_{i_1} = s'_{j_1}$. Since the path $s'_{i_1} s'_{i_1+1} \ldots s'_{j_1}$ forms a closed path on m'_i, there exists (unless $j_1 = |w|$, in which case $m = 1$) a state pair s'_{i_2} and s'_{j_2}, $i_1 < i_2 < j_1 < j_2$, such that $s'_{i_2} = s'_{j_2}$ and $s'_{j_2} \neq s'_t$, for any $t > j_2$. Repeating this argument, we see that there exists a sequence of state pairs which meets the conditions above.

Let $(s_{i_1}, s_{j_1}), (s_{i_2}, s_{j_2}), \ldots (s_{i_m}, s_{j_m})$ be the sequence of state pairs in M that corresponds to the sequence on the substring graph. For each pair of states (s'_{i_n}, s'_{j_n}), $1 \leq n \leq m$, from the sequence, consider the path $P'_n = s'_{i_n-k}\, s'_{i_n-k+1} \ldots\, s'_{i_n} \ldots\, s'_{j_n}$. Recall that, in a substring graph of order k, each state which is not on the initial transient path has a unique input span of length k (Lemma 7-(a)). Let w_n be the span that corresponds to the path P'_n, $1 \leq n \leq m$. The fact that $s'_{i_n} = s'_{j_n}$ implies that $w_n = x_n y_n = z_n x_n$, where $|x_n| = k$, $\delta'(s'_{i_n-k}, x_n) = s'_{i_n}$, $\delta'(s'_{i_n}, y_n) = s'_{j_n}$, $\delta'(s'_{i_n-k}, z_n) = s'_{j_n-k}$, $\delta'(s'_{j_n-k}, x_n) = s'_{j_n}$.

Now, for each pair of states (s'_{i_n}, s'_{j_n}), $1 \leq n \leq m$, on the substring graph, consider the pair (s_{i_n}, s_{j_n}) on the automaton M. We have $\delta(s_{i_n-k}, x_n y_n) = \delta(s_{i_n}, y_n) = s_{j_n}$. By Lemma 5,

(a) s_{j_n} is in an SCC m_n and TS-equivalent to s_{i_n} in the reaching component from s_{i_n} to m_n, and

(b) s_{i_n} and s_{j_n} have as a common input span x_n of length k.

Now, by Lemma 6, s_{i_m} is in an SCC m_m and there is a state t_x in m_m which is TS-equivalent to s_{i_1}. Both t_x and s_{i_1} have x_1 as an input span. Notice that, by the TS-equivalence property, since $\delta(s_{i_1}, w) = s_{i_m}$ is in m_m, $\delta(t_x, w)$ is also in m_m. Let $t_1 t_2, ..., t_{|w|}$ be the path corresponding to the transition $\delta(t_x, w)$, where $t_1 = t_x$. Since t_1 and s'_{i_1} have the same input span of length k, t_n and s_{i_n} have the same input span of length k. All SCC's in M are s-local. Hence, no two states in m_m have the same input span of length k by Lemma 2-(c). Recall that $\delta'(s'_{i_1}, w)$ covers all edges in SCC m'_i of the substring graph. This implies that for every transition $\delta'(s'_i, a) = s'_j$ in m'_i, there exists a unique transition $\delta(s_i, a) = s_j$ in SCC m_m in M. It follows that $\delta(t_x, w')$ is in m_m. Actually, $\delta(t_x, w) = \delta(t_x, w')$ because $\delta'(s'_{i_1}, w') = \delta'(s'_{i_1}, w)$.

Now, by the same argument with w′ from y, we get the following:

(a) There is an SCC m_n in M which has a state t_y such that s_{i_1} and t_y are TS-equivalent in the reaching component from s_{i_1} to m_n, and have the same input span of length k as that of s'_{i_1} in m'_i.

(b) $\delta(s_{i_1}, w') = s_{|w'|}$ is in m_n, and

(c) $\delta(t_y, w')$ and $\delta(t_y, w)$ are in m_n.

Now, since M is TS-local, m_m and m_n must be the same SCC. Otherwise, $\delta(s_{i_1}, w') = s_{|w'|}$ is not in the reaching component from s_{i_1} to m_m while $\delta(t_x, w')$ is in m_m which is in the reaching component. The states t_x and s_{i_1} cannot be TS-equivalent. By Lemma 5-(b) and (c), it can be shown that $s_{|w|}$ and $s_{|w'|}$ have the same input span of length k. Hence, if m_n and m_m are the same SCC, $s_{|w|}$ and $s_{|w'|}$ must be the same state. Otherwise, m_m is not s-local because it has two states which have the same input span of length $k \geq |Q|^2$. □

A fast algorithm to decide on simple grammars equivalence

Didier CAUCAL

IRISA , Campus de Beaulieu , F35042 Rennes Cedex , France

Abstract . We present an algorithm to decide on simple grammars equivalence. Its complexity in time and space is polynomial in the valuation and the length of the description of the compared grammars, and exponential if we only take the last parameter into account. From this algorithm, we deduce an optimal upper bound of the number of parallel derivations to be applied to decide on equivalence.

Introduction

The equivalence problem for two classes C and D , and only one if C is equal to D , of context-free grammars is : can we decide for each pair (G,H) of CxD , if the language generated by G is equal to the language generated by H ? The equivalence problem for context-free grammars class is undecidable. On the other hand, the problem is open for twenty years [Gi-Gr 66] for strict deterministic grammars class. But nevertheless, there are numerous algorithms, called equivalence ones, to decide on equality of deterministic context-free grammars subfamilies. We distinguish two families of equivalence algorithms. First one is the Valiant algorithms family (see among others [Va 74] , [Va-Pa 75] , [Oy-Ho 78] , [Oy-Ho-In 80]). For each pair of automata to be compared, these algorithms build a third one which simulates the behaviour of the two others, and does not accept any word if the languages accepted by the first two are equal. The other equivalence algorithms family is that of branching algorithms or Korenjak and Hopcroft algorithms (see among others [Ko-Ho 66] , [Ol-Pn 77] , [Ha-Ha-Ye 79] , [To 84]). For each pair of grammars (or automata) to be compared, these algorithms build a finite tree, rooted by the pair of axioms to be compared, and of which each label is a pair of sets of nonterminal strings. Although branching algorithms method is direct and determines in a good way the studied equivalence properties, Valiant method is preferred. Time and space complexity of an equivalence algorithm generally is double exponential in the length of the description n_G of $G = G_1 \cup G_2$ (for (G_1,G_2) to be compared), and at least simple exponential in n_G and in the valuation v_G of G . We recall that the valuation of a nonterminal string α of G is the shortest length of the terminal strings generated by G from α , and v_G is the greatest valuation of nonterminals of G .

In this paper, we apply the notion of self-proving relation, introduced by Courcelle [Co 83] , to reduce the complexity of Korenjak and Hopcroft algorithm [Ko-Ho 66] . The complexity of the obtained branching algorithm is polynomial in n_G and v_G , and as v_G exponentially depends of n_G , is exponential in n_G . From this algorithm, we deduce an optimal upper bound of the number of parallel derivations, in the number of nonterminals of G and in the maximal valuation of the right members of the rules of G , to be applied on the pair of axioms on which we want to decide on equivalence.

1. Preliminaries

Let E be a set, 2^E is the set of subsets of E, and $\#E$ is the cardinality of E. Let A be a set of 2^F, $\cup A = \{ x \in B \mid B \in A \}$ is the set of F union of A. Let R be a relation included in $E{\times}F$, we would note $x\,R\,y$ instead of $(x,y) \in R$, and the reverse relation of R is $R^{-1} = \{ (y,x) \mid x\,R\,y \}$. By abuse of notation, we would write x instead of $\{x\}$. The image $R(A)$ of a set A by R is $\{y \mid \exists\, x \in A, x\,R\,y \}$. We write $Dom(R) = R^{-1}(F)$ for the domain of R and $Im(R) = R(E)$ for the image of R. The identity relation on E is denoted by $id_E = \{ (x,x) \mid x \in E \}$. The composition of a relation R of $E{\times}F$ with a relation S of $F{\times}G$ is the relation $R \circ S = \{ (x,y) \mid \exists\, z \in F\, (x\,R\,z \wedge z\,S\,y) \}$. The kernel of a relation R is defined by the relation $Ker(R) = R \circ R^{-1} = \{ (x,y) \mid R(x) \cap R(y) \neq \emptyset \}$. A relation R is functional (or is a function) if $Ker(R^{-1}) \subseteq id_F$, that is the image $R(x)$ of an element x of $Dom(R)$ is a singleton; we would write $R : E \rightarrow F$. A function R from E to F is total if $Dom(R) = E$. The reflexive and transitive closure of $R \subseteq E{\times}E$ for the operation $\circ$ is the relation $R^* = \cup\{ R^i \mid i \geq 0 \}$ with $R^0 = id_E$ and $R^{i+1} = R^i \circ R$. The set of positive integers (resp. strict positive) is denoted by $\mathbb{N}$ (resp. by $\mathbb{N}_+$). For each non-null integer n, we have $[n] = \{1,\ldots,n\}$.

Let $(X^*,.)$ be the free monoid generated by a set X of symbols, called letters. Its neutral element is noted ε. A word on X is an element of X^* and a language on X is a set of X^*. Let u be a word on X, we note $|u|$ the length of u, that is the number of occurrences of letters of u. A word v is a prefix of a word u if it exists a word w such that $u = vw$. A language $L \subseteq X^*$ is prefix if each word of L does not admit any other prefix than itself.

Let us consider a binary relation R on X^*. A word u of X^* is irreducible according to R if $R(u) = \emptyset$. We say that R is canonical if it is noetherian (there is no infinite sequence $(u_n)_{n\geq 0}$ such that $u_n\,R\,u_{n+1}$) and confluent ($(R^{-1})^* \circ R^* \subseteq R^* \circ (R^{-1})^*$), which means that for each word u of X^*, there is a single word $u{\downarrow}R$ irreducible such that $u\,R^*\,u{\downarrow}R$. We define the relation $\xrightarrow[R]{}$ equal to $\{ (uxv,uyv) \mid x\,R\,y \}$ of one-step rewriting according to R and $\xleftrightarrow[R]{} = \xrightarrow[R]{} \cup (\xrightarrow[R]{})^{-1}$ of one-step rewriting according to $R \cup R^{-1}$. We say that R is compatible (for the operation .) if $\xrightarrow[R]{}$ is equal to R, and that R is a congruence if R is a compatible equivalence relation.

The relation $\xrightarrow[R]{*} = (\xrightarrow[R]{})^*$ is the rewriting according to R , that is the smallest compatible preorder including R . The smallest congruence including R is the relation $\overset{*}{\underset{R}{\longleftrightarrow}} = (\underset{R}{\longleftrightarrow})^*$. We note

$$|R|_1 = \max\{ |u| \mid u \in Dom(R) \} \quad \text{and} \quad |R|_2 = \max\{ |u| \mid u \in Im(R) \} .$$

A tree T over a set F is a function from $(\mathbb{N}_+)^*$ to F such that the set Dom(T) of nodes of T is closed under prefix (if $uv \in Dom(T)$ then $u \in Dom(T)$) and all the left neighbours (or brothers) of a node exist (if $ui \in Dom(T)$ and $i \in \mathbb{N}$ then $uj \in Dom(T)$ for each j of [i]). The image Im(T) of a tree T is the set of the labels of T . A tree is said to be finite if its domain is so. A node u of T is said to be terminal if $u1 \notin Dom(T)$.

A context-free grammar G over X is a finite relation of XxX^* . We note $N_G = Dom(G)$ the set of nonterminals of G , and $T_G = \{ u(i) \in X\text{-}N_G \mid u \in Im(G) \wedge i \in [|u|] \}$ the set of terminals of G . The sets $N_G{}^*$ and $T_G{}^*$ are respectively the sets of nonterminal strings and terminal strings of G . The language generated by G from a word u of X^* is the set $L(G,u) = \xrightarrow[G]{*}(u) \cap (T_G)^*$. We note $\xrightarrow[G\ g]{*} = (\xrightarrow[G\ g]{})^*$ the left rewriting according to G with $\xrightarrow[G\ g]{} = \{ (uxv,uyv) \mid x\ G\ y \wedge u \in T_G{}^* \}$. The valuation of $u \in X^*$ according to G is $\tau_G(u) = \min(\{ |v| \mid v \in L(G,u) \} \cup \{\infty\})$. The valuation v_G of G is the greatest valuation of nonterminals of G , that is $v_G = \max\{\tau_G(A) \mid A \in Dom(G) \}$. We say that a context-free grammar is under Greibach normal form if $Im(G) \subseteq T_G(N_G)^*$. A grammar G is reduced [resp. prefix] if for each nonterminal A of G , $L(G,A) \neq \emptyset$ [resp. L(G,A) is prefix].

2. Simple grammars equivalence

A simple grammar is a LL(1)-grammar under Greibach normal form.

Definition. A *simple grammar* G over X is a context-free grammar over X under Greibach normal form such that

$$(A\ G\ a\alpha \ \wedge\ A\ G\ a\beta \ \wedge\ a \in X) \Rightarrow (\alpha = \beta) .$$

We recall [Ha 78] that each simple grammar is prefix. To each simple grammar G , we associate the binary relation $\equiv_G$ over $(N_G)^*$ defined by

$$\alpha \equiv_G \beta \quad \text{iff} \quad L(G,\alpha) = L(G,\beta) .$$

The relation $\equiv_G$ is a congruence.

The simple grammars equivalence problem is to decide on the equality $L(G_1,A_1) = L(G_2,A_2)$ for all simple grammars G_1 et G_2 and for all nonterminals A_1 and A_2 of respectively G_1 and G_2 . This problem is inter-reducible to the decidability of $\equiv_G$ for each simple grammar G . Actually, for the necessary condition, we may , after renaming, suppose that $Dom(G_1)$ and $Dom(G_2)$ are distinct and so we take $G = G_1 \cup G_2$. The other way round, to test if $\alpha \equiv_G \beta$, we exclude the trivial case where α or β is the empty word, and we take $G_1 = G \cup \{ (A_1,\gamma) \mid \alpha \underset{G}{\longrightarrow}_g \gamma \}$ and $G_2 = G \cup \{ (A_2,\gamma) \mid \beta \underset{G}{\longrightarrow}_g \gamma \}$ with $A_1,A_2 \in X - (N_G \cup T_G)$. Moreover, we may restrict the study of $\equiv_G$ to each reduced simple grammar G . In fact and for each simple grammar G , we can determine the set $M = \{ A \in Dom(G) \mid L(G,A) \neq \emptyset \}$ of finitely valuated nonterminals of G , and build the reduced simple grammar $H = \{ (A,\alpha) \in G \mid L(G,\alpha) \neq \emptyset \}$ in $O(\#N_G.\#T_G.|G|_2)$; and we have

$$(\alpha \equiv_G \beta) \Leftrightarrow ((\alpha,\beta \notin M^*) \vee (\alpha,\beta \in M^* \wedge \alpha \equiv_H \beta)) .$$

Henceforth G is a reduced simple grammar and we don't precise anymore the symbol G in notations. A pair (α,β) of nonterminal strings is incomparable, and we note it $\alpha <> \beta$, if $L(G,\alpha)$ differ from $L(G,\beta)$ at a first terminal letter, that is

$$\alpha <> \beta \ \text{ iff } \ \exists\, a \in T ,\ \neg((\exists\, \gamma, \alpha \longrightarrow_g a\gamma) \Leftrightarrow (\exists\, \delta, \beta \longrightarrow_g a\delta)) .$$

The most natural means to study equivalence of a pair of nonterminal strings is to test whether the pair is incomparable and either to transfer this test to pairs obtained by left parallel derivation. The one-step left parallel derivation with incomparability test is a mapping T_A from N^*xN^* into $2^{N^*xN^*}$, introduced by Harrison [Ha 78] and defined for all α and β of N^* by

$$T_A(\alpha,\beta) = \{(\varepsilon,\varepsilon)\} \text{ if } \alpha=\beta=\varepsilon$$
$$T_A(\alpha,\beta) = \emptyset \text{ if } \alpha <> \beta$$
$$T_A(\alpha,\beta) = \{ (\gamma,\delta) \mid \exists a \in T, (\alpha \longrightarrow_g a\gamma \wedge \beta \longrightarrow_g a\delta) \} \text{ in other cases.}$$

The mapping T_A is valid in the sense that

$$\alpha \equiv \beta \Leftrightarrow \emptyset \neq T_A(\alpha,\beta) \subseteq \equiv .$$

To iterate the mapping T_A, we extend it to each set E of N^*xN^* as follows

$$T_A(E) = \emptyset \text{ if } \exists (\alpha,\beta) \in E,\ T_A(\alpha,\beta) = \emptyset$$
$$T_A(E) = \cup\{ T_A(\alpha,\beta) \mid (\alpha,\beta) \in E \} \text{ else.}$$

The study of equivalence of a pair by iterating T_A finds expression in the following proposition.

Proposition 2.1 . $\alpha \equiv \beta \Leftrightarrow \forall n,\ T_A{}^n(\alpha,\beta) \neq \emptyset$.

Proof.

$\Rightarrow$: By induction and because of the validity of T_A.

$\Leftarrow$: If α is not equivalent to β then there exists a word u of minimal length belonging to only one of the languages $L(G,\alpha)$ and $L(G,\beta)$. By symmetry of α and β, we may assume that $u \in L(G,\alpha) - L(G,\beta)$. Let v be the greatest prefix of u such that $\exists \delta \in N^*,\ \beta \xrightarrow{*}_g v\delta$. According to the definition of u, there exists $(\gamma,\delta) \in T_A{}^{|v|}(\alpha,\beta)$ with $\alpha \xrightarrow{*}_g v\gamma$. According to the definition of v, $\gamma <> \delta$ so $T_A{}^{|v|+1}(\alpha,\beta) \neq \emptyset$. ♦

Proposition 2.1 gives a semi-decision procedure of the non-equivalence. Therefore, it implies that all pairs of nonterminal strings of a set closed by T_A are equivalent.

Definition. A binary relation R over N^* is *closed by T_A* if

$$\emptyset \neq T_A(R) \subseteq R .$$

Corollary 2.2 . If R is closed by T_A then R is included in $\equiv$.

Proof.

If $\emptyset \neq T_A(R) \subseteq R$ then by induction on n, $\emptyset \neq T_A{}^n(R) \subseteq R$ and by prop. 2.1, $R \subseteq \equiv$. ♦

A more general condition than closure by T_A has been given by Courcelle [Co 83] .

Definition. A binary relation R over N^* is *self-proving* if

$$\emptyset \neq T_A(R) \subseteq \overset{*}{\underset{R}{\longleftrightarrow}} .$$

Before we extend corollary 2.2 to self-proving relations, we establish that each element of T_A, this last one applied to rewriting according to R, is obtained by rewriting according to $R \cup T_A(R)$.

Lemma 2.3 . Given a relation R such that $T_A(R) \neq \emptyset$, we have

$$\emptyset \neq T_A(\overset{*}{\underset{R}{\longrightarrow}}) \subseteq \overset{*}{\underset{S}{\longrightarrow}} \text{ with } S = R \cup T_A(R) .$$

Proof.

For $T_A(R) \neq \emptyset$ and $S = R \cup T_A(R)$, we verify by induction on n that

$$\emptyset \neq T_A(\overset{n}{\underset{R}{\longrightarrow}}) \subseteq \overset{*}{\underset{S}{\longrightarrow}} .$$

♦

It follows that the self-provability of a relation R corresponds to the closure by T_A of the smallest congruence including R.

Proposition 2.4 . A relation R is self-proving if and only if $\overset{*}{\underset{R}{\longleftrightarrow}}$ is closed by T_A.

Proof.

$\Rightarrow$: Let us consider a self-proving relation R, that is $\emptyset \neq T_A(R) \subseteq \overset{*}{\underset{R}{\longleftrightarrow}}$.

As $T_A(R^{-1}) = (T_A(R))^{-1}$, and by lemma 2.3 , we have

$$\emptyset \neq T_A(\overset{*}{\underset{R}{\longleftrightarrow}}) \subseteq \overset{*}{\underset{S}{\longleftrightarrow}} \text{ with } S = R \cup T_A(R) .$$

So $\overset{*}{\underset{S}{\longleftrightarrow}} = \overset{*}{\underset{R}{\longleftrightarrow}}$, hence $\overset{*}{\underset{R}{\longleftrightarrow}}$ is closed by T_A.

$\Leftarrow$: Immediate. ♦

From corollary 2.2 and proposition 2.4 results the following corollary.

Corollary 2.5 . Every self-proving relation is included in $\equiv$.

We will decide on $\alpha \equiv \beta$ by extraction of a self-proving relation R such that $\alpha \overset{*}{\underset{R}{\longleftrightarrow}} \beta$, and the optimization of the decision is obtained by extraction of such a relation, minimal in regard to inclusion.

Let us take a mapping Val from N into T^* which, to each nonterminal A associates a word Val(A) of L(G,A) of minimal length, that is such that $|Val(A)| = \tau(A)$. Research of a self-proving relation, of which closure by congruence includes the pair to be studied, is obtained by splitting of the pairs of nonterminal strings. This splitting is carried out by the mapping T_B from $N^+ x N^+$ into $2^{N^* x N^*}$ defined for all A and B of N and for all α and β of N^* by

$$T_B(A\alpha,B\beta) = (T_B(B\beta,A\alpha))^{-1} \text{ if } \tau(A) < \tau(B)$$
$$T_B(A\alpha,B\beta) = \{ (A,B\gamma) , (\gamma\alpha,\beta) \} \text{ if } (\tau(B) \le \tau(A)) \wedge (A \overset{*}{\underset{g}{\longrightarrow}} Val(B)\gamma) \wedge (\gamma \in N^*)$$
$$T_B(A\alpha,B\beta) = \emptyset \text{ in other cases.}$$

As T_A , the mapping T_B is valid, that is, for all α and β of N^+

$$\alpha \equiv \beta \Leftrightarrow \emptyset \neq T_B(\alpha,\beta) \subset \equiv .$$

The decision algorithm of $\equiv$, defined in the following section, extracts self-proving relations under the following form :

Definition. A binary relation R over N^* is said *fundamental* if it verifies the three following conditions :

(i) $Dom(R) \subseteq N$ and $Im(R) \subseteq (N\text{-}Dom(R))^*$

(ii) R is functional : if $A\,R\,\alpha$ and $A\,R\,\beta$ then $\alpha = \beta$

(iii) $R \subseteq Ker(\tau)$: if $A\,R\,\alpha$ then $\tau(A) = \tau(\alpha)$.

Lemma 2.6 . Given a fundamental relation R , we have

$$\#R < \#N \text{ and } \underset{R}{\longrightarrow} \text{ is canonical .}$$

Proof.

Let R be a fundamental relation. From (i) , Dom(R) is included in N and from (ii) , $\#R < \#N$. By (i) , every rewriting according to R from $\alpha \in N^*$ is at most $|\alpha|$ long, so $\underset{R}{\longrightarrow}$ is noetherian. As Dom(R) $\subseteq$ N and R is functional, $\underset{R}{\longrightarrow}$ is confluent. Finally, $\underset{R}{\longrightarrow}$ is canonical. ♦

3. Equivalence algorithm

We define a branching algorithm to decide on $\equiv$. For each (α,β) of N^*xN^*, this algorithm expands from T_A or T_B and in lexical order a tree over N^*xN^* rooted by (α,β). Every reflexive pair labeled node is terminal. A pair cannot be expanded if its image by T_A or T_B is empty, or its words have different valuations. The construction of the tree halts if a pair to be expanded cannot be expanded, and in this case α is not equivalent to β, or if all terminal nodes of the tree are labelled by reflexive pairs, and in this other case α is equivalent to β. A pair is expanded by means of T_A if one of its words is a letter. Before expanding, every pair (λ,μ) is reduced in $(\lambda{\downarrow}R,\mu{\downarrow}R)$ by the relation R of pairs which have been expanded by T_A.

Given nonterminal strings α and β, we put down

$$(T_{\alpha,\beta},u_{\alpha,\beta},R_{\alpha,\beta}) = \mathrm{Const}(\{(\varepsilon,(\alpha,\beta))\},\varepsilon,\emptyset)$$

where Const is a recursive procedure, input-output parameter of which is (T,u,R), with T a tree, u a node of T, and R a fundamental relation over N^*. The procedure Const will be now defined. We precise that the symbol $\leftarrow$ is the assignment one and that the procedure Stop halts execution.

Procedure Const(T,u,R)

a) If the words of the current label have different valuation then we halt.

$(\lambda,\mu) \leftarrow T(u)$
If $\tau(\lambda) \neq \tau(\mu)$ **Then** Stop **Endif**

b) We compute the pair of normal forms of $T(u)$ according to R and we suppress the greatest common left factor. If the resulting pair differs from $T(u)$, we add it to the tree.

If $\lambda \neq \mu$ **Then**
    $(\lambda',\mu') \leftarrow (\lambda{\downarrow}R,\mu{\downarrow}R)$
    If $\lambda' \neq \mu'$ ***Then*** $(\gamma\rho,\gamma\eta) \leftarrow (\lambda',\mu')$ with $|\gamma|$ max. ; $(\lambda',\mu') \leftarrow (\rho,\eta)$ ***Endif***
    If $(\lambda',\mu') \neq (\lambda,\mu)$ ***Then***
        $T \leftarrow T \cup \{(u1,(\lambda',\mu'))\}$; $u \leftarrow u1$; $(\lambda,\mu) \leftarrow (\lambda',\mu')$
    Endif
Endif

c) If the current label T(u) is reflexive then its node is terminal and we pass to the following node (in lexical order) if it exists.

If $\lambda = \mu$ **Then**

If u has no successor in lexical order

Then Stop

Else Const(T,v,R) for v the following node of u in lexical order

Endif

Endif

d) If the words of T(u) are not letters then we apply T_B to T(u) else we add T(u) to R and apply T_A to T(u). If the mapping fails we halt else we go on with the leftmost "daughter" pair.

If $\min(|\lambda|,|\mu|) > 1$ **Then** $E \leftarrow T_B(\lambda,\mu)$ **Else**

$E \leftarrow T_A(\lambda,\mu)$

If $|\lambda| > 1$ *Then* $(\lambda,\mu) \leftarrow (\mu,\lambda)$ *Endif*

$R \leftarrow \{ (A,\alpha\downarrow\{(\lambda,\mu)\}) \mid A\, R\, \alpha \} \cup \{(\lambda,\mu)\}$

Endif

If $E = \emptyset$ **Then** Stop **Else**

$T \leftarrow T \cup \{ (ui,e_i) \mid i \in [\#E] \}$ with $\{e_1,\ldots,e_{\#E}\} = E$

Const(T,u1,R)

Endif

End of procedure

We state in the next chapter that this algorithm always halts and that $\alpha \equiv \beta$ if and only if $T_{\alpha,\beta}(u_{\alpha,\beta})$ is a reflexive pair. Figures 3.1 and 3.2 give some applications of the algorithm ; operations T_A, T_B and reduction one are respectively represented by a vertical bar, a double vertical bar and an arrow.

Let the following simple grammar : G = { (A,a) , (A,bABBBA) , (B,aA) , (B,bBBBAB) } .

To apply the algorithm to (AB,BA) gives the following tree $T_{AB,BA}$:

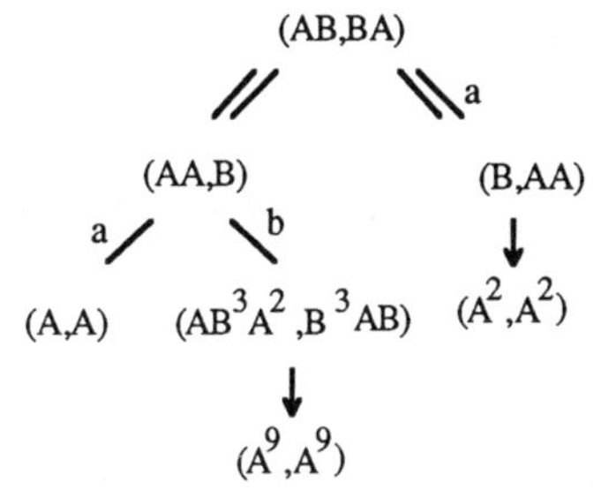

(AB,BA)
// ; \\ a

(AA,B) ; (B,AA)
a / ; \ b ; ↓

(A,A) ; (AB^3A^2, B^3AB) ; (A^2,A^2)

↓

(A^9,A^9)

Moreover $u_{AB,BA} = 21$ and $R_{AB,BA} = \{ (B,AA) \}$.

So $T_{AB,BA}(u_{AB,BA}) = (A^2,A^2)$, hence AB ≡ BA .

Fig 3.1 . A case of equivalence

Let the following simple grammar :

G = { (A,a) , (A,bACB) , (A,cBCAB) , (B,a) , (B,bBCA) , (B,cADB) , (C,aB) , (D,aC) } .

To apply the algorithm to (A,B) gives the following tree $T_{A,B}$:

(A,B)
a / ; | b ; \ c

(ε,ε) ; (ACB,BCA) ; (BCAB,ADB)

↓ ; ↓

(BCB,BCB) ; (CBB,DB)

// ; \\ aa

(CB,D) ; (BB,BB)

| a

(BB,C)

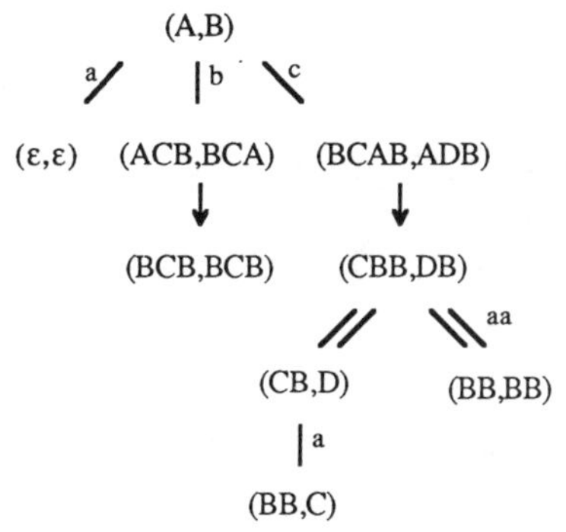

Moreover $u_{A,B} = 3111$ and $R_{A,B} = \{ (A,B) , (D,CB) \}$.

So $T_{A,B}(u_{A,B}) = (BB,C)$, hence A is not equivalent to B .

Fig 3.2 . A case of non-equivalence .

4. Validity and complexity

We need to verify that the recursive function Const is valid, always halts, and that $\alpha \equiv \beta$ if and only if all labels of leaves of $T_{\alpha,\beta}$ are reflexive.

Lemma 4.1 . For each α and β of N^*, $T_{\alpha,\beta}$ exists and we have

$$\alpha \equiv \beta \iff T_{\alpha,\beta}(u_{\alpha,\beta}) \in id_{N^*} .$$

Proof.

Let us consider the sequence $(T_i,u_i,R_i)_{i\geq 0}$ of successive call parameters of Const with

$$T_0 = \{ (\varepsilon,(\alpha,\beta)) \} \ , \ u_0 = \varepsilon \text{ and } R_0 = \emptyset .$$

a) We verify by induction on i that the relation R_i is fundamental. By lemma 2.6 and for each i, $\#R_i < \#N$. Therefore the number of nodes u_i, labels of which have been expanded by T_A is finite, and we deduce that the sequence $(T_i,u_i,R_i)_{i\geq 0}$ is finite.

b) If $\alpha \equiv \beta$ then by validity of T_A and of T_B, we show by induction on $|u|$ for $u \in Dom(T_{\alpha,\beta})$, that $T_{\alpha,\beta}(u) \in \equiv$. From the halt conditions of the algorithm, we deduce that $T_{\alpha,\beta}(u_{\alpha,\beta})$ is a reflexive pair.

c) If $T_{\alpha,\beta}(u_{\alpha,\beta}) \in id_{N^*}$ all terminal nodes of $T_{\alpha,\beta}$ have reflexive labels. Let R be the set of labels of $T_{\alpha,\beta}$ already expanded by T_A. By induction on $i \geq 0$, we have $R_i \subseteq \overset{*}{\underset{R}{\longleftrightarrow}}$. Consequently, for each node u of $T_{\alpha,\beta}$ and by inverse induction on $|v|$ such that $uv \in Dom(T_{\alpha,\beta})$, we have $T_{\alpha,\beta}(uv) \in \overset{*}{\underset{R}{\longleftrightarrow}}$. Thus $\forall\, u \in Dom(T_{\alpha,\beta})$, $T_{\alpha,\beta}(u) \in \overset{*}{\underset{R}{\longleftrightarrow}}$. Especially $\alpha \overset{*}{\underset{R}{\longleftrightarrow}} \beta$ and $\emptyset \neq T_A(R) \subseteq \overset{*}{\underset{R}{\longleftrightarrow}}$, that is R is self-proving. By corollary 2.5, $R \subseteq \equiv$ hence $\overset{*}{\underset{R}{\longleftrightarrow}} \subseteq \equiv$, so $\alpha \equiv \beta$. ◆

We have yet to evaluate the complexity of the algorithm. From lemma 2.6, the number of nodes, labels of which have been expanded by T_A is at most #N and similarly for the nodes expanded by T_B. So the size $\#Dom(T_{\alpha,\beta})$ of the equivalence tree $T_{\alpha,\beta}$ is in $O(\#T.\#N)$. The construction of $R_{\alpha,\beta}$ is in $O(v.(\#N)^2)$ where v is the grammar valuation. As $\{ \tau(\alpha) \mid \alpha \in Im(G)\} \leq v.|G|_2$, the maximal valuation of a label of $T_{\alpha,\beta}$ is in $O(\max(\tau(\alpha),\tau(\beta),v.|G|_2))$. The mapping Val can be defined in $O(\#G.|G|_2 + \#N.v)$ and the cost of a mapping T_B is in $O(v)$. As v is in $O(|G|_2^{\#N})$, we finally get the following theorem.

Theorem 4.2 . *Simple grammars equivalence problem is decidable and there is a decision algorithm with a polynomial, in the length of description and the valuations of compared grammars, and exponential, in the length of description only, time and space complexity.*

We recall that the algorithms of [Ho-Ko 66] and [Ha 78] to decide on the simple grammars equivalence have exponential, in the length of description and the valuations of compared grammars, and double exponential, when you do not take the valuation into account, time and space complexity. We remark that the theorem may be establish with the removing of the reduction operation and, on the opposite, with the strenghthening of the halt condition of the pair expanding : a pair is a leaf if it is reflexive or if it (or its symmetric) has already been expanded in the tree.

5. Optimal bound of decision

The algorithm gives a way to determine from the length of description of G and the length of the nonterminal strings α and β, a bound $b_{\alpha,\beta}$ of the number of T_A to be applied to (α,β) to decide on the equivalence of α and β, that is to say

$$\alpha \equiv \beta \iff T_A^{b\alpha,\beta}(\alpha,\beta) \neq \emptyset .$$

The total function Div : $N^* x N^* \to \mathbb{N} \cup \{\infty\}$, called divergence function, is defined by

$$\mathrm{Div}(\alpha,\beta) = \min(\{ n \mid T_A^{n}(\alpha,\beta) = \emptyset \} \cup \{\infty\}) .$$

We would like to determine an optimal upper bound $b_{\alpha,\beta}$ of $\mathrm{Div}(\alpha,\beta)$ when α is not equivalent to β, from #N, $e = \max\{ \tau(\alpha) \mid \alpha \in \mathrm{Im}(G) \} - 1$ and $\tau(\alpha,\beta) = \min(\tau(\alpha),\tau(\beta))$.

Example 5.1 . Let us consider an integer $n \geq 2$ and the following simple grammar G :

$$G = \{ (A_i,a) \mid i \in [n] \} \cup \{ (A_i,b\lambda A_{i+1}) \mid i \in [n-1] \} \cup \{ (A_n,b\lambda) \}$$

with $\lambda \in N^* = \{A_1,\ldots,A_n\}^*$. We let $\alpha = A_1\gamma$ and $\beta = A_2\gamma$ with $\gamma \in N^*$. We have $\#N = n$, $e = \tau(\lambda)+1$ and $\tau(\alpha) = \tau(\beta) = \tau(\gamma)+1$. We verify :

$$\mathrm{Div}(\alpha,\beta) = (1+\tau(\lambda))(n-1) + \tau(\gamma) + 1 = e(\#N-1) + \tau(\alpha,\beta) .$$

First, we give the basic properties of Div .

Lemma 5.2 . For all α, β, γ, δ of N^* such that $\gamma \equiv \delta$, we have

(i) $1 \leq \mathrm{Div}(\alpha,\beta) \leq \mathrm{Div}(\alpha\gamma,\beta\delta) \leq \mathrm{Div}(\alpha,\beta) + \tau(\gamma) = \mathrm{Div}(\gamma\alpha,\delta\beta)$

(ii) $\min(\mathrm{Div}(\alpha,\beta),\mathrm{Div}(\beta,\gamma)) \leq \mathrm{Div}(\alpha,\gamma)$.

The proof is of no difficulty (we will remind that G is prefix). From this lemma, we deduce the divergence of the mapping T_B .

Lemma 5.3 . For all $(\lambda,\mu) \in T_B(\alpha,\beta)$ with $\tau(\alpha) = \tau(\beta)$, we have

$\tau(\lambda,\mu) \leq \tau(\alpha,\beta)$ and $\mathrm{Div}(\alpha,\beta) \leq \mathrm{Div}(\lambda,\mu) + \tau(\alpha,\beta) - \tau(\lambda,\mu)$.

We extend the divergence to all binary relation R over N^* by

$$\mathrm{Div}(R) = \min\{ \mathrm{Div}(\alpha,\beta) \mid \alpha R \beta \} ,$$

and similarly, we deduce the divergence of the reduction.

Lemma 5.4 . For each binary relation R over N^* and for all $\alpha \overset{*}{\underset{R}{\longleftrightarrow}} \lambda$ and $\beta \overset{*}{\underset{R}{\longleftrightarrow}} \mu$, we have : $\quad \mathrm{Div}(R \cup \{(\alpha,\beta)\}) \leq \mathrm{Div}(\lambda,\mu)$.

Proof.

We extend the one-step rewriting $\underset{R}{\longrightarrow}$ over $(N^*)^2 x (N^*)^2$ by

$$(\alpha,\beta) \underset{R}{\longrightarrow} (\lambda,\mu) \ \text{ if } \ (\alpha \underset{R}{\longrightarrow} \lambda \wedge \beta = \mu) \vee (\alpha = \lambda \wedge \beta \underset{R}{\longrightarrow} \mu) ,$$

and by induction on $n \geq 0$ and for $(\alpha,\beta) \overset{n}{\underset{R}{\longleftrightarrow}} (\lambda,\mu)$, we show that

$$\mathrm{Div}(R \cup \{(\alpha,\beta)\}) \leq \mathrm{Div}(\lambda,\mu) .$$

♦

From now on, α and β are nonterminal strings, and we note $T(u)$ instead of $T_{\alpha,\beta}(u)$.

Let $(T_i,u_i,R_i)_{0\leq i\leq p}$ be the finite sequence of parameters of the recursive procedure Const with $T_0 = \{ (\varepsilon,(\alpha,\beta)) \}$, $u_0 = \varepsilon$ and $R_0 = \emptyset$. We define the set

$$U = \{ u_i \mid 0 \leq i \leq p \wedge \#R_i < \#R_{i+1} \}$$

of nodes of $T_{\alpha,\beta}$, labels of which have been expanded by T_A , and for each node u of $T_{\alpha,\beta}$ and for the lexical order $\leq_{lex}$ over $\mathbb{N}^*$, we let

$$R_u = \{ v \in U \mid v \leq_{lex} u \} ,$$

and $\quad D_u = \{ uv \in \mathrm{Dom}(T_{\alpha,\beta}) \mid \forall w ((1 \leq_{lex} w \leq_{lex} v) \wedge (1 \neq w \neq u) \wedge (uw \notin U) \}$.

From the two previous lemmas, we get an upper bound of the divergence of every partial tree of $T_{\alpha,\beta}$ of which we have obtained the nodes by T_B and by reduction.

Lemma 5.5 . $\forall u \in \mathrm{Dom}(T_{\alpha,\beta}) - U$, $\forall v \in D_u$,

$\tau(T(v)) \leq \tau(T(u))$ and $\mathrm{Div}(R_u \cup \{T(u)\}) \leq \mathrm{Div}(T(v)) + \tau(T(u)) - \tau(T(v))$.

Proof.

By induction on $|v|-|u| \geq 0$.

♦

Lemma 5.6 . $\forall\, u \in U$, $\forall\, v \in D_u - \{u\}$,

$\tau(T(v)) \le e$ and $Div(R_u) \le Div(T(v)) + e - \tau(T(v))$.

Proof.

We let $v = wv'$ with $|w| = |u|+1$. As $T(w) \in T_A(T(u))$ and $u \in U$, we have $\tau(T(w)) \le e$ and $Div(T(u)) \le Div(T(w)) - 1$. The two following cases occur :

Either $w \in U$. From the definition of D_u , we have $v = w$ and we deduce the inequalities of the lemma.

Or $w \notin U$. So $v \in D_w$ and the lemma 5.6 comes from lemma 5.5 . ◆

Lemma 5.6 can be extended to every tree $T_{\alpha,\beta}$ when $\min(|\alpha|,|\beta|) = 1$.

Lemma 5.7 . If $T(\varepsilon) \in U$ then for each $u \in Dom(T_{\alpha,\beta}) - \{\varepsilon\}$, we have

$Div(\alpha,\beta) \le Div(T(u)) + (\#N-1)e - \tau(T(u))$.

Proof.

Let $\{u_1,\ldots,u_p\} = U$ with $p = \#U$ and $u_i \le_{lex} u_{i+1}$ for $1 \le i < p$. With lemma 5.6, we establish by induction on $i \in [p]$ and for each u of $D_{u_i} - \{u_i\}$, that

$Div(\alpha,\beta) \le Div(T(u)) + (\#R_{u_i})e - \tau(T(u))$.

Lemma 5.7 comes from lemma 5.6 and the following equality :

$Dom(T_{\alpha,\beta}) - \{\varepsilon\} = \bigcup\{ D_{u_i} - \{u_i\} \mid i \in [p] \}$. ◆

As a function of #N , e and $\tau(\alpha,\beta)$, the decision bound $b_{\alpha,\beta}$ is the following :

Definition. For all $\alpha , \beta \in N^*$, we define

$$b_{\alpha,\beta} = \tau(\alpha,\beta) + 1 \quad \text{if } \tau(\alpha) \ne \tau(\beta)$$

$$b_{\alpha,\beta} = (\#N-1)e + \tau(\alpha) \quad \text{if } \tau(\alpha) = \tau(\beta) .$$

For all α and β of N^+ , we have $Div(\alpha,\alpha\beta) = \tau(\alpha) + 1 = b_{\alpha,\alpha\beta}$. So the optimality of the decision bound results from example 5.1 .

Proposition 5.8 . If α is not equivalent to β then $Div(\alpha,\beta) \leq b_{\alpha,\beta}$.

Proof.

Let α and β be non-equivalent non-terminals .

a) Either $\tau(\alpha) \neq \tau(\beta)$: By symmetry of α and β , we may assume that $\tau(\alpha) < \tau(\beta)$. Let u be the greatest prefix of $Val(\alpha)$ such that $\exists\, \gamma \in N^*\ (\beta \xrightarrow{*}_g u\gamma)$. Then

$$Div(\alpha,\beta) \leq \min(\tau(\alpha),|u|) + 1 \leq b_{\alpha,\beta} .$$

b) Or $\tau(\alpha) = \tau(\beta)$ and $\min(|\alpha|,|\beta|) = 1$: As $T(u_{\alpha,\beta})$ is a pair which either cannot be expanded by T_A or T_B , or has words of distinct valuations, we establish

$$Div(T(u_{\alpha,\beta})) \leq \tau(T(u_{\alpha,\beta})) + 1 .$$

If $u_{\alpha,\beta} = \varepsilon$ then $Div(\alpha,\beta) = 1 \leq b_{\alpha,\beta}$, and else we may apply lemma 5.7 and get

$$Div(\alpha,\beta) \leq (\#N-1)e + Div(T(u_{\alpha,\beta})) - \tau(T(u_{\alpha,\beta})) \leq (\#N-1)e + 1 \leq b_{\alpha,\beta} .$$

c) From a) and b) , we set by induction on $\tau(\alpha,\beta)$: $Div(\alpha,\beta) \leq b_{\alpha,\beta}$. ♦

From the definition of Div and proposition 5.8 , we are able to establish the following semi-decision procedure :

$$\alpha \equiv \beta \iff T_A^{\,b_{\alpha,\beta}}(\alpha,\beta) \neq \emptyset ,$$

that is to say

Theorem 5.9 . *Given nonterminal strings u and v of a reduced simple grammar G , and the number $b_{u,v}$ of the previous definition, G generates the same terminal language from u and v if and only if the result of the parallel derivation T_A , applied $b_{u,v}$ times from (u,v) , is not the empty set.*

To compare this bound $b_{\alpha,\beta}$ to already known ones, we must restrict our study to the case where $|G|_2 \leq 3$, $|\alpha| = |\beta| = 1$ and $\tau(\alpha) = \tau(\beta)$. In that case, we have $Div(\alpha,\beta) \leq 2v(\#N)$ instead of $v(\#N^{(v+1)(v+3)})$ in [Ko-Ho 66] , $v(\#N^{2(v+1)})$ in [Wo 73] , $2v(\#N)^2$ in [Bu 73] .

Conclusion

The algorithm, described in this paper, has been generalised [Ca 87] to decide on the equivalence of deterministic stateless pushdown automata which accept on stack letters. The complexity of this algorithm remains polynomial in function of the length of description and the valuation, and exponential if we only take the length of description into account, and this instead the double exponential complexity of [Oy-Ho 78] .

References

Bu 73 — P. Butzbach "Sur l'équivalence des grammaires simples" , Actes des premières journées d'informatique théorique. Langages algébriques. Bonascre. pp 223-245 .

Ca 87 — D. Caucal "How to improve branching algorithms for deciding on grammars equivalence" , Report INRIA 618 .

Co 83 — B. Courcelle "An axiomatic approach to the KH algorithms" , Math. Systems Theory 16 , pp 191-231 .

Gi-Gr 66 — S. Ginsburg , S.A. Greibach "Deterministic context-free languages" , Information and Control 9 , pp 602-648 .

Ha 78 — M.A. Harrison "Introduction to formal language theory" , Addison-Wesley .

Ha-Ha Ye 79 — M.A. Harrison , I.M. Havel , A. Yeduhaï "On equivalence of grammars through transformation trees" , TCS 9 , pp 191-231 .

Ko-Ho 66 — A.J. Korenjak , J.E. Hopcroft "Simple deterministic languages" , Seventh Annual IEEE Switching and Automata Theory Conference , pp 36-46 .

Ol-Pn 77 — T. Olshansky , A. Pnueli "A direct algorithm for checking equivalence of LL(k) grammars" , TCS 4 , pp 321-349 .

Oy-Ho 78 — M. Oyamaguchi , N. Honda "The decidability of equivalence for deterministic stateless pushdown automata" , Information and Control 38 , pp 367-376 .

Oy-Ho In 80	M. Oyamaguchi , N. Honda , Y. Inagaki "The equivalence problem for real-time strict deterministic languages" , Information and Control 45 , pp 90-115 .
To 84	E. Tomita "An extended direct branching algorithm for checking equivalence of deterministic pushdown automata" , TCS 32 , pp 87-120 .
Va 74	L.G. Valiant "The equivalence problem for deterministic finite-turn pushdown automata" , Information and Control 25 , pp 123-153 .
Va-Pa 75	L.G. Valiant , M.S. Paterson "Deterministic one-counter automata" , JCSS 10 , pp 340-350 .
Wo 73	D. Wood "Some remarks on the KH algorithms for s-grammars" , BIT 13 , pp 476-489 .

COMPLEXITY OF THE PARALLEL GIVENS FACTORIZATION ON SHARED MEMORY ARCHITECTURES

Michel COSNARD, El Mostafa DAOUDI and Yves ROBERT[1]

LIP - IMAG
Ecole Normale Supérieure de Lyon
46, allée d'Italie
69364 - Lyon Cedex 07, France
e-mail: cosnard@frensl61.bitnet

ABSTRACT

We study the complexity of the parallel Givens factorization of a square matrix of size n on shared memory multicomputers with p processors. We show how to construct an optimal algorithm using a greedy technique. We deduce that the time complexity is equal to:

$$T_{opt}(p) = \frac{n^2}{2p} + p + o(n) \quad \text{for } 1 \leq p \leq \frac{n}{2+\sqrt{2}}$$

and that the minimum number of processors in order to compute the Givens factorization in optimal time T_{opt} is equal to $p_{opt} = \frac{n}{2+\sqrt{2}}$.

These results complete previous analysis presented in the case where the number of processors is unlimited.

Keywords: Parallel linear algebra, complexity of parallel algorithms, orthogonal factorization, Givens rotations

1. INTRODUCTION

Computing the orthogonal decomposition of a mxn matrix is a classical problem in scientific computation. Two well known methods are available for solving such a problem: Householder reduction and Givens rotations. The parallelization of Givens' method gives rise to very interesting algorithmic problems.

Such a factorization requires mn - n(n+1)/2 steps on a sequential machine, each step being the time necessary to achieve a Givens rotation. This O(m*n) number of steps to obtain the factorization motivates its parallelization.

Section 2 is devoted to some definitions and the short description of previous results mainly in

1 Support by the French GRECO C3 is aknowledged.

the case when $p = \lfloor m/2 \rfloor$, together with a brief review of existing algorithms. In section 3, we study the generalization of the greedy algorithm in the case of limited parallelism. In section 4 we derive a lower bound and construct in section 5 an optimal algorithm for p=n/4. The last part is devoted to the presentation of a general theorem.

2. DEFINITIONS AND PREVIOUS RESULTS

Throughout the paper, A is an m by n rectangular matrix. We assume that p Givens rotations can be performed simultaneously. We denote lim f(x) the limit of a function f as x goes to infinity, and we recall that f(x) = o(x) means that f(x)/x tends to zero as x goes to infinity. We let R(i,j,k), $i \neq j$, $1 \leq i,j \leq m$ and $1 \leq k \leq n$, denote the rotation in plane (i,j) which annihilates the element A(i,k).

Indeed, parallel versions for SIMD or MIMD computers of the Givens transformations algorithm have been introduced: see Lord et al. [8], Sameh and Kuck [11] and the survey papers of Heller [6]. In this paper we shall concentrate on synchronous parallel algorithms: we call time step the duration of a Givens rotation and assume that such a transformation takes a time independent of the size of the vectors.

Clearly at each step no more than $\lfloor m/2 \rfloor$ elements can be simultaneouly annihilated. A rather complete analysis of the case $p=\lfloor m/2 \rfloor$ can be found in the literature. In this section we shall make a review of the existing results. An algorithm M requiring T steps to achieve the Givens factorization of A is represented by M = (M(1),...,M(T)) where for $1 \leq t \leq T$, M(t) is a group of r(t) disjoint rotations ($r(t) \leq p$), which can be performed simultaneously. M is used in order to construct a sequence A(t) such that : A(0)=A and for $1 \leq t \leq T$, A(t) is obtained by applying in parallel the rotations in M(t) to A(t-1). For short we use (M,T,R) where R = r(1)+...+r(T).

To describe informally the parallel algorithms, let us fix m = 13 and n = 6. The table of figure 1 illustrates Sameh and Kuck's annihilation scheme [11]. An integer r is entered when zeros are created at the r-th step. We do not specify completely each rotation. A zero can be created in a row using any other row with the same number of annihilated elements. For instance at step 3, it can perform simultaneously the rotations R(11,10,1) and R(13,12,2). R(13,12,2) is the only possible choice, but R(11,10,1) can be replaced by R(11,x,1) for any $x \leq 10$. Sameh and Kuck's scheme is easy to program and to analyze: a possible choice is to use R(i,i-1,k) in order to annihilate the element in position (i,k). Clearly, the total number of steps is m+n-2 if m>n (17 in the example) and 2n-3 otherwise.

Cosnard and Robert [1] and independently Modi and Clarke [9] have introduced a greedy algorithm which performs at each step as many rotations as possible, annihilating the elements in each column from bottom to top and in each row from left to right. This scheme is depicted in the second table of figure 1: it begins very fast since it performs 6 rotations from step 1 to step 2, 5 rotations from step 4 to step 8, 4 rotations at the steps 9 and 10, but it terminates slowly with only 2 rotations at the steps 11 and 12 and one at the last two steps.

Sameh and Kuck's							Greedy					
*							*					
12	*						4	*				
1	13	*					3	6	*			
10	12	14	*				3	5	8	*		
9	11	13	15	*			2	5	7	10	*	
8	10	12	14	16	*		2	4	7	9	12	*
7	9	11	13	15	17		2	4	6	8	11	14
6	8	10	12	14	16		1	3	6	8	10	13
5	7	9	11	13	15		1	3	5	7	9	12
4	6	8	10	12	14		1	3	5	7	9	11
3	5	7	9	11	13		1	2	4	6	8	10
2	4	6	8	10	12		1	2	4	6	8	10
1	3	5	7	9	11		1	2	3	5	7	9

Sameh and Kuck's Greedy

Figure 1: Annihilation schemes

Numerical experiments show that the greedy algorithm takes appreciably fewer stages than Sameh and Kuck's scheme, which is confirmed by an asymptotic theoretical analysis: Modi and Clarke obtain the approximation

$$\log m + (n-1)\log\log m$$

which is valid if m goes to infinity, n fixed (when not stated explicitly the basis of the logarithms will be 2). In their paper, Modi and Clarke introduce another class of algorithms, namely the Fibonacci schemes, and discuss in detail their performances. However, they observe that these schemes seem to be less efficient than the greedy algorithm.

The Fibonnacci annihilation scheme of order 1 is derived as indicated in figure 2. We fill up the first column from bottom to top. There is a single zero, below are $u_2=2$ successive copies of -1, then $u_3=3$ copies of -2, and so on (there are $u_k=u_{k-1}+1$ values of -(k-1), with $u1=1$). The second column is like the first except that all the entries are increased by 2, and the whole column moved down one place. Other columns are filled up using the same rule. Now, to get an annihilation scheme, just add u+1 to all integers in the lower part of the matrix, where u is the integer in the left hand bottom corner. It is not straightforward to ensure that this scheme is actually a Givens annihilation scheme: we would need to check carefully whether it is possible to annihilate the elements of the matrix in such a way. However, this can be done owing to the very systematic construction of the scheme.

*							*					
0	*						5	*				
-1	2	*					4	7	*			
-1	1	4	*				4	6	9	*		
-2	1	3	6	*			3	6	8	11	*	
-2	0	3	5	8	*		3	5	8	10	13	*
-2	0	2	5	7	10		3	5	7	10	12	15
-3	0	2	4	7	9		2	5	7	9	12	14
-3	-1	2	4	6	9		2	4	7	9	11	14
-3	-1	1	4	6	8		2	4	6	9	11	13
-3	-1	1	3	6	8		2	4	6	8	11	13
-4	-1	1	3	5	8		1	4	6	8	10	13
-4	-2	1	3	5	7		1	3	6	8	10	12

Figure 2: Fibonnacci scheme of order 1 (Add 5 to get the annihilation scheme)

The Fibonacci scheme of order 2 is obtained in a similar way, replacing the relation $u_k=u_{k-1}+1$ by $u_k=u_{k-1}+u_{k-2}+1$ (with $u_1=1$, $u_2=2$), and adding 3 instead of 2 to the entries of column j to get the entries of column j+1. The number of steps for the Fibonnacci scheme of order 1 is equal to k+1+2(n-1), where k is the least integer such that $k(k+1)/2 \geq m-1$ (leading to 15 in our example). The number of steps for the scheme of order 2 is 19 in our example. Rather than providing the exact expression (there is no simple formula), we recall the asymptotic value $\log_a m + 3n$, where $a = (1+\sqrt{5})/2$ given in [9].

All the algorithms presented by Sameh and Kuck [11] and Modi and Clarke [9] are Givens sequences, that is sequences of Givens rotations in which zeros once created are preserved, i.e. for any couple of distinct rotations R(i,j,k) and R(i',j',k') in M we have $(i,k) \neq (i',k')$.. The question whether temporarily annihilating elements and introducing zeros that will be destroyed later on can lead to any additional parallelism is a nice example of specific questions that parallelism can rise.

In the remaining of this section, we describe the results obtained by Cosnard and Robert [2] and Cosnard, Muller and Robert [3]. (M,T,R) is called a Standard Parallel Givens Sequence (SPGS for short) if it is a Givens sequence which reduces A to upper triangular form and annihiiates elements from left to right and from bottom to top, i.e. if $R(i,j,k) \in M(t)$ and $R(i',j',k') \in M(t')$ where $t \leq t'$, then $k<k'$ or ($k=k'$ and $i \geq i'$). Clearly, all the algorithms introduced so far are SPGS, in particular the Greedy algorithm.

After showing that in the set of optimal algorithms there exists a SPGS, Cosnard and Robert [3] prove that, for any value of m and n (with $m \geq n$), and for $p = \lfloor m/2 \rfloor$, the greedy SPGS is an optimal algorithm. Hence the complexity of the QR decomposition using plane rotations is the number of steps OPT(m,n) of the greedy algorithm and the asymptotic efficiency ASE(m,n) could be easily deduced. The proof of the following results can be found in [2] and [3] (4. is a new but easy consequence of these proofs):

Theorem 1 : Let $p = \lfloor m/2 \rfloor$.
1. If n = m (A is a square matrix), then OPT(m,n) = 2n + o(n) and ASE(m,n) = 1/2.
2. If n is fixed, then OPT(m,n) = log2m + o(log2m) and ASE(m,n) = 2n/logm.
3. If $m = o(n^2)$, then OPT(m,n) = 2n + o(n) and ASE(m,n) = 1.
4. If $m = o(n^k)$, with $k \geq 3$, then $2n + o(n) \leq$ OPT(m,n) $\leq 3n + o(n)$ and ASE(m,n) $\geq 2/3$.

An interesting consequence [2], [3] of this asymptotic analysis is that it allows to select among various parallel algorithms those which are aymptotically optimal. This is important since the computation of the indices of the rows involved in plane rotations at a given step of the greedy algorithm is somewhat difficult. Hence it must be avoided. This can be done by replacing the greedy SPGS by another algorithm which, although not optimal, will have good performances and could be implemented in a simple way.

Corollary 2: Let $p = \lfloor m/2 \rfloor$.
1. If m = n, then Sameh and Kuck's scheme is asymptotically optimal.
2. If $m = o(n^2)$, then the Fibonacci scheme of order 1 is asymptotically optimal.
3. If $m = o(n^k)$ and $m \geq n^2$, then the Fibonacci scheme of order 2 is of asymptotic efficiency 2/3.

When n is fixed and m goes to infinity, none of the previous algorithms but the greedy is asymptotically optimal.

3. WITH P ROTATIONS SIMULTANEOUSLY

In this and the following sections, we shall limit the number of plane rotations which can be performed simultaneously to a given integer $p \leq \lfloor m/2 \rfloor$. The first following results are straightforward extensions of results presented in [2] and [3]. For any algorithm (M,T,R), call r(i,t) the number of zeros in column i introduced at time t and s(i,t) the total number of zeros in column i after step t. Hence for Givens Sequences we have s(i,t) = r(i,1) + ... + r(i,t). A simple proof (see [1]) enables us to characterize the SPGS among the Givens Sequences and to show that there exists an optimal SPGS:

Lemma 3: Let $p \leq \lfloor m/2 \rfloor$ and M be a Givens Sequence. M is a SPGS if and only if :

* $s(0,t) = m$ for $t \geq 0$ * $s(i,0) = 0$ for $1 \leq i \leq m$

* $r(i,t) \leq \min \{ p - \sum_{j=1}^{i-1} r(j,t) , \lfloor (s(i-1,t-1) - s(i,t-1))/2 \rfloor \}$ for $i,t \geq 1$.

Lemma 3 means that the total number of zeros introduced at time t cannot exceed p and that, just as before, no more zeros than half the difference between the number of zeros in column i and i-1 at step t-1 can be created in column i at step t. We shall see later on that these constraints make the analysis of the problem much more difficult.

Lemma 4: In the set of optimal algorithms there exists a SPGS.

A great difference with the case $p=\lfloor m/2 \rfloor$ is the fact that there is not a unique greedy SPGS. The greedy algorithms can be unformally described in the following manner: at each step do the maximum number of rotations.

Definition: Let M be a SPGS. M is a greedy algorithm if and only if:

$$\sum_{j=1}^{n} r(j,t) = \min \{ p , \sum_{j=1}^{n} \lfloor (s(j-1,t-1) - s(j,t-1))/2 \rfloor \}$$

Clearly this can be achieved with different strategies. To illustrate this fact we present two algorithms: the vertical greedy, which fills up the columns, and the horizontal greedy, which fills up the rows.

Vertical greedy:

- $svg(0,t) = m$ for $t \geq 0$
- $svg(i,0) = 0$ for $1 \leq i \leq m$
- $rvg(i,t) = \min \{ p - \sum_{j=1}^{i-1} rvg(j,t) \ , \ \lfloor (svg(i-1,t-1) - svg(i,t-1))/2 \rfloor \}$ for any $i,t \geq 1$.

Horizontal greedy:

- $shg(0,t) = m$ for $t \geq 0$
- $shg(i,0) = 0$ for $1 \leq i \leq m$
- $rhg(i,t) \leq \min \{ p - \sum_{j=i+1}^{n} rhg(j,t) + \lfloor (shg(i-1,t-1) - shg(i,t-1))/2 \rfloor \}$ for any $i,t \geq 1$.

The problem of optimality will be solved in the following sections. Remark that none of the two preceding greedy algorithms is optimal. However for particular choices of m, n or p both can be optimal.

Theorem 5: Given m, n and p, there exists an optimal algorithm in the set of greedy SPGS.

Proof: Given a SPGS, we construct a greedy SPGS performing the decomposition in less steps. This is done by induction on t.

Assume that $s(i,t) \leq sg(i,t)$ and let us prove it for t+1. Recall that $s(i,t+1) = s(i,t) + r(i,t)$ and define $rg1(i,t+1) = r(i,t+1) - (sg(i,t) - s(i,t))$. From Lemma 3 we deduce that :

$$r(i,t+1) \leq \lfloor (s(i-1,t) - s(i,t))/2 \rfloor$$

which implies that:

$$\begin{aligned} rg1(i,t+1) &\leq \lfloor (s(i-1,t) - s(i,t))/2 \rfloor - sg(i,t) + s(i,t) \\ &\leq \lfloor (s(i-1,t) + s(i,t))/2 \rfloor - sg(i,t) \\ &\leq \lfloor (sg(i-1,t) + sg(i,t))/2 \rfloor - sg(i,t) \\ &\leq \lfloor (sg(i-1,t) - sg(i,t))/2 \rfloor \end{aligned}$$

Moreover

$$rg1(i,t+1) \leq r(i,t+1) \leq p - \sum_{j=1}^{i-1} r(j,t+1) \leq p - \sum_{j=1}^{i-1} rg1(j,t+1)$$

Hence we obtain:

$$rg1(i,t+1) \leq \min \{ p - \sum_{j=1}^{i-1} rg1(j,t+1) , \lfloor (sg(i-1,t) - sg(i,t))/2 \rfloor \}$$

Lemma 3 implies that this scheme is a SPGS. It is now easy to derive a greedy SPGS.

If

$$\sum_{j=1}^{n} rg1(j,t+1) < \min \{ p , \sum_{j=1}^{n} \lfloor (s(j-1,t) - s(j,t))/2 \rfloor \}$$

saturate the inequality by adding as many zeros as possible, for instance from left to right.
We then have:

$$s(i,t+1) = s(i,t) + r(i,t) \leq sg(i,t) + rg1(i,t) \leq sg(i,t) + rg(i,t) = sg(i,t+1)$$

which concludes the proof.

From the theorem, it can be easily seen that, although there exists an optimal greedy algorithm for any m, n and p, we do not know how to construct it. Figure 3 shows that this optimal greedy SPGS can be different from the vertical or the horizontal greedy. We choose m=n=12 and p=3.

*											
5	*										
4	8	*									
3	7	11	*								
3	7	10	14	*							
3	6	10	13	16	*						
2	6	9	12	15	18	*					
2	6	9	12	15	17	20	*				
2	5	9	12	14	17	19	22	*			
1	5	8	11	14	16	18	21	23	*		
1	4	8	11	13	16	18	20	22	24	*	
1	4	7	10	13	15	17	19	21	23	25	*

Vertical greedy

*											
15	*										
13	16	*									
11	14	17	*								
5	12	15	18	*							
4	7	13	16	19	*						
3	6	9	14	17	20	*					
2	5	8	10	15	18	21	*				
2	4	7	9	11	16	19	22	*			
1	3	6	8	10	12	17	20	23	*		
1	3	5	7	9	11	13	18	21	24	*	
1	2	4	6	8	10	12	14	19	22	25	*

Horizontal greedy

*											
7	*										
6	10	*									
5	9	13	*								
4	8	12	15	*							
4	7	11	14	17	*						
2	6	10	13	16	19	*					
2	5	9	12	15	18	20	*				
2	4	8	11	14	17	19	21	*			
1	3	7	10	13	16	18	20	22	*		
1	3	6	9	12	15	17	19	21	23	*	
1	3	5	8	11	14	16	18	20	22	24	*

Better greedy (optimal, see later)

Figure 3: Various greedy algorithms

4. A LOWER BOUND

In this section, we shall derive a lower bound for the complexity of the parallel Givens factorization in the case m=n. Hence in the remaining, we shall assume that m=n=pq, where p is the number of processors. We call $T_{opt}(p)$ the optimal time for computing the Givens factorization with p processors and T_{opt} the optimal time when an unlimited number of processors is available (in fact we deduce from the preceding section that $T_{opt}=T_{opt}(p)$ for $p \geq \lfloor n/2 \rfloor$). The following relations are straightforward:

* $T_{opt} = T_{opt}(\lfloor \frac{n}{2} \rfloor) \leq T_{opt}(p)$ for all p * $T_{opt} = 2n-o(n)$

Theorem 6: For $1 \le p \le n/2$, we have: $T_{opt}(p) \ge B_{inf}(p) = \frac{n(n-1)}{2p} + p - 1$.

Proof: Let : - $ac(k)$ be the number of active processors at step k

- $id(k)$ be the number of idle processors at step k
- $T(p)$ be the execution time with p processors

The following relation holds: (1) $id(k)=p-ac(k)$

Define $$P_{ac} = \sum_{k=1}^{T(p)} ac(k) \quad \text{and} \quad P_{id} = \sum_{k=1}^{T(p)} id(k)$$

Using (1) we obtain:

(2) $$T(p)=\frac{P_{id}+P_{ac}}{p}$$

$\sum_{k=1}^{T(p)} ac(k)$ is the number of annihilated elements equal to $\frac{n(n-1)}{2}$

then we have $$T(p) = \frac{q(n-1)}{2} + \frac{P_{id}}{p}$$

Now we show that $P_{id} \ge p(p-1)$ therefore we have

$$T(p) \ge \frac{n(n-1)}{2p} + p - 1 = B_{inf}(p)$$

Since the elements are annihilated from bottom to top and from left to right then the execution time is the time necessary to annihilate the elements of the subdiagonal, i.e the elements in positions $(i+1,i)$ for $i=1,\ldots,n$. Let t_i be the step at which the element in position $(i+1,i)$ is annihilated. Since for annihilating one element we use two rows, at the most $\min(p, \lfloor\frac{n-i+1}{2}\rfloor)$ elements can be annihilated simultaneously with the element in position $(i+1,i)$, then:

$$ac(t_i) \le \min(p, \lfloor\tfrac{n-i+1}{2}\rfloor) \quad \text{and} \quad id(t_i) \ge p - \min(p, \lfloor\tfrac{n-i+1}{2}\rfloor)$$

For $i \ge n-2p+1$, $\min(p, \lfloor\frac{n-i+1}{2}\rfloor) = \lfloor\frac{n-i+1}{2}\rfloor$ then $id(t_i) \ge p - \lfloor\frac{n-i+1}{2}\rfloor$

So $$\sum_{i=n-2p+1}^{n-1} id(t_i) \ge \sum_{i=n-2p+1}^{n-1} \left(p - \lfloor\tfrac{n-i+1}{2}\rfloor\right) = p(p-1)$$

Therefore we have

$$P_{id} \ge p(p-1)$$

which concludes the proof of theorem 6.

A direct consequence of the theorem is that the optimal number of processors in order to obtain the optimal time is bounded below by the value of p for which $B_{inf}(p)=2n$.

Corollary 7: Let p_{opt} be the minimum number of processors for computing the Givens factorization of a dense square matrix of size n in T_{opt}. Then asymptotically we have:

$$\frac{n}{2+\sqrt{2}} \leq p_{opt} \leq \frac{n}{2}$$

5. AN OPTIMAL ALGORITHM FOR $p=\frac{n}{4}$

In this section we shall prove that the preceding lower bound is tight for $p=\frac{n}{4}$. For this we first define a new class of algorithms: the modified Fibonacci algorithms of order i [4]. They are constructed in two phases (recall that p=m/2 for this construction). An example is shown in figure 4.

First phase: we fill in the first column of A using the same method as for the standard Fibonacci algorithm of order i, but we begin in position (3,1). The kth column is obtained by adding (i+1) to the elements of the (k-1)th column after a shift of two positions.
Second phase: we use Sameh and Kuck scheme.

```
 *                                      *
 +  *                                  13  *
 0  +  *                                4 14  *
-1  +  +  *                             3 13 15  *
-1  2  +  +  *                          3  6 14 16  *
-2  1  +  +  +  *                       2  5 13 15 17  *
-2  1  4  +  +  +  *                    2  5  8 14 16 18  *
-2  0  3  +  +  +  +  *                 2  4  7 13 15 17 19  *
-3  0  3  6  +  +  +  +  *              1  4  7 10 14 16 18 20  *
-3  0  2  5  +  +  +  +  +  *           1  4  6  9 13 15 17 19 21  *
-3 -1  2  5  8  +  +  +  +  +  *        1  3  6  9 12 14 16 18 20 22  *
-3 -1  2  4  7  +  +  +  +  +  +  *     1  3  6  8 11 13 15 17 19 21 23  *
```

Figure 4: Modified Fibonacci scheme of order 1 with n=m=12 and p=6

Lemma 8: For $p=\frac{n}{4}$ we have: $T_{opt}(\frac{n}{4})=B_{inf}(\frac{n}{4}) + o(n)=\frac{9n}{4} +o(n)$.

Proof: We subdivide the matrix into two blocks: the first one is composed with the 2p first rows and the second one is composed with the 2p last rows, as indicated figure 5. We construct an algorithm in four phases. Each phase consists in executing one or several regions numbered in figure 5 by I, I', II, III and IV.

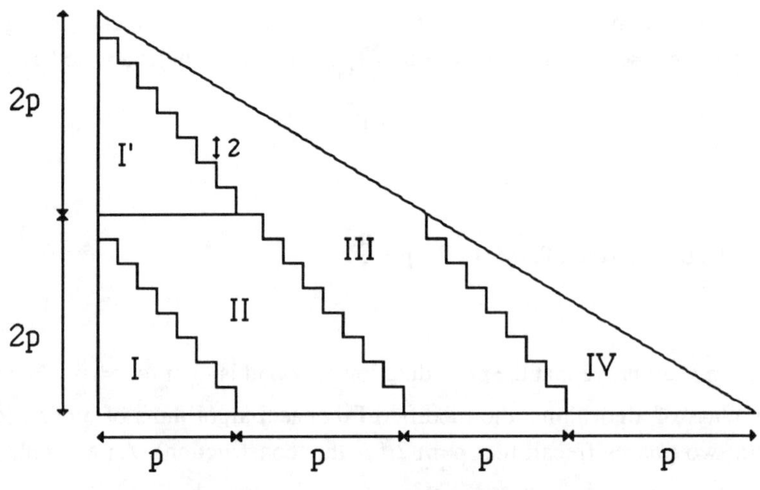

Figure 5: Decomposition of the matrix

First phase: The elements in regions I and I' are annihilated asymptotically with full efficiency. The execution time is $T_1(\frac{n}{4})=\frac{n}{2}+o(n)$.

Proof:

- We assign all the processors to execute region I' with the first phase of the modified Fibonacci algorithm of order 1. Hence the execution time is $\lfloor\sqrt{n-2}\rfloor+\frac{n}{2}-4=\frac{n}{2}+o(n)$. Remark that from step $\lfloor\sqrt{n-2}\rfloor$ up to the end of this phase, the number of idle processors increases, at least, by one at each two steps.
- The processors which are released from region I' are assigned to region I for annihilating with Sameh and Kuck algorithm starting at step $\lfloor\sqrt{n-2}\rfloor$. In fact, when the elimination progresses in region I the number of elements which can be immediately annihilated increases by one at each 2 steps. Hence the execution time of region I is $\lfloor\sqrt{n-2}\rfloor+\frac{n}{2}-2=\frac{n}{2}+o(n)$.

Second phase: The elements in region II are annihilated with full efficiency. The execution time is equal to: $T_2(\frac{n}{4})=\frac{n}{2}$.

Proof: At the end of the execution of regions I and I', the numbers of annihilated elements in two consecutive columns differ by 2 (staircase with steps of size 2). The elimination in this phase consists in annihilating one element in each of p consecutive columns of region II (translating the staircase from left to right). Since the number of elements to be annihilated is $2p^2$, we deduce that $T_2(\frac{n}{4})=\frac{n}{2}$.

```
*                                *
+  *                             18 *
3  +  *                          3  20 *
2  +  +  *                       2  18 21 *
2  5  +  +  *                    2  5  20 23 *
1  4  +  +  +  *                 1  4  18 22 25 *
1  4  7  +  +  +  *              1  4  7  20 23 26 *
1  3  6  +  +  +  +  *           1  3  6  18 22 25 27 *
11 13 15 17 +  +  +  +  *        11 13 15 17 20 24 26 29 *
10 12 14 16 +  +  +  +  +  *     10 12 14 16 19 22 25 28 +  *
9  11 13 15 17 +  +  +  +  +     9  11 13 15 17 21 24 27 29 +  *
8  10 12 14 16 +  +  +  +  +     8  10 12 14 16 19 22 25 28 +  +  *
7  9  11 13 15 17 +  +  +  +     7  9  11 13 15 17 21 24 27 29 +  +  *
6  8  10 12 14 16 +  +  +  +     6  8  10 12 14 16 19 23 26 28 +  +  +
5  7  9  11 13 15 17 +  +  +     5  7  9  11 13 15 17 21 24 27 29 +  +
4  6  8  10 12 14 16 +  +  +     4  6  8  10 12 14 16 19 23 26 28 +  +
```

Figure 6: The first three phases of the optimal algorithm, n=16 and p=4.
(second phase starts from step 10, and third phase starts from step 18)

Third phase: The elements in region III are annihilated with full efficiency. The execution time is equal to: $T_3(\frac{n}{4})=\frac{3n}{4}$.
Proof: At the end of the second phase, the number of annihilated elements of two consecutive columns among the first 2p columns differ by 2. This staircase is translated from left to right in time $T_3(\frac{n}{4})=\frac{3n}{4}$.

Fourth phase: The elements in region IV are annihilated in optimal time, the execution time is equal to: $T_4(\frac{n}{4})=\frac{n}{2}$ -1.
Proof: After execution of the first three phases, the first 2p columns are annihilated and the next p columns have the shape of a staircase with steps of size 2. Once again we translate the staircase until we reach the last column in time $T_4(\frac{n}{4})=\frac{n}{2}$ -1.

The execution time of the algorithm is the sum of the execution times of the four phases:

$$T(\frac{n}{4})=T_1(\frac{n}{4})+T_2(\frac{n}{4})+T_3(\frac{n}{4})+T_4(\frac{n}{4})=B_{inf}(\frac{n}{4})+o(n)=\frac{9n}{4}+o(n).$$

*															
18	*														
3	20	*													
2	18	21	*												
2	5	20	23	*											
1	4	18	22	25	*										
1	4	7	20	23	26	*									
1	3	6	18	22	25	27	*								
11	13	15	17	20	24	26	29	*							
10	12	14	16	19	22	25	28	30	*						
9	11	13	15	17	21	24	27	29	31	*					
8	10	12	14	16	19	22	25	28	30	32	*				
7	9	11	13	15	17	21	24	27	29	31	33	*			
6	8	10	12	14	16	19	23	26	28	30	32	34	*		
5	7	9	11	13	15	17	21	24	27	29	31	33	35	*	
4	6	8	10	12	14	16	19	23	26	28	30	32	34	36	*

Figure 7: The optimal algorithm, n=16 and p=4.

6. A GENERAL THEOREM

In fact, we can prove that the preceding lower bound can be reached for $p \leq \frac{n}{2+\sqrt{2}}$. Hence we deduce that the asymptotic complexity of the Givens decomposition algorithm is equal to $B_{inf}(p)$.

Theorem 9: For $1 \leq p \leq \frac{n}{2+\sqrt{2}}$, $T_{opt}(p) = \frac{n^2}{2p} + p + o(n)$.

Proof: The proof is long and tedious see [4]. We only give a sketch of the proof. We first consider the case where $p = \frac{n}{2+\sqrt{2}}$.

CASE $p = \frac{n}{2+\sqrt{2}}$

We subdivide the matrix into five regions. The algorithm is constructed in three phases.

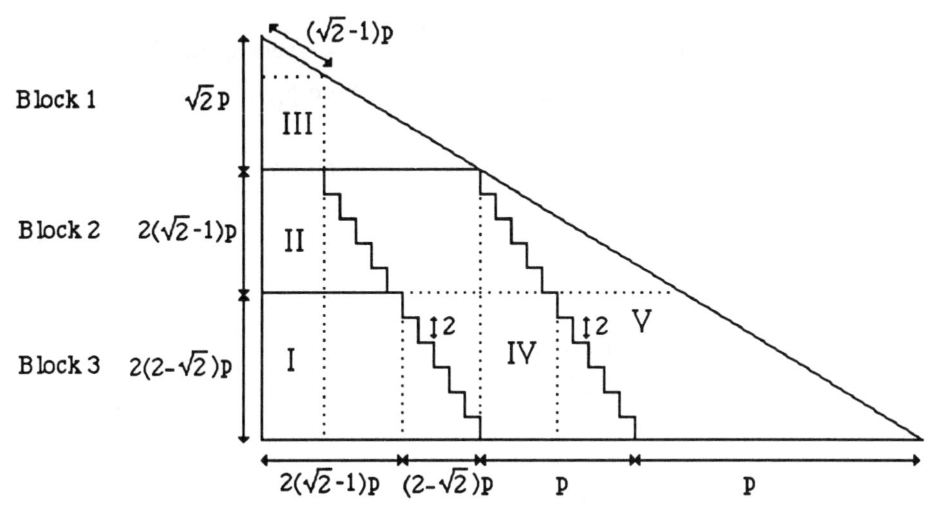

Figure7: Decomposition of the matrix in regions

First phase: The elements in regions I, II and III are annihilated with $p=\frac{n}{2+\sqrt{2}}$ processors in full asymptotic efficiency. The execution time is equal to:

$$T_1\left(\frac{n}{2+\sqrt{2}}\right)=2\sqrt{2}p+o(p)=2(\sqrt{2}-1)n+o(n)$$

Proof: The elimination progresses in two stages:

Stage 1: We annihilate the elements in subregions I_1, II_1 and III_1 as indicated in figure 7.a with $p=\frac{n}{2+\sqrt{2}}$ processors in full asymptotic efficiency. The execution time is equal to:

$$t_1(p)=2(2-\sqrt{2})p+o(p).$$

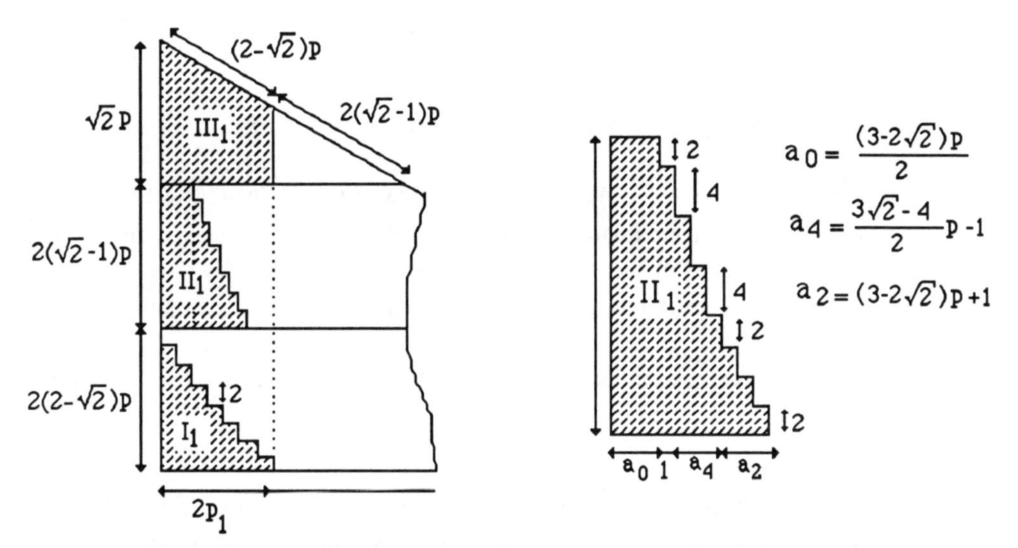

Figure 7.a: Decomposition into subregions

1. We annihilate in the subregion I_1 with $p_1=\frac{2-\sqrt{2}}{2}p$ processors using the two phases of algorithm §.5. The execution time is equal to: $t_1(p)=2(2-\sqrt{2})p+o(p)$.

2. We annihilate in subregion III_1 using Fibonacci algorithm of order 1 with the rest of the processors $p_2=p-p_1=\frac{\sqrt{2}}{2}p$. The execution time is equal to $2(2-\sqrt{2})p+o(p)$ and the first $(2-\sqrt{2})p$ columns are annihilated.

3. The processors which are released from the subregion III_1 (asymptotically, one processor is released at each 4 steps) are assigned to the subregion II_1. The annihilation process in II_1 is divided in two stages:

3.a. During the first $2(\sqrt{2}-1)p+o(p)$ steps we annihilate by column, from bottom to top and from left to right. We annihilate in position (i,k) with R(i,i-1,k) as soon as the elements in positions (i,k-1), (i-1,k-1), (i-2,k-1), (i-3,k-1) are annihilated. At the end of this stage the elements in subregions II'_1 and III'_1 are annihilated as indicated in figure 7.b.

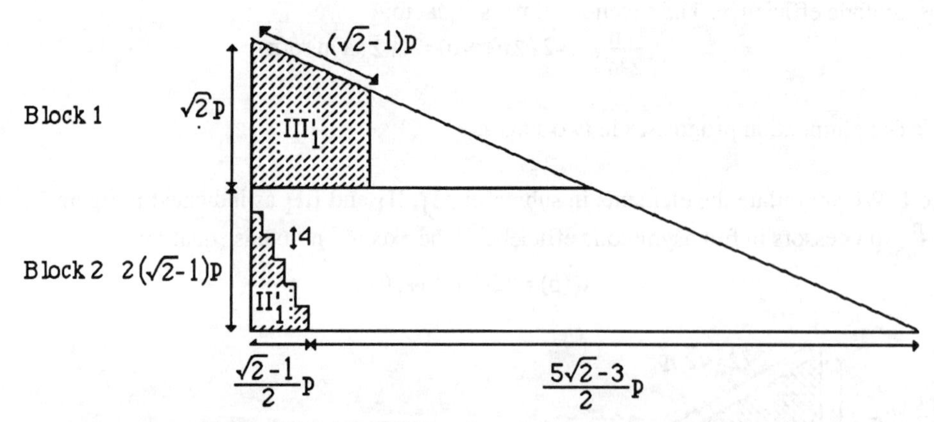

Figure 7.b: Annihilation in subregions II'_1 and III'_1

3.b. This stage starts at time $2(\sqrt{2}-1)p+o(p)$ and finishes at time $t_1(p)=2(2-\sqrt{2})p+o(p)$. The elimination process consists in annihilating, at each step, one element on each column (column of Block 2) on which we can annihilate some element in order to:

(i) Increase by 1 at each 4 steps the number of columns on which we can annihilate some element.

(ii) Decrease in the staircase the number of steps of length 4 by 1 and increase the number of steps of length 2 by 2 at each 4 steps of the algorithms.

Let $t=2(\sqrt{2}-1)p+o(p)$ and $q=\frac{\sqrt{2}-1}{2}p$. In figure 7.c we show the evolution of the elimination during the 4 first steps. It is easy to show by induction that conditions (i) and (ii) are satisfied.

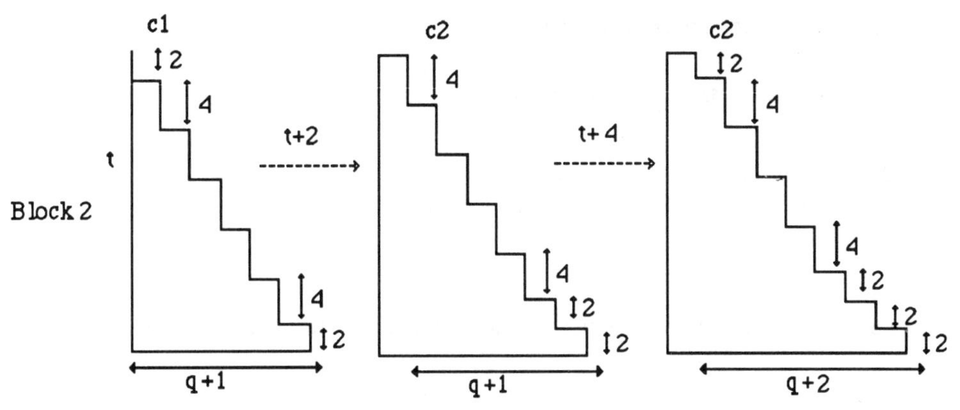

Figure 7.c: Evolution of the elimination in subregion II_1

Stage 2: We annihilate in subregions I_2, II_2 and III_2 as indicated in figure 7.d with $p=\frac{n}{2+\sqrt{2}}$ processors in full asymptotic efficiency. The execution time is equal to:

$$t_2(p)=4(\sqrt{2}-1)p+o(p).$$

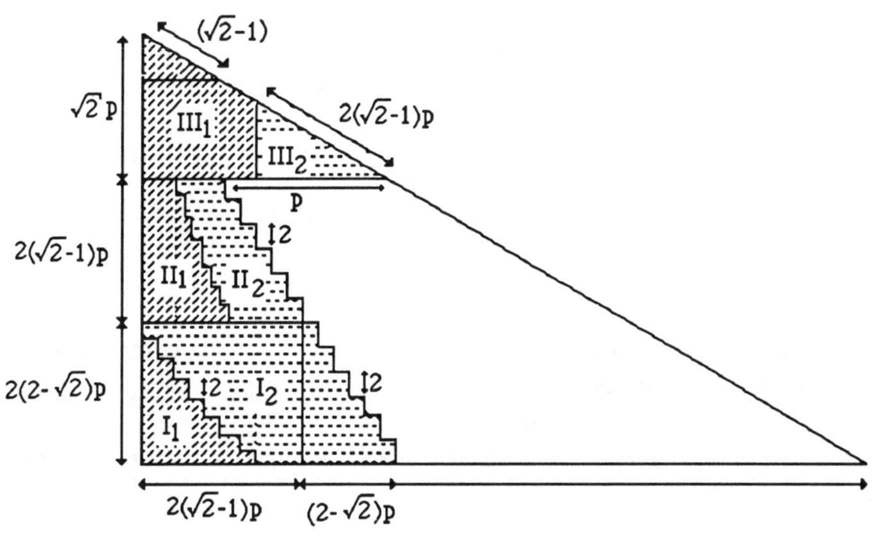

Figure 7.d: Execution of subregions I_2, II_2 et III_2

1. We annihilate in subregion I_2 with $p'_1=(2-\sqrt{2})p$ using the second phase of algorithm § 5. The execution time is equal to: $4(\sqrt{2}-1)p$ steps.

2. We annihilate in subregion III_2 with $p'_2=p-p'_1=(\sqrt{2}-1)p$ using the Fibonacci algorithm of order 1. The execution time is equal to: $4(\sqrt{2}-1)p+o(p)$

3. We annihilate in subregion II_2 using the processors which are released from subregion III_2 (

asymptotically, one processor is released at each 4 steps) by a cyclic annihilation.

The execution time of regions I, II and III is equal to: $t_1(p)+t_2(p)=2\sqrt{2}p+o(p)$

Second phase: The elements in region IV are annihilated in full efficiency using the second phase of algorithm § 5. The execution time is equal to:

$$T_2\left(\frac{n}{2+\sqrt{2}}\right)=(2-\sqrt{2})n$$

Third phase: The elements in region V are annihilated in optimal time. The execution time is equal to:

$$T_3\left(\frac{n}{2+\sqrt{2}}\right)=(2-\sqrt{2})n-1.$$

The execution time of the algorithm is the sum of the execution times of the three phases:

$$T_{opt}\left(\frac{n}{2+\sqrt{2}}\right)=2(\sqrt{2}-1)n+(2-\sqrt{2})n+(2-\sqrt{2})n+o(n)=2n+o(n)=B_{inf}\left(\frac{n}{2+\sqrt{2}}\right)+o(n).$$

A straightforward consequence of the previous analysis is the value of $p_{opt}=\frac{n}{2+\sqrt{2}}$.

Corollary 10: The minimum number of processors in order to obtain the Givens decomposition of a square matrix of size n in asymptotically optimal time, $T_{opt}=2n-o(n)$, is equal to:

$$p_{opt}=\frac{n}{2+\sqrt{2}}$$

GENERAL CASE

(i) If p is greater than p_{opt} then we have only to use the algorithm constructed for p_{opt} to construct the algorithm asymptotically in T_{opt}.

(ii) If p is lower than p_{opt} then we decompose n in $(2+\sqrt{2})p+rp$. We present only the case where r is greater than 2, $r\geq 2$.

The matrix is subdivided in four regions as indicated in figure 8. We construct the algorithm in four phases.

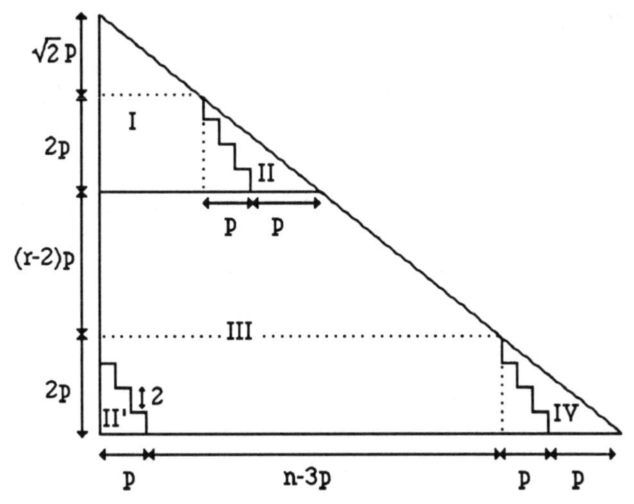

Figure 8: Decomposition of the matrix in regions

<u>First phase:</u> The elements in region I are annihilated using the two phases of the preceding algorithm with p processors in full asymptotic efficiency.

<u>Second phase:</u> We annihilate simultaneously the elements in regions II and II' in the following way:

- We annihilate in region II in optimal time.
- We annihilate in region II' with Sameh and Kuck's algorithm using the processors which are released from the region II (the number of processors released increases by 1 at each 2 steps).

<u>Third phase:</u> The elements in region III are annihilated with p processors using a cyclic elimination in full asymptotic efficiency.

<u>Fourth phase:</u> The elements in region IV are annihilated in optimal time.

7. CONCLUDING REMARKS

Several conjectures were made concerning the complexity of the parallel Givens factorization with a limited number of processors.

First it was conjectured that m/2 processors were necessary to solve the problem in asymptotically optimal time. In this paper, we proved that this is false: the number is equal to $p_{opt} = \frac{n}{2+\sqrt{2}}$ in the case m=n. Remark that the general problem (rectangular matrix) is still open.

It was conjectured that the greedy algorithm would be optimal. We have proven that:

1. there exist various greedy algorithms with different running time.
2. in the set of the greedy algorithms, there exists an optimal algorithm
3. we have constructed an asymptotically optimal algorithm for any p provided m=n.

Notice that the algorithm that we construct in section 5 and 6 is not a greedy algorithm (but it is not very far...)

Evaluating the complexity of the problem in the general rectangular case is still open. However, we do think that our proof and construction will help solving it.

The asymptotically optimal algorithm that we construct is rather difficult to describe. However, we think that, since we now know the time complexity of the square problem, it would be easier to obtain simpler constructions. In particular is it possible to modify Sameh an Kuck scheme in order to reach asymptotically the bound (as in the case of unlimited parallelism)?

REFERENCES

[1] COSNARD M., ROBERT Y., Complexité de la factorisation QR en parallèle, C. R. Acad. Sc. Paris, 297, A, 549-552 (1983)

[2] COSNARD M., ROBERT Y., Complexity of parallel QR decomposition, J. ACM 33 (4), 712-723 (1986)

[3] COSNARD M., MULLER J.M., ROBERT Y., Parallel QR decompostion of a rectangular matrix, Numerische Mathematik 48, 239-249 (1986)

[4] DAOUDI E.M., Etude de la complexité de la décomposition orthogonale d'une matrice sur plusieurs modèles d'architectures parallèles, Thèse de l'INP Grenoble, mai 1989.

[5] GENTLEMAN W.M., Least squares computations by Givens transformations without square roots, J. Inst. Maths. Applics. 12, 329-336 (1973)

[6] HELLER D., A survey of parallel algorithms in numerical linear algebra, SIAM Review 20, 740-777 (1978)

[7] KUMAR S.P., Parallel algorithms for solving linear equations on MIMD computers, PhD. Thesis, Washington State University (1982)

[8] LORD R.E., KOWALIK J.S., KUMAR S.P., Solving linear algebraic equations on an MIMD computer, J. ACM 30 (1), 103-117 (1983)

[9] MODI J.J., CLARKE M.R.B., An alternative Givens ordering, Numerische Mathematik 43, 83-90 (1984)

[10] SAMEH A., Solving the linear least squares problem on a linear array of processors, Proc. Purdue Workshop on algorithmically-specialized computer organizations, W. Lafayettte, Indiana, September 1982

[11] SAMEH A., KUCK D., On stable parallel linear system solvers, J. ACM 25 (1), 81-91 (1978)

Optimal Bounds on the Dictionary Problem

Arne Andersson
Department of Computer Science
Lund University
Lund, Sweden

Abstract

A new data structure for the dictionary problem is presented. Updates are performed in $\Theta(\log n)$ time in the worst case and the number of comparisons per operation is $\lceil \log n + 1 + \epsilon \rceil$, where ϵ is an arbitrary positive constant.

1 Introduction

One of the fundamental and most studied problems in computer science is the *dictionary problem*, that is the problem of how to maintain a set of data during the operations search, insert and delete. It is well known that in a comparison-based model the lower bound on these operations is $\lceil \log(n+1) \rceil$ comparisons both in the average and in the worst case. This bound can be achieved by storing the set in an array or in a perfectly balanced binary search tree. However, for both these data structures the overhead cost per update is high, $\Theta(n)$ in the worst case.

An efficient dynamic data structure for the dictionary problem should have a worst case cost of $\Theta(\log n)$ per operation. The first efficient solution was presented by Adelson-Velski and Landis [1]. Their data structure, the AVL-tree, requires $1.44 \log n$ comparisons per operation and allows updates in logarithmic worst case time. Other data structures for the dictionary problem with good worst case performance are symmetric binary B-trees [3] and trees of bounded balance [5], which require $2 \log n$ comparisons. A close approximation to the optimal bound was given by Mauer et. al. [4]. They presented the k-neighbour tree which requires $\lfloor (1+\epsilon) \log n + 2 \rfloor$ comparisons where ϵ is an arbitrary positive constant. Similar results are obtained by van Leeuven's and Overmars's stratified balanced trees [6] and by Andersson's generalized symmetric binary B-trees [2]. For all these structures there is a tradeoff between the number of comparisons and the maintenance cost in such a way that a lower value of ϵ corresponds to a higher cost for restructuring during updates.

Although a bound of $(1+\epsilon)\log n$ is sufficiently low for practical purposes there is still a gap between this and the optimal bound. Recently, Andersson [2] showed that there exists a data structure, which is based on the generalized symmetric binary B-tree, requiring at the most

$$\left\lfloor \left(1 + \frac{1}{\log \log n}\right) \log n \right\rfloor + 1 = \log n + o(\log n) \tag{1}$$

comparisons per search and $\Theta(\log n)$ amortized cost per update. This bound is optimal in the leading term but the $o(\log n)$-term is quite large.

Here we show that the difference between the upper and lower bound on the dictionary problem is only a small additive constant. We present the *ϵ-tree*, which require

$$\lceil \log n + 1 + \epsilon \rceil \tag{2}$$

comparisons per search or update for an arbitrary small value of ϵ. We may choose ϵ in such a way that mostly we make only one, and never more than two extra comparisons compared to the lower bound.

The paper is organized in the following way: In section 2 we give a short description of the k–neighbour tree [4], on which our result is based. We show how to improve the insertion algorithm by adding some information to the nodes in the tree. In section 3 we introduce the ϵ-tree. This data structure is a k-neighbour tree in which we let the value of the tuning parameter k change as the number of stored elements changes. The ϵ-tree can be maintained with an amortized cost of $\Theta\left(\frac{\log n}{\epsilon}\right)$ per update. Using two ϵ-trees we improve the amortized bound into a worst case bound.

We use the same terminology regarding trees as used in [4]. The height of a tree containing n elements is denoted H_n and the number of comparisons required to search among n elements is denoted C_n. log denotes the logarithm to the power of 2 and ln denotes the natural logarithm.

2 k–neighbour trees

The k-neighbour trees presented by Mauer et. al. [4] are unary-binary trees where there are at least k binary nodes between two unary nodes on the same level. We have the following definition:

Definition 1 *A binary tree is called a k-neighbour tree iff*

1. *All leaves have the same depth.*
2. *If a node v has only one child then*
   (a) *v has at least one right neighbour*
   (b) *if v has at least k right neighbours then the k nearest right neighbours have two children otherwise all the right neighbours have two children.*

An example of a k-neighbour tree is given in Figure 1. The tree is a leaf-oriented search tree which represents an ordered set in the following way:

1. All elements are stored in the leaves in sorted order from left to right.
2. Each internal node contains the value of its smallest child.

The tree has a maximum height of

$$H_n \leq \left\lfloor \frac{\log n}{\log(2 - \frac{1}{k+1})} + 1 \right\rfloor \tag{3}$$

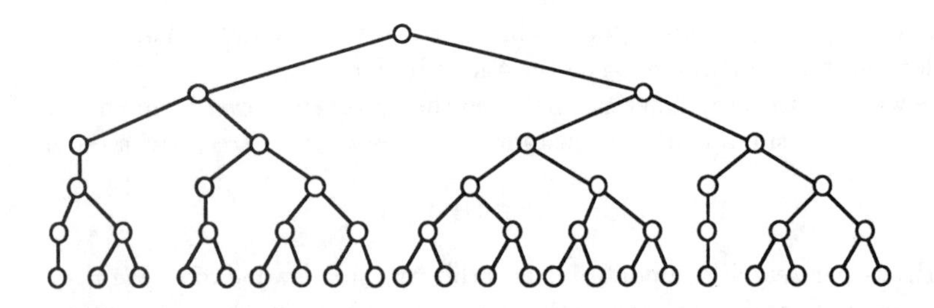

Figure 1: *A 3-neighbour tree.*

Although the elements are not ordered in the same way as in an ordinary binary search tree there is an efficient search algorithm such that

$$C_n = H_n + 1 \tag{4}$$

Mauer e. al. gave maintenance algorithms with $\Theta(\log n)$ worst case performance. Here we are concerned only with the insertion algorithm, which works as follows:

1. Follow a search path down the tree and locate the node p below which a new leaf is to be inserted.

2. Create a new leaf as a child of p and call the procedure INSERT(p).

The recursive procedure INSERT(p) works as follows:

Case 1: p has two children. The insertion is completed.

Case 2: p has three children. Two subcases occur:

Case 2.1: p has a neighbour q at a distance $\leq k$ which has only one child. Make both p and q binary by moving all leaves between them one step.

Case 2.2: p has no neighbour q as described above. Create a new unary node p' and let its only child be the leftmost child of p. Then

Case 2.2.1: p has no parent. Create a new root of the tree to be the common parent of p and p'.

Case 2.2.2: p has a father. Insert p' as a new child of p's father and call INSERT(p's parent).

The largest amount of restructuring work is made during case 2.1 when a number of nodes are moved between p and q. The worst case cost for this is $\Theta(k + \log n)$. Since the algorithm terminates in case 2.1 this work is performed only once. The search for a unary neighbour of p in case 2 takes $\Theta(k)$ time. Since case 2.2 may occur on each level of the tree we get a worst case cost of $\Theta(k \log n)$ for insertion.

In order to obtain our result we have to improve the complexity of the insertion algorithm for k-neighbour trees. This can be made by adding some information to the nodes, as shown in Lemma 1 below.

Lemma 1 *An insertion into a k-neighbour tree requires $\Theta(\log n + k)$ time in the worst case.*

Proof: To each internal node we add a boolean variable telling whether the node has a unary descendant at the distance of $\lfloor \log(k+1) \rfloor$ or not. From the definition of k-neighbour trees follows that each node has at the most one unary descendant at this distance.

Insertions are performed as in the algorithm described in section 2 with the following modification: Instead of making an explicit search for a unary neighbour of p in case 2 we check the ancestor at the distance $\lfloor \log(k+1) \rfloor$ as well as its left and right neighbours. If any of the three nodes has a unary descendant q at the same level as p we proceed with case 2.1 even if the distance between p and q is greater than k. Using this algorithm, the time spent locating the node q in case 2 is $O(1)$ at each level and the total time is $O(\log n)$.

Thus the dominating cost is the one for moving nodes in case 2.1. This cost is depending on the longest possible distance nodes are moved between p and q. Since the ancestors of p and q at distance $\lfloor \log(k+1) \rfloor$ are neighbours (or maybe even the same node) the distance between p and q is less than

$$2 \cdot 2^{\lfloor \log(k+1) \rfloor} \leq 4k + 2 = \Theta(k)$$

Thus the cost for insertion is $\Theta(\log n + k)$ in the worst case and the proof is completed. □

An example of the insertion algorithm is given in Figure 2.

3 ϵ-trees

The result in Lemma 1 allows us to let k take a value of $\Theta(\log n)$, still having a logarithmic cost per insertion. However, the nature of k-neighbour trees does not allow us to change the value of k dynamically. This problem is solved by changing k at repeated intervals.

In this way we achieve a tree with a maximum height of $\rceil \log n + \epsilon \rceil$. The resulting data structure, called the *ϵ-tree* due to its low height, is defined below.

Definition 2 *An ϵ-tree is an k-neighbour tree with the following modifications*

1. *The stored elements are of two types: present and deleted.*
2. $k \geq \left\lceil \frac{3 \log n}{\epsilon} \right\rceil$, *$n$ is the number of present elements.*
3. *The number of deleted elements is less than $\frac{\epsilon}{3} n$*

where ϵ is an arbitrary constant greater than zero.

Queries are performed in the following way:

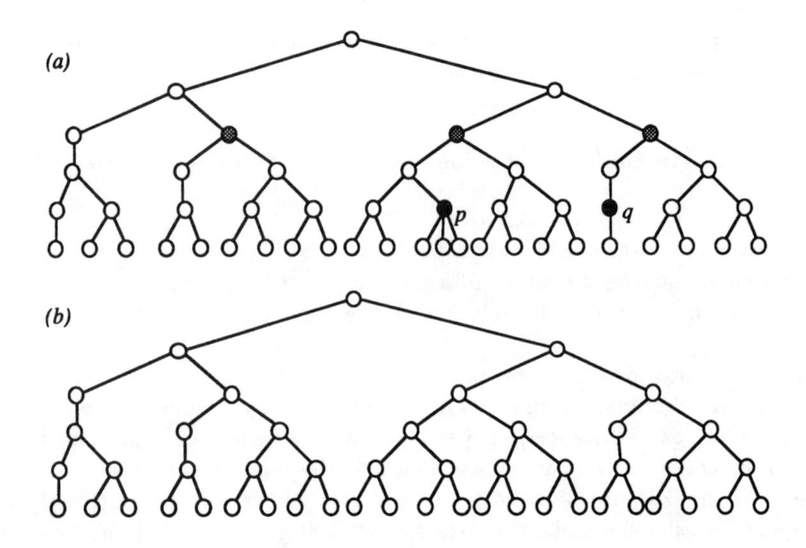

Figure 2: *A insertion in a 3-neighbour tree. (a) A new leaf is inserted below p. The ancestor at a distance of* $\lfloor \log(k+1) \rfloor$ *and its neighbours (gray-coloured) are examined. One of them has a unary descendant q, which is a neighbour of p.*
(b) The tree after moving nodes between p and q.

Insertion: Insert the element into the tree and mark it as present.

Deletion: Locate the element and mark it as deleted.

Search: Locate the element. If the element is found and marked as present the search is successful.

In order to satisfy the definition we rebuild the tree repeatedly. When the number of updates since the latest rebuilding exceeds $\frac{\epsilon}{3}n$ all deleted elements are removed and the tree is rebuilt. In this way the number of inserted elements is always less than $\frac{\epsilon}{3}n$. Let n_0 denote the value of n at the latest rebuilding of the tree. We have

$$n_0 \geq \left(1 - \frac{\epsilon}{3}\right) n \tag{5}$$

When the tree is rebuilt we set

$$k = \left\lceil \frac{3}{\epsilon} \log \left(\frac{n}{1 - \frac{\epsilon}{3}} \right) \right\rceil \tag{6}$$

Equations (5) and (6) gives that

$$k = \left\lceil \frac{3}{\epsilon} \log \left(\frac{n_0}{1 - \frac{\epsilon}{3}} \right) \right\rceil \geq \left\lceil \frac{3 \log n}{\epsilon} \right\rceil \tag{7}$$

In Lemma 2 and 3 below we analyze the height and maintenance cost of an ϵ-tree.

Lemma 2 *The following is true for an ϵ-tree:*

$$H_n \leq \lceil \log n + \epsilon \rceil$$

Proof: The height of the three depends on k and the total number of elements (present and deleted), denoted N. From equation (3) we have

$$H_n \leq \left\lfloor \frac{\log N}{\log\left(2 - \frac{1}{k+1}\right)} + 1 \right\rfloor \tag{8}$$

From the definition of the ϵ-tree we know that the number of deleted elements is less than $\frac{\epsilon}{3}n$, which implies that $N < (1 + \frac{\epsilon}{3})n$. This together with the fact that $k \geq \left\lceil \frac{3\log n}{\epsilon} \right\rceil$ gives that

$$H_n \leq \left\lfloor \frac{\log\left(1 + \frac{\epsilon}{3}\right) n}{\log\left(2 - \frac{1}{\left\lceil \frac{3\log n}{\epsilon} \right\rceil + 1}\right)} + 1 \right\rfloor$$

$$= \left\lfloor \frac{\log n}{\log\left(2 - \frac{1}{\left\lceil \frac{3\log n}{\epsilon} \right\rceil + 1}\right)} + \frac{\log\left(1 + \frac{\epsilon}{3}\right)}{\log\left(2 - \frac{1}{\left\lceil \frac{3\log n}{\epsilon} \right\rceil + 1}\right)} + 1 \right\rfloor \tag{9}$$

For small values of ϵ we have that

$$\log\left(1 + \frac{\epsilon}{3}\right) < \frac{\epsilon}{3\ln 2} < 0.5\epsilon \tag{10}$$

and

$$\log\left(2 - \frac{1}{\left\lceil \frac{3\log n}{\epsilon} \right\rceil + 1}\right) \approx 1 \tag{11}$$

which implies that

$$\frac{\log\left(1 + \frac{\epsilon}{3}\right)}{\log\left(2 - \frac{1}{\left\lceil \frac{3\log n}{\epsilon} \right\rceil + 1}\right)} < 0.6\epsilon < \left(1 - \frac{1}{6\ln 2}\right)\epsilon \tag{12}$$

Combining equation (9) and (12) gives that

$$H_n \leq \left\lceil \frac{\log n}{\log\left(2 - \frac{1}{\left\lceil \frac{3\log n}{\epsilon} \right\rceil + 1}\right)} + \left(1 - \frac{1}{6\ln 2}\right)\epsilon \right\rceil \tag{13}$$

The proof follows from the inequality

$$\frac{\log n}{\log\left(2 - \frac{1}{\frac{3}{\epsilon}\log n + 1}\right)} + \left(1 - \frac{1}{6\ln 2}\right)\epsilon < \log n + \epsilon \ , \quad 2 < n < \infty \tag{14}$$

The inequality (14) is not obvious but can be shown by the substitution

$$t = 2 - \frac{1}{\frac{3\log n}{\epsilon} + 1} \tag{15}$$

This gives

$$\frac{\frac{1}{2-t}-1}{3\log t}\epsilon < \frac{\epsilon}{3}\left(\frac{1}{2-t}-1\right)+\frac{\epsilon}{6\ln 2} \quad , \quad 1<t<2$$

$$\frac{\frac{1}{2-t}-1}{\log t} < \frac{1}{2-t}-1+\frac{1}{2\ln 2} \quad , \quad 1<t<2$$

$$\log t - \frac{t-1}{\left(1-\frac{1}{2\ln 2}\right)t-1+\frac{1}{\ln 2}} > 0 \quad , \quad 1<t<2 \tag{16}$$

By setting

$$f(t) = \log t - \frac{t-1}{\left(1-\frac{1}{2\ln 2}\right)t-1+\frac{1}{\ln 2}} \tag{17}$$

the inequality (16) is equivalent to

$$f(t) \geq 0 \quad , \quad 1 \leq t \leq 2 \tag{18}$$

Derivation gives that

$$f'(t) = \frac{(4(\ln 2)^2 - 4\ln 2 + 1)t^2 + (12\ln 2 - 10(\ln 2)^2 - 4)t + 4(\ln 2)^2 - 8\ln 2 + 4}{((2\ln 2 - 1)t - 2\ln 2 + 2)^2 t \ln 2} \tag{19}$$

Both $f(t)$ and $f'(t)$ are continuous in the interval $1 \leq t \leq 2$. The equation

$$f'(t) = 0 \tag{20}$$

has two solutions, namely

$$t_1 = \frac{2(\ln 2)^2 - 4\ln 2 + 2}{4(\ln 2)^2 - 4\ln 2 + 1} \approx 1.262 \tag{21}$$

and

$$t_2 = 2 \tag{22}$$

Inspection of $f'(t)$ gives that

$$f'(t) > 0 \quad , \quad 1 \leq t < t_1 \tag{23}$$

$$f'(t) < 0 \quad , \quad t_1 < t < 2 \tag{24}$$

Thus $f(t)$ has a local maximum at t_1. This together with the fact that $f(1) = f(2) = 0$ gives that

$$f(t) \geq 0 \quad , \quad 1 \leq t \leq 2 \tag{25}$$

which completes the proof. □

Lemma 3 *The amortized cost per update in an ϵ-tree is $\Theta\left(\frac{\log n}{\epsilon}\right)$.*

Proof: From Lemma 1 and equation (7) we get the cost for a single update to be

$$\Theta(\log n + k) = \Theta\left(\log n + \frac{3\log n}{\epsilon}\right) = \Theta\left(\frac{\log n}{\epsilon}\right) \tag{26}$$

The rebuilding of a tree requires $\Theta(n)$ time. This is amortized over $\Theta(\epsilon n)$ updates which implies that the amortized cost per update is $\Theta\left(\frac{1}{\epsilon}\right)$. Thus the dominating cost is the one given in equation (26) which completes the proof. □

The result of Lemma 2 and 3 allows us to achieve a very low upper bound on the dictionary problem at a logarithmic amortized cost as shown in Theorem 1 below.

Theorem 1 *For any value of $\epsilon, \epsilon > 0$, there is a data structure such that*

$$C_n \leq \lceil \log n + 1 + \epsilon \rceil$$

and the amortized cost per update is

$$\Theta\left(\frac{\log n}{\epsilon}\right)$$

Proof: The proof follows from equation (4) and Lemma 2 and 3. □

Finally, we show how to improve the result of Theorem 1 into a worst case result. BY keeping two ϵ-trees we do not have to stop making queries while we are rebuilding a tree since there will always be one tree in which the queries can be performed.

Theorem 2 *For any value of $\epsilon, \epsilon > 0$, there is a data structure such that*

$$C_n \leq \lceil \log n + 1 + \epsilon \rceil$$

and the worst case cost per update is

$$\Theta\left(\frac{\log n}{\epsilon}\right)$$

Proof: We use a data structure consisting of two ϵ-trees. In each node of the trees we store the number of present descendants. The data structure is maintained in the following way:

1. When one of the trees is rebuilt the rebuilding work is distributed over the updates in such a way that the maximum time spent per update is $\Theta(\log n)$.

2. The two trees are not being rebuilt at the same time.

3. Updates are performed in the following way:

   (a) The update is made in a tree which is not being rebuilt. From the stored information about the number of present descendants we compute the cardinality of the inserted/deleted element.

(b) If the other tree is not being rebuilt an update is also made in that tree. For this second update we use the cardinality of the element and therefore no comparisons are required.

(c) If the other tree is being rebuilt we store the update in a queue to be performed at the end of the rebuilding. The cardinality of the element is stored in the queue to avoid comparisons when the update is to be made in the tree.

A rebuilding consist of two phases: construction of a new tree and performance of the queued updates. The first step requires $\Theta(n)$ time and the second one requires $\Theta(n \log n)$ time. Therefore, the entire construction can be distributed over a linear number of updates at a worst case cost of $\Theta(\log n)$ per update. Although each update is performed in both trees we make element-comparisons only the first time. The rest of the proof is similar to the proof of Theorem 1. □

4 Comments

We have shown that there exist a data structure such that we can perform the dictionary queries insert, search and delete in logarithmic time, making at the most two comparisons more than the optimal number.

The method to improve complexity by using a "varying constant" (k in the ϵ-tree) is not restricted to search trees and it might also be used to improve complexity in other applications.

References

[1] G. M. Adelson-Velski and E. M. Landis. An algorithm for the organization of information. *Dokladi Akademia Nauk SSSR*, 146(2), 162.

[2] A. Andersson. Binary search trees of almost optimal height. *tech. report, Department of Computer Science, Lund University*, 1988.

[3] R. Bayer. Symmetric binary B-trees: Data structure and maintenance algorithms. *Acta Informatica*, 1(4), 1972.

[4] H. A. Mauer, Th. Ottman, and H. W. Six. Implementing dictionaries using binary trees of very small height. *Information Processing Letters*, 5(1), 1976.

[5] J. Nievergelt and E. M. Reingold. Binary trees of bounded balance. *SIAM Journal on Computing*, 2(1), 1973.

[6] J. van Leeuwen and M. H. Overmars. Stratified balanced search trees. *Acta Informatica*, 18, 1983.

OPTIMAL CONSTANT SPACE MOVE-TO-REAR LIST ORGANIZATION†

B. John Oommen and **David T.H. Ng**
School of Computer Science,
Carleton University,
Ottawa, K1J 5B6,
CANADA.

ABSTRACT

We consider the problem of adaptively organizing a list whose elements are accessed with a fixed but unknown probability distribution. We present a strategy which has constant additional space requirements and which achieves the reorganization by performing a data restructuring operation on an element **exactly once**. The scheme, which is stochastically absorbing, and is of a Move-to-Rear flavour is shown to be asymptotically optimal. In other words, by suitably performing the Move-to-Rear operation the probability of converging to the optimal arrangement can be made as close to unity as desired. Considering all of these features, this strategy is probably the most ideal list organization strategy reported in the literature. Simulation results demonstrating the power of the scheme have been included. The paper also includes a hybrid data reorganization scheme in which an absorbing Move-To-Rear rule and an ergodic rule are used in conjunction with each other.

I. INTRODUCTION

The most elementary and commonly used data structure in computer science is the linear list. The structure has a myriad of applications and is favoured primarily because of the ease in analysing its behaviour in a variety of applications and also because of the ease in implementing it. The latter is particularly true because each element of the list points only to the next element, and thus pointer manipulations are rendered extremely simple.

When a set of elements are to be organized as a list, one would like to arrange the list in such a way that the data retrieval process is optimal. Here, of cause, the data organization is dependent on the users' access probabilities because we would like the more frequently accessed elements to be towards the front of the list in comparison with the less frequently used ones. Indeed, if the access probabilities of the elements are known, it is a trivial exercise to show that the list is optimally arranged if the elements are arranged in the

† Partially supported by the Natural Sciences and Engineering Research Council of Canada.

descending order of their access probabilities.

The problem is however far more complex when the access probabilities of the elements are unknown *a priori*. One can then, of cause, envision a scheme which estimates these probabilities and which arranges the elements in the descending order of these estimates. As opposed to this, a large amount of research has gone into the study of having the list organize itself. This paper deals with the problem of adaptively organizing the list **without** estimating the access probabilities of the elements.

To be more specific, let $\mathfrak{R} = \{R_1, ..., R_N\}$ be a list of elements where each R_i is accessed with a probability s_i. $\mathfrak{S} = \{s_1, ..., s_N\}$ is called the user's access distribution and in all the literature available this distribution is assumed to be stationary, and thus $\mathfrak{S}$ is time-invariant. We intend to arrange the elements of $\mathfrak{R}$ as a list so that the average access time is asymptotically minimized. To render the problem non-trivial, $\mathfrak{S}$ is assumed unknown.

Various adaptive strategies have been described in the literature. A few of them are listed below for the case when the accessed element is not at the front of the list.

(i) The Move-To-Front Rule : The accessed element is moved to the front of the list.

(ii) The Transposition Rule : The accessed element is moved towards the front of the list by interchanging it with its preceding element.

(iii) The Move-k-Ahead Rule : The accessed element is moved k positions forward toward the front of the list unless it is found in the first k positions, in which case, it is moved to the front of the list itself.

(iv) The POS-k Rule : The accessed element is moved to position k of the list if it is in positions k+1 through N. It is transposed with its preceding element if it is in positions 2 through k.

In all of the above cases the list is unchanged if the accessed element is at the front of the list. The properties of the various schemes are found in the respective references and the surveys [1, 3, 4, 6, 13, 15].

Observe that all of the above schemes have an ergodic Markovian representation. Thus, even in the limit, the list can be in any one of its N! configurations. Consequently, a list which was completely organized can be significantly disorganized by the Move-To-Front (or POS-k or Move-k-Ahead) rule by a **single** request of the most infrequently accessed element. Observe that after this unfortunate occurrence, it will take a long time for the list to be organized again - i.e., for this element to dribble its way to the tail of the list. In spite of this drawback, ergodic schemes are generally to be desired if the environment is non-stationary.

The literature reports the following **absorbing** list organizing methods :

(i) A stochastic Move-To-Front rule in which at time 'n' the accessed element is moved to the front of the list with a probability $f(n) = a^n$. The scheme is expedient [11] if $f(0) = 1$ and $0 < a < 1$.

(ii) A stochastic Move-To-Rear rule in which at time 'n', if the accessed element is R_j, the element is moved to the **rear** of the list with a probability $g(n) = a^{Z_j(n)}$, where $Z_j(n)$ is the number of times R_j has been accessed. The scheme is asymptotically optimal as $a \rightarrow 1$ [11].

(iii) A deterministic Move-To-Rear rule which moves the accessed element to the **rear** of the list if it has been accessed k times. The scheme is asymptotically optimal as $k \to \infty$ and performs exactly N data reorganization operations. The drawback of the scheme however is that it requires an extra linear amount of memory [12].

(iv) A deterministic Move-To-Rear rule which moves the accessed element to the **rear** of the list if it has been accessed k consecutive times [12]. This scheme was proven to be expedient but conjectured optimal as $k \to \infty$.

In this paper, we shall present a new Move-To-Rear rule. In one sense it is but a modified version of the one described in (iv) above. But in another sense, it is a rule in its own right and is distinct from all the other rules examined in the literature. The scheme requires a **constant** amount of additional memory locations and performs **exactly** N data reorganization operations. Finally, and most importantly, the scheme is absorbing and asymptotically optimal. It is thus the ideal list organizing strategy being advantageous in terms of the speed (i.e., the number of data movement operations is the minimum possible), in terms of accuracy and in terms of Markovian behaviour. The paper proves some of the theoretical properties of the scheme. A hybrid heuristic in which the MTR rule and an ergodic rule work simultaneously to enhance the data retrieval is alluded to in the paper. Such a hybrid system is completely novel to the literature.

II. THE IDEAL MOVE-TO-REAR STRATEGY

Let $\mathfrak{V}$ be the set of elements that have not been moved from their initial positions. Note that $\mathfrak{V}$ is initially set to be $\mathfrak{R}$ and is repeatedly being shrunk until it is the null set. When an element R_i is accessed, if R_i is not in $\mathfrak{V}$ (i.e., if R_i has already been moved from its initial position) although the contents of the record R_i are presented to the user, the data reorganization strategy completely ignores the access and thus the list organizing strategy is by-passed. We shall merely now consider the case when $R_i \in \mathfrak{V}$.

Let M be any fixed integer and **Z** be a vector memory location having two integer fields, Z_1 and Z_2. Whenever the element $R_i \in \mathfrak{V}$ is accessed, if Z_1 is zero, it is set to i and Z_2 is set to unity. Furthermore, if Z_1 was previously set to i, Z_2 is incremented. Otherwise the values of both Z_1 and Z_2 are reset to zero. Whenever the value of Z_2 is exactly M, the element R_i (i.e. the element whose index is stored in Z_1) is moved to the **rear** of the list and $\mathfrak{V}$ is updated by deleting from it the element R_i. Subsequently, the element R_i is never moved. This describes the list organization strategy completely. Observe that essentially all that this scheme does is that it moves a record $R_i \in \mathfrak{V}$ to the rear of the list after it has been accessed M **consecutive** times (although in between these accesses, elements of $\mathfrak{R} - \mathfrak{V}$ may have been accessed). We define the condition of having $Z_1 = Z_2 = 0$ as the state of maximum uncertainty.

The above technique is given formally in Algorithm Ideal_MTR below. We shall now prove some of the properties of the scheme.

Theorem I : The Algorithm Ideal_MTR described above is absorbing. Furthermore, the total number of list reorganizing operations done is exactly N.

Proof : The second assertion is obvious. The absorbing nature of the chain follows from the fact that after every element has been accessed M consecutive times (ignoring the elements which were previously accessed M consecutive times) the list remains in its final configuration, and no more list reorganizing is performed. •••

ALGORITHM Ideal_MTR

Input : A sequence of accesses on the list $\mathfrak{R}$. R_i is accessed with an unknown access probability s_i. This is read in using procedure ReadInput.

Output : A reorganized list.

Memory Requirements :

(i) An extra memory location **Z** with two integer fields Z_1 and Z_2. Z_1 stores the index of the last accessed record in $\mathfrak{V}$ and Z_2 records the number of times the record whose index is Z_1 has been consecutively accessed if accesses of elements in $\mathfrak{R} - \mathfrak{V}$ are ignored.

(ii) An integer constant M.

Method

$\mathfrak{V} \leftarrow \mathfrak{R}$;

$Z_1 \leftarrow 0$; $Z_2 \leftarrow 0$ /* Initialize the algorithm to the state of Maximum Uncertainty */

Repeat

ReadInput (R_i)

If $R_i \in \mathfrak{V}$ **Then**

If ($Z_1 = i$) **Then** $Z_2 \leftarrow Z_2 + 1$

Else If $Z_1 \neq 0$ **Then** $Z_1 \leftarrow 0$; $Z_2 \leftarrow 0$ /*Reset to Maximum Uncertainty */

Else $Z_1 \leftarrow i$; $Z_2 \leftarrow 1$

Endif

Endif

If ($Z_2 = M$) **Then** /* R_i has been accessed M consecutive */

Move-To-Rear (R_i) /* times ignoring accesses in $\mathfrak{R} - \mathfrak{V}$ */

$\mathfrak{V} \leftarrow \mathfrak{V} - \{R_i\}$

Endif

Endif

Forever

End Algorithm Ideal_MTR.

Note that it is not just the **expected** number of move operations which is finite as in [12]. Here the number of moves performed is exactly equal to the size of the list.

The following results concerning the probability of being absorbed in the optimal configuration is more powerful and is the main result of the paper. We shall show that the scheme is asymptotically optimal, i.e., the probability of being absorbed in the optimal configuration can be made as close to unity as desired. To aid the proof, we now introduce the following notations. The access probability vector $\mathfrak{S}$ is an N by 1 probability vector satisfying :

$$\mathfrak{S} = [s_1, s_2, \ldots s_N]^T, \text{ where,} \tag{2}$$

$$\sum_{i=1}^{N} s_i = 1. \tag{3}$$

Let $\mathfrak{V}$ be any subset of $\mathfrak{R}$. We define the conditional access probability vector $\mathfrak{S}|\mathfrak{V}$ as the vector of normalized probabilities in which only the quantities corresponding to the elements of the set $\mathfrak{V}$ have non-zero values. Thus, $\mathfrak{S} \mid \mathfrak{V}$ consists of the set $\{s_i'\}$, where,

$$s_i' = 0 \qquad \text{if } R_i \notin \mathfrak{V} \tag{4}$$

$$= \frac{s_i}{\sum_{R_i \in \mathfrak{V}} s_i} \qquad \text{otherwise.} \tag{5}$$

Observe that the vector $\mathfrak{S}$ defined by (2) and (3) is exactly equivalent to $\mathfrak{S} \mid \mathfrak{R}$, and thus these above normalized probabilities are conditional probabilities appropriately conditioned.

The main result in Markov chains which we shall use to compute the probability of absorption into an absorbing barrier is given below [8].

Lemma 0 : Let F be the transition matrix of an irreducible homogeneous Markov chain, where, $F_{i,j}$ is the transition probability associated with the transition from state i to state j. Let $x_{i,A}$ be the probability of being absorbed into an absorbing state A, where $i \in \tau$, the set of transient states. Then $x_{i,A}$ obeys the following recursive equation :

$$x_{i,A} = F_{i,A} + \sum_{j \in \tau} (F_{i,j} \cdot x_{j,A}) \tag{6}$$

Theorem II : Let $\mathfrak{S}$ be the vector of access probabilities. Then, the probability that R_u precedes R_v in the final list obtained by using Algorithm Ideal_MTR is exactly $s_u^M/(s_u^M+s_v^M)$.

Proof : Let R_u and R_v be any two distinct elements with indices $u, v \in \{1, 2, \ldots, N\}$, and let their corresponding access probabilities be s_u and s_v respectively. Since the theorem is trivial for the case when either (or both) these probabilities are zero, with no loss of generality we assume that both s_u and s_v are positive. It is required to prove that in this case, the probability of being absorbed into a list in which R_u precedes R_v is $s_u^M / (s_u^M + s_v^M)$. We shall prove the theorem by performing a computation of the probability conditioned on the length of the list starting from the first element that is moved to the rear.

For the indices $u, v \in \{1, 2, \ldots, N\}$, let $\xi_{u,v}(\mathfrak{V})$ be the event that either R_u or R_v is the element which is selected from $\mathfrak{V}$ to be moved to the rear, where $\mathfrak{R} \supseteq \mathfrak{V}$. Also, let $\xi_{u,v}(i)$ be the event that the ith element moved contains the first appearance of R_u or R_v. Thus, by virtue of the scheme, R_u ultimately precedes R_v if it is selected to be moved **before** R_v, since if that is the case, R_u is the ith element moved, and R_v will be the jth element moved where $i < j$.

Since $\mathfrak{V}$ is the set of elements which have not been moved to the rear of the list, by definition, the conditional access probabilities are defined by (5). Let

$$p = s_u / \sum_{R_i \in \mathfrak{V}} s_i \quad \text{, and,} \quad q = s_v / \sum_{R_i \in \mathfrak{V}} s_i \ .$$

Then p is the conditional access probability of accessing R_u, q is the conditional access probability of accessing R_v and 1-p-q is the conditional access probability of accessing any element in $\mathfrak{V} - \{R_u, R_v\}$.

We now compute the probability of R_u preceding R_v given $\xi_{u,v}(\mathfrak{V})$. To do this we describe the following homogeneous irreducible Markov chain. The chain has a set of states, and it is in state $\phi(n)$ defined below :

$$\phi(n) = U_i \quad \text{if } Z_1 = u \text{ and } Z_2 = i \tag{7}$$
$$= V_i \quad \text{if } Z_1 = v \text{ and } Z_2 = i \tag{8}$$
$$= \theta_0 \quad \text{if } Z_1 = 0 \text{ and } Z_2 = 0 \tag{9}$$
$$= \theta_t \quad \text{otherwise.} \tag{10}$$

Clearly, the Markov chain represents the contents of the memory locations Z_1 and Z_2 given $\xi_{u,v}(\mathfrak{V})$, where θ_0 is the state of maximum uncertainty, U_i is the state when R_u has been accessed i consecutive times if accesses of elements in $\mathfrak{R} - \mathfrak{V}$ are ignored, V_i is the state when R_v has been accessed i consecutive times if accesses of elements in $\mathfrak{R} - \mathfrak{V}$ are ignored and θ_t is the state when any element in $\mathfrak{V} - \{R_u, R_v\}$ has been accessed at least once (but not M consecutive times). Note that given $\xi_{u,v}(\mathfrak{V})$, the chain from θ_t leaves θ_t with probability one. Thus the transition probabilities of the Markov chain are given by the stochastic matrix F, where, for all $i \leq i \leq M\text{-}1$,

$$F_{U_i,U_{i+1}} = p \tag{11}$$
$$F_{U_i,\theta_0} = 1 - p \tag{12}$$
$$F_{V_i,V_{i+1}} = q \tag{13}$$
$$F_{V_i,\theta_0} = 1 - q \tag{14}$$
$$F_{\theta_0,U_1} = p \tag{15}$$
$$F_{\theta_0,V_1} = q \tag{16}$$
$$F_{\theta_0,\theta_t} = 1 - p - q \tag{17}$$
$$F_{\theta_t,\theta_0} = 1 \tag{18}$$

Observe that since U_M and V_M are absorbing states,

$$F_{U_M,U_M} = F_{V_M,V_M} = 1. \tag{19}$$

The transition map of the Markov chain is given diagramatically in [16]. Additional properties of the map are also found in [16] but are omitted here for the sake of brevity.

Let a_i be the probability of converging to U_M given $\xi_{u,v}(\mathfrak{V})$ and that the starting state is U_i. Similarly let b_i be the probability of converging to U_M given $\xi_{u,v}(\mathfrak{V})$ and that the starting state is V_i. Also let Δ_0 be the probability of converging to U_M given $\xi_{u,v}(\mathfrak{V})$ and that the starting state is θ_0 and let Δ_t be the probability of converging to U_M given $\xi_{u,v}(\mathfrak{V})$ and that the starting state is θ_t. Observe that Δ_0 is the quantity we aim to compute. Using Lemma 0, the quantities $\{a_i\}$, $\{b_i\}$, Δ_0 and Δ_t satisfy :

$$a_M = 1; \tag{20}$$
$$a_i = p\, a_{i+1} + (1 - p)\, \Delta_0, \quad \text{for } 1 \leq i \leq M\text{-}1; \tag{21}$$
$$b_M = 0; \tag{22}$$
$$b_i = q\, b_{i+1} + (1 - q)\, \Delta_0, \quad \text{for } 1 \leq i \leq M\text{-}1. \tag{23}$$

Finally, since θ_t is a reflecting barrier given $\xi_{u,v}(\mathfrak{V})$, $\Delta_t = \Delta_0$, and thus,

$$\Delta_0 = p\, a_1 + q\, b_1 + (1\text{-}p\text{-}q)\, \Delta_0. \tag{24}$$

Solving the difference equation (21) using (20) as a boundary condition yields the value of a_1 as :

$$a_1 = p^{M-1} + \left[\sum_{j=0}^{M-2} p^j\right] (1\text{-}p)\, \Delta_0. \tag{25}$$

Similarly, solving the difference equation (23) using (21) as a boundary condition yields the value of b_1 as :

$$b_1 = \left[\sum_{j=0}^{M-2} q^j\right] (1\text{-}q)\, \Delta_0. \tag{26}$$

Substituting for a_1 and b_1 in (24) yields the equation for Δ_0 as :

$$\Delta_0 = p^M + \left[\left\{\sum_{j=1}^{M-1} p^j\right\} (1\text{-}p) + \left\{\sum_{j=1}^{M-1} q^j\right\} (1\text{-}q)\right] \Delta_0 + (1 - p - q)\, \Delta_0,$$

whence, $\Delta_0 = \frac{p^M}{\mathrm{Den}}$ (27)

where the Denominator Den is :

$$\mathrm{Den} = (p + q) - \left[\sum_{j=1}^{M-1} p^j\right] (1\text{-}p) - \left[\sum_{j=1}^{M-1} q^j\right] (1\text{-}q). \tag{28}$$

Summing the two geometric series in p^j and q^j yields Den = $p^M + q^M$.
But $p = s_u / \sum_{R_i \in \mathfrak{V}} s_i$ and $q = s_v / \sum_{R_i \in \mathfrak{V}} s_i$.

Since the denominators for p and q are the same, we simplify Δ_0 to yield :

$$\Delta_0 = \frac{s_u^M}{s_u^M + s_v^M}.$$

Since this is true for all $\mathfrak{V} \supseteq \{R_u, R_v\}$, we have thus proved that for all i, obeying $1 \le i \le M1$,

$$\Pr\left[R_u \text{ ultimately precedes} \mid \xi_{u,v}(\mathfrak{V})\right] = \frac{s_u^M}{s_u^M + s_v^M}.$$

To complete the proof we note that since $\xi_{u,v}(1), \ldots, \xi_{u,v}(M\text{-}1)$ are mutually exclusive and collectively exhaustive events, using the laws of total probability,

$$\Pr(R_u \text{ precedes } R_v) = \sum_i \Pr[R_u \text{ precedes } R_v \mid \xi_{u,v}(i)] \,.\, \Pr[\xi_{u,v}(i)]$$

Since $\Pr[R_u \text{ precedes } R_v \mid \xi_{u,v}(i)]$ is the same for all i,

$$\Pr(R_u \text{ precedes } R_v) = \Pr[R_u \text{ precedes } R_v \mid \xi_{u,v}(i)] \,.\, \sum_i \Pr[\xi_{u,v}(i)] = \frac{s_u^M}{s_u^M + s_v^M}$$

and the result follows. • • •

The final theorem regarding the asymptotic optimality of the scheme is now proved.

Theorem III : Let $\mathfrak{R} = \{R_1, R_2, \ldots, R_N\}$ be the list of elements with distinct access probabilities $\{s_1, \ldots, s_N\}$. Then, the probability of the list converging to the optimal arrangement converges to unity as $M \to \infty$.

Proof : For every distinct pair u, v $\{1, \ldots, N\} \times \{1, \ldots, N\}$, we have shown that the probability of R_u ultimately preceding R_v is :

$$\Pr[R_u \text{ ultimately precedes } R_v] = \frac{s_u{}^M}{s_u{}^M + s_v{}^M}.$$

As M tends to ∞, we see that if $s_u > s_v$,

$$\lim_{M\to\infty} \Pr\left[R_u \text{ ultimately precedes } R_v\right] = \lim_{M\to\infty} \frac{s_u{}^M}{s_u{}^M + s_v{}^M} = \lim_{M\to\infty} \frac{1}{1 + \left(\frac{s_v}{s_u}\right)^M} = 1.$$

Thus R_u ultimately precedes R_v w.p. 1 as $M\to\infty$.

Let P* be the probability that the list converges to optimal arrangement. Since the records are independently drawn according to the distribution $\{s_i\}$, the probability P* is greater than or equal to the probability that **every single pair** is in the decreasing order of the access probabilities. Thus,

$$P^* \geq \prod_{u\neq v} \Pr\left[R_u \text{ ultimately precedes } R_v \mid s_u > s_v\right]$$

The result that P* tends to unity follows since every quantity on the product term of the R.H.S. does so as $M \to\infty$. •••

Apart from proving that the probability that R_u ultimately precedes R_v in the final list converges to unity as $M\to\infty$ (if $s_u > s_v$), the above results can also give us a closed form expression for the probability of any final list arrangement. The detailed expression for this probability is given in [16].

The difference between the Algorithm Deterministic-MTR-CSpace [12] and the algorithm Ideal_MTR described in this paper are found in [16]. However, there is one fundamental difference between the algorithm Ideal_MTR and **all** of the other algorithms reported in the literature. In all the algorithms reported in the field of data reorganization, operations are executed based on the probability of the accessed element being requested by the user. However, in the case of Algorithm Ideal_MTR, the data reorganization is based on the **conditional** access probability vector $\mathfrak{S} \mid \mathfrak{V}$. Thus, although accesses of elements in the set $\mathfrak{R} - \mathfrak{V}$ can occur and that, even with a high probability, the occurrence of such accesses **does not** place an impediment on the convergence of the scheme. Since elements in $\mathfrak{R} - \mathfrak{V}$ have been placed in their appropriate positions they are merely reported to the user on being accessed. But elements in $\mathfrak{V}$ (the ones which are still in their "unabsorbed states") are the only elements which are involved in any data reorganization computations. The computational impact of this strategy is illustrated in [16].

III. IMPLEMENTATION DETAILS

The scheme that we have introduced, Ideal-MTR has been described in terms of certain set inclusion operations. Thus, we have chosen to move a record R_i if it is in a set $\mathfrak{V}$, which is a subset of $\mathfrak{R}$, the set of all records. Furthermore, the Move-To-Rear operation was performed to the tail of the list which requires that the most frequently accessed element will seem to linger at the tail of the list until all the elements found their places. Finally, the algorithm was also described in terms of transition to the state of maximum uncertainty and the transition depended critically on the question of whether R_i and R_j (where R_j is the record interrupting a string of R_i's) are elements of $\mathfrak{V}$. Clearly, implementing the algorithm can be clumbersome if the implementation involved only the set operations. However, we shall now show that the implementation is very straight forward if the list is maintained using pointers (as in usually the case) and an additional pointer is used to point of the element of the list which was first moved to the rear.

Let FrontOld be the pointer to the head of the list. This pointer always points to the front of the original list prior to any data reorganization. A second pointer, FrontNew, initially points to the tail of the list. (i.e., it is initialized to NIL). Subsequently as the elements of the list are moved to the rear, they are appended to the tail of the list pointed at by FrontNew. Thus, the elements which have found their place lie between FrontNew and the tail of the list and the elements which are still learning their place lie between FrontOld and FrontNew. Whenever the user request a record R_i, the search for R_i starts from FrontNew and continues towards the tail of the list. If R_i is found prior to encountering the null pointer, this indeed represent the case when R_i is contained in the set $\mathfrak{R}$ - $\mathfrak{V}$. In such a case the counters Z_1 and Z_2 are unchanged.

However, if the accessed element R_i has not been found by the time the null pointer is encountered, the search for R_i continues progressing from FrontOld all the way down to FrontNew. This traversal could result in a data reorganization operation after the corresponding changes to Z_1 and Z_2 have been made.

The implementation of the scheme is shown in Figure I for a list of five elements namely the list $\mathfrak{R} = \{R_1, R_2, \ldots R_5\}$ in which $s_1 > s_2 > \ldots > s_5$. In Figure I, we have considered a case when the elements are reorganized (i.e. attain the value M) in the order (1, 2, 3, 4, 5). Initially, on R_1 being accessed M consecutive times, it is moved to the rear of the list. FrontNew now points to R_1. Subsequently, on R_2 being accessed M consecutive times (i.e. ignoring the accesses on R_1), **it** is moved to the rear of the list. As time proceeds whenever R_3 is accessed M consecutive times (i.e. ignoring the accesses of R_1 and R_2), R_3 is moved to the rear of the list. However, to catalyze the accessing process, the list pointers FrontOld and FrontNew point to the reorganized list.

Observe that whenever R_i lies between FrontOld and FrontNew, no data reorganization is performed. It is very easy to generalize this to yield a hybrid system in which the elements between FrontOld and FrontNew are reorganized based on an ergodic heuristic until they have finally found their place at the tail of the list. Thus, although R_i has not been accessed M consecutive times, (i.e. ignoring access of $\mathfrak{R}$ - $\mathfrak{V}$) it need not remain static in its original position. Whenever R_i is accessed, it can be moved forwards towards FrontOld by a heuristic such as the Transposition rule, and thus we are not only guarantied that the cost will be optimal when the list has converged to its final absorbing configuration but we are also guarantied that in between the time instances when a MTR operation is done, the elements which are still in their transient configuration will migrate towards FrontOld so as to minimize the retrieval cost. Such a hybrid system is completely novel to the field of data retrieval.

The algorithmic details of the hybrid scheme and the pointer manipulations involved can be found in [16]. They are omitted here for the sake of brevity. The latter reference [16] includes extensive simulation results involving the scheme introduced here for various families of random distributions.

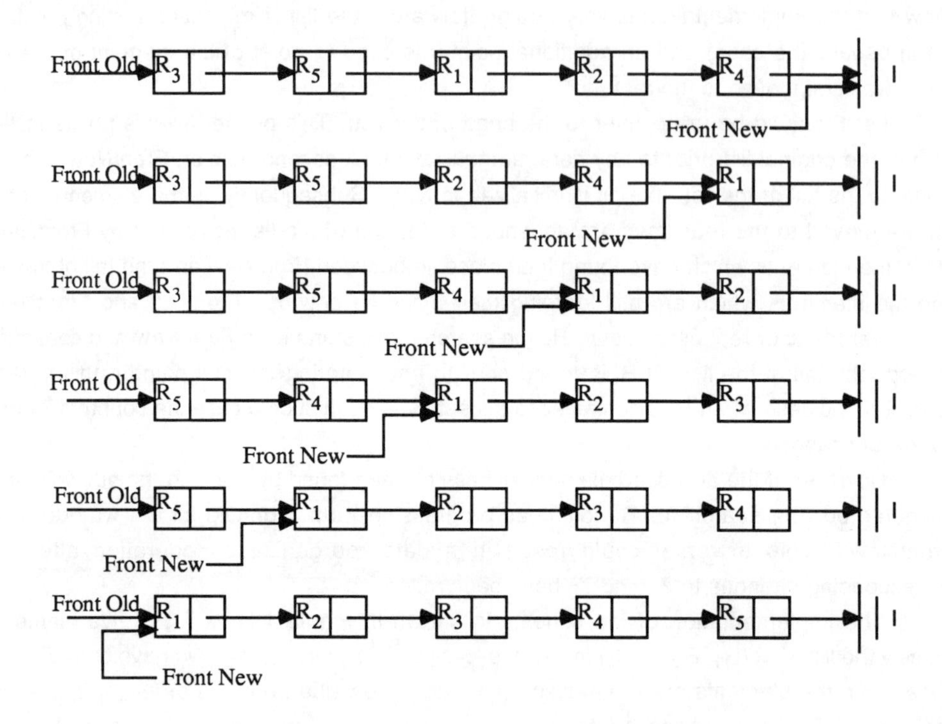

Figure I. The operations of MTR on a list of elements R_1, R_2, ..., R_5 for the case when R_1 is accessed M consecutive times, then R_2 is accessed M consecutive times (ignoring accesses on R_1), and so on for R_3, R_4 and R_5, until all 5 elements are reorganized.

IV. CONCLUSIONS

In this paper, we have considered the problem of organizing the linear list of N elements in which the elements are accessed with a fixed but unknown probability distribution. We have presented a strategy which requires a constant number of extra memory locations and which achieves the data reorganization by performing exactly N restructuring operations. The scheme is stochastically absorbing and of Move-To-Rear flavour, and we have shown that the probability of the scheme converging to the optimal arrangement can be made as close to unity as desired. The algorithm is fundamentally distinct from **all** of the other algorithms reported in the literature because, unlike the contemporary methods, operations are not executed based on the probability of the accessed element being requested by the user, but they are done based on a **conditional** access probability vector, the details which have been described in the paper. The paper also suggests a hybrid data reorganization scheme in which an absorbing Move-To-Rear rule and an ergodic rule are used in conjunction with each other to optimize the data retrieval process.

REFERENCES

[1] Arnow, D.M. and Tenebaum, A.M., "An Investigation of the Move-Ahead-k Rules", Congressus Numerantium, Proc. of the Thirteenth Southeastern Conference on Combinatorics, Graph Theory and Computing, Florida, February 1982, pp. 47-65.

[2] Bitner, J.R., "Heuristics That Dynamically Organize Data Structures", SIAM J. Comput., Vol.8, 1979, pp. 82-110.

[3] Burville, P.J. and Kingman, J.F.C., "On a Model for Storage and Search", J. Appl. Probability, Vol.10, 1973, pp. 697-701.

[4] Gonnet, G.H., Munro, J.I. and Suwanda, H., "Exegesis of Self Organizing Linear Search", SIAM J. Comput., Vol.10, 1981, pp.613-637.

[5] Hendricks, W.J., "An Extension of a Theorem Concerning an Interesting Markov Chain", J. App. Probability, Vol.10, 1973, pp.231-233.

[6] Hester, J.H. and Hirschberg, D.S., "Self-Organizing Linear Search", ACM Computing Surveys, Vol. 17, 1985, pp.295-311.

[7] Kan, Y.C. and Ross, S.M., "Optimal List Order Under Partial Memory Constraints", J. App. Probability, Vol.17, 1980, pp. 1004-1015.

[8] Isaacson, D.L. and Madsen, R.W., "Markov Chains : Theory and Applications" New York, John Wiley & Son, 1976.

[9] Knuth, D.E., "The Art of Computer Programming, Vol.3, Sorting and Searching", Addison-Wesley, Reading, MA., 1973.

[10] McCabe, J., "On Serial Files With Relocatable Records", Operations Research, Vol.12, 1965, pp.609-618.

[11] Oommen, B.J. and Hansen, E.R., "List Organizing Strategies Using Stochastic Move-to-front and Stochastic Move-to-Rear Operations", to appear in SIAM Journal on Computing. Vol. 16, (1987), 705-716.

[12] Oommen, B.J., Hansen, E.R. and Munro, J.I., "Deterministic Optimal and Expedient Move-To-Rear List Organizing Strategies", to appear in Theoretical Computer Science (Preliminary abridged version was published in the Proc. of the 25th Annual Allerton Conference, Urbana, Illinois, Sept/Oct. 1987, pp.54-63).

[13] Rivest, R.L., "On Self-Organizing Sequential Search Heuristics", Comm. ACM, Vol.19, 1976, pp.63-67.

[14] Sleator, D. and Tarjan, R., "Amortized Efficiency of List Update Rules", Proc. of the Sixteenth Annual ACM Symposium on Theory of Computing, April 1984, pp. 488-492.

[15] Tenenbaum, A.M. and Nemes, R.M., "Two Spectra of Self-Organizing Sequential Search Algorithms", SIAM J. Comput., Vol.11, 1982, pp-557-566.

[16] Oommen, B.J., Ng, D.T.H., "Ideal List Organization for Stationary Environments", Submitted for publication. Also available as a Technical Report from the School of Computer Science, Carleton University, Ottawa K1S 5B6, CANADA.

IMPROVED BOUNDS ON THE SIZE OF SEPARATORS OF TOROIDAL GRAPHS

L. G. Aleksandrov and H. N. Djidjev

Center of Informatics and Computer Technology
Bulgarian Academy of Sciences
Acad. G. Bonchev. Str. bl. 25-A, Sofia, Bulgaria

Abstract. It is known that the set of vertices of any toroidal graph (graph of orientable genus 1) can be divided into two edge-disjoint sets of size no greater than 2/3 times the size of the original graph by deleting no more than $\sqrt{18}\sqrt{n}$ vertices [2]. The paper improves the constant before $\sqrt{n}$ in the above theorem to $\sqrt{12}$ by using the structure *separation graph* and gives a lower bound on the optimal constant that can replace $\sqrt{12}$.

1. Introduction

Let G be an n-vertex graph of orientable genus g and a spanning tree T. There exists a set of no more than $\beta\sqrt{(g+1)n}$ vertices of G whose removal divides G into components of no more than $2n/3$ vertices each [1,3]. In this paper we address the problem of estimating the optimal constant that can replace β in the above separator theorem. Suppose that G is connected and let T be a spanning tree of G. A *T-cycle* of G will be called any simple cycle of G that contains exactly one edge not in T. The previous proof of the above separator theorem uses the fact that there exists a set of no more than $2g+1$ T-cycles of G whose removal divides G into components of no more than $2n/3$ vertices each. Intuitively, removal of $2g$ cycles makes the graph "almost planar" (i.e. such that each of the remaining cycles divides the embedding of the graph) and one additional cycle is needed to divide the resulting graph. Any essential improvement of the number $2g+1$ of these cycles will lead to an improvement the constant β in the separator theorem. It is not known what is the lowest value of $k(g)$ that can replace the constant 2 before g. In the paper we show that for some small values of g, i.e. for $g=1$ (the case of toroidal graphs), $k(g)=1$ and we prove the corresponding separator theorem. Our result reduces the constant before $\sqrt{n}$ in the size of the separator from $\sqrt{18}\approx 4,2426$ to $\sqrt{12}\approx 3,4641$.

To obtain the claimed improvement we use the structure *separation graph* of G, denoted by $S(G)$, as defined in [2]. $S(G)$ has f vertices and $f+2g-1$ edges, where f is the number of the faces and g is the genus of G. Using the separation graph we reduce the problem of partitioning a graph of small genus to the problem of partitioning a sparse graph. For example, if G is planar then $S(G)$ is a tree and if G is toroidal then $S(G)$ is a tree plus two extra edges. The separation graphs have been exploited in [3] to obtain an $O(n)$ algorithm for finding a so-called ε*-separator*, an improvement over the previous $O(n \log n)$ algorithm.

A technique for establishing a lower bound on the size of the separating set of vertices for toroidal graphs is proposed. It is shown that the lowest constant that can replace the leading constant 3,4641 in the size of the separator for toroidal graphs is not smaller than 1,9046. These results restrict the interval containing the optimal constant from the best previous [1,5550, 4,2426] to [1,9046 , 3,4641].

The paper is organized as follows. In Section 2 we introduce the notion of a separation graph and establish a relation between the problem of dividing a graph and the problem of dividing its separation graph. In Section 3 we prove the separator theorem for toroidal graphs and in Section 4 we obtain a lower bound on the size of the separator.

2. Separation graphs

Let G be connected graph with spanning tree T with a fixed root embedded in a surface Z of genus g such that each face is a triangle. Let each vertex v of G has a nonnegative weight $w(v)$. For any connected subgraph K of G, by $w(K)$ we denote the sum of the weights of the vertices of K. Following [3], we shall define a *separation graph* $S=S(G,T)$ of G with respect of T that can be used to find a small set of T-cycles that form a separator of G. We define vertices of S to be the faces of G and edges between all pairs of vertices whose corresponding faces share a common nontree edge. Thus S is a subgraph of the *dual* graph of G whose edges correspond to the nontree edges of G. Since the embedding of G is a triangulation, the degree of each vertex of S is at most three. Obviously S can be constructed in $O(|V(G)|+|E(G)|)$ time given the embedding of G. As shown below, if g is small in comparison with $|V(G)|$, then S is much simpler than G. Moreover, if weights are assigned to the edges of S in a proper way, we can find separators consisting of T-cycles by considering S instead of G. First we generalize a claim from [3].

LEMMA 2.1. Let n and m denote the number of vertices and the number of edges of S respectively. Then $m-n = 2g-1$.

PROOF: Let m', n' and f' be the number of edges, vertices, and faces of G. By Euler formula we have $n'-m'+f' = 2-2g$. By the definition of S it holds $n = f'$ and $m = m'-(n'-1)$. Thus $m-n = m'-n'+1-f' = 2g-1$. q.e.d.

Let $v \in V(G)$. Denote by $E(v,0)$ the set of all non-tree edges of G incident to v and by $E_0(v,1)$ the set of all non-tree edges of G whose both endpoints are adjacent to v. As there are no simple cycles in the spanning tree of G, $E_0(v,1)\neq\emptyset$. Moreover, if $E(v,0)=\emptyset$, one can easily see that there exists an endpoint of an edge in $E_0(v,1)$ whose parent in T is v. Denote by $E(v,1)$ the set of all such edges. As we showed above, $E(v,0) \cup E(v,1)\neq\emptyset$.

Give to the non-tree edges of G (or, equivalently, to the edges of S) different numbers from 1 to $|E(S)|$. We are going to define for each edge of G (of S) a set $vert(e)$ of vertices of G such that the sets $vert(e)$, $e \in E(G)$, form a partition of $V(G)$. Initially let $vert(e)=\emptyset$ for each $e \in E(G)$. Consider the following cases for any $v \in V(G)$.

i) $E(v,0)\neq\emptyset$. Let e_1 be the edge in $E(v,0)$ with the minimum number. Then update $vert(e_1):=vert(e_1) \cup \{v\}$.

ii) $E(v,0)=\emptyset$. Then $E(v,1)\neq\emptyset$ and let e_2 be the edge in $E(v,1)$ with the minimum number. Then update $vert(e_2):=vert(e_2) \cup \{v\}$.

We repeat the above procedure consecutively for all vertices of G. Then for any $e \in E(S)$ we define the *weight* of e

$$w(e) := \sum_{v \in vert(e)} w(v).$$

It is easy to construct a linear time algorithm that computes the weights of the edges of S.

From the definition it follows directly:

LEMMA 2.2. $\sum_{e \in E(S)} w(e) = \sum_{v \in V(G)} w(v)$.

THEOREM 2.1. Let M be a set of edges that divide S into components $S_1,\ldots,S_k$. Then the corresponding set of cycles $C(M)$ divides G into (possibly some empty) components $G_1,\ldots,G_k$ such that

$$(2.2) \qquad w(G_i) \leq w(S_i) \;, i=1,..,k.$$

PROOF: Let the curve corresponding to the set of cycles $C(M)$ divides the surface Z into connected regions $Z_1,\ldots,Z_{k'}$. Denote by E'_i, $i=1,\ldots,k'$, the set of all non-tree edges of G embedded in Z_i and the corresponding sets of edges in S by S'_i, $i = 1,\ldots,k'$. It is easy to see that S'_i induces a connected graph. Moreover for $i\neq j$, and any $e' \in S'_i$, $e'' \in S'_j$, edges e' and e'' share no common endpoint since $Z_i \neq Z_j$. We showed that $M \cup S'_1 \cup \ldots \cup S'_{k'} = E(S) = M \cup E(S_1) \cup \ldots \cup E(S_k)$ and S'_i induces a connected graph. Then obviously $k' = k$ and $S'_i = E(S_{perm(i)})$, $i=1,\ldots,k$, where *perm* is a permutation of $1,\ldots,k$. Denote by V_i the set of all endpoints of edges in $S_{perm(i)}$ minus the set of vertices on cycles from $C(M)$ and denote by G_i the subgraph of G induced by V_i. We shall prove that (2.2) holds for G_i.

Let $v \varepsilon V(G_i)$ and e be the edge of $E(S)$ such that $v \varepsilon vert(e)$. Then e is either incident, or has an endpoint adjacent to v and, from the definition of S_i, $e \varepsilon S_i \cup M$. On the other hand it is not possible that $e \varepsilon M$ because this will lead to the contradiction that v belongs to a cycle from $C(M)$. Thus (2.2) follows from the definition of $w(S_i)$. q.e.d.

3. A separator theorem for toroidal graphs

Known proofs of separator theorems for graphs of fixed genus [2,3] consist of the following three basic steps:

1. Reducing the radius of the graph by deleting and contracting sets of vertices on certain levels.

2. Reducing the genus of the graph to zero by deleting the sets of vertices of certain $2g$ T-cycles, where T is a spanning tree of the original graph.

3. Constructing a separator for the resulting graph by deleting the vertices on one additional cycle.

Thus, for the case of toroidal graphs ($g=1$), we have to delete three T-cycles in order to divide the graph of reduced radius. Our goal in this section is to show, by using the separation graphs defined in Section 2, that two T-cycles suffice to divide the graph.

LEMMA 3.1. Let S be an n-vertex, $(n+1)$-edge connected graph of degree three with weights $w(3)$ on its edges. There exist a pair e_1, e_2 of edges whose deletion divides S into components of weight not exceeding $(2/3)w(S)$ each.

PROOF: If there is an edge, such that $w(e) \geq (1/3)w(S)$ then delete that edge. Assume that all edges of S have weights less than $(1/3)w(S)$. Since S contains exactly two T-cycles, there exists one pair x_1, x_2 of vertices whose removal divides S into exactly three binary trees T_1, T_2 and T_3 (Fig. 3.1). Let the vertices x_1 and x_2, along with the connecting edges, belong to all their adjacent trees.

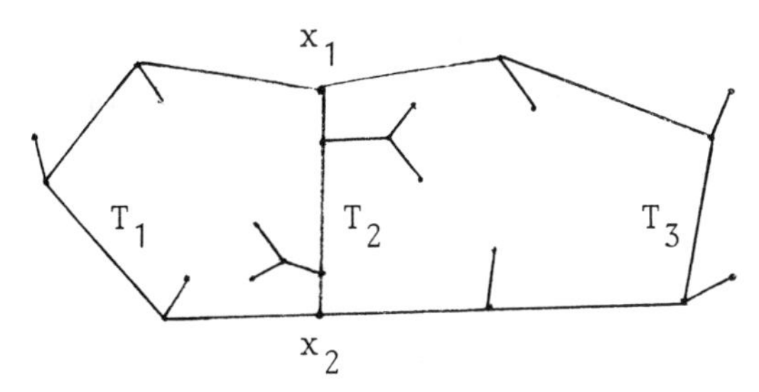

Figure 3.1. The decomposition of S

Suppose that T_1 has the maximum weight. Then $w(T_1) \geqq (1/3)w(S)$. If $w(T_1) \leqq (2/3)w(S)$, then the removal of the two edges of T_1 incident to x_1 and x_2 proves the claim.

Let $w(T_1) > (2/3)w(S)$. Let p be the path between x_1 and x_2 in T_1. Consider the set A of all subtrees of T_1 resulting after the deletion of the edges of p. If a tree T' in A exists whose weight exceeds $(1/3)w(S)$, then delete the edge of T' incident with p. If the remainder of T' has weight exceeding $(2/3)w(S)$, then we need to delete only one more edge to divide that tree into subtrees of weights less or equal to $(2/3)w(S)$. Now suppose that no tree in A has weight exceeding $(1/3)w(S)$. Contract each tree in A to a vertex with a weight equal to the weight of the tree. Then the weight of each vertex is less than $(1/3)w(S)$ and the sum of the weights of all vertices and edges on the resulting path p^* (which is equal to $w(T_1)$) exceeds $(2/3)w(S)$. Clearly p^* and p have the same set of edges. There exist two edges e_1 and e_2 in p whose removal leaves a subpath that does not contain neither x_1 nor x_2 and has weight between $(1/3)w(S)$ and $(2/3)w(S)$. Obviously e_1 and e_2 satisfy the lemma. q.e.d.

Next we present a straightforward generalization of a result from [1].

Let G be a connected n-vertex graph whose vertices are divided into levels according to their distance to some vertex t. For any pair l',l'' of levels denote by $C(l',l'')$ the set of vertices on levels l' and l''. Remove $C(l',l'')$ from G and distribute the resulting components (more precisely their vertices) in sets $A(l',l'')$ and $B(l',l'')$ such that

$$\max\{|A(l',l'')|,\ |B(l',l'')|\} \text{ is minimized.}$$

Let G be as above, $k > 0$, and let $L(l)$ denote the number of vertices on level l. Let for $\alpha \leqq 1$, l_α denote the lowest level such that

$$\sum_{l=0}^{l_\alpha} L(l) \geqq \alpha n.$$

LEMMA 3.2 (see [1,2]). There exist levels l' and l'', $l' \leqq l_{1/3} \leqq l_{2/3} \leqq l''$, such that either

(3.1) $\max\{|A(l',l'')|,|B(l',l'')|\} \leqq (2/3)n, |C(l',l'')| \leqq \sqrt{6kn}$,

or

(3.2) $L(l')+L(l'') + 2k(l''-l'-1) \leqq \sqrt{6kn}$.

THEOREM 3.1. Let G be an n-vertex toroidal graph. The vertices of G can be divided into three sets A, B, C such that no edge joins A and B, $\max(|A|,|B|) \leqq (2/3)n$ and $|C| \leqq \sqrt{12n}$.

PROOF: Without a loss of generality assume that G is connected. Choose a vertex t and assign levels to the vertices of G according to their distance to t. Choose k=2. If a pair l',l'' of levels exists satisfying (3.1) then the theorem obviously holds. Assume that no such pair exists. Then a pair l',l'' satisfying (3.2) exists. As $l' \leqq l_{1/3}$ and $l'' \geqq l_{2/3}$, then the graph induced by the set of the vertices on levels below l' and above l'' has no component of more than $(2/3)n$ vertices. Contract the subgraph induced by the set of vertices on levels $\leqq l'$ to a single vertex t^* and remove all vertices on levels l'' and above. If the resulting graph K^* has no more than $(2/3)n$ vertices, then the theorem holds. Assume $|K^*| \geqq (2/3)n$. Embed K^* on the torus and add new edges to make the embedding a triangulation. Let T be a breath-first spanning tree of K^* with root t^*. Assign weights $1/|K^*|$ to the vertices of K^*. Construct the separation graph S of K^*. Find the pair of edges dividing S as in Lemma 3.1. By Theorem 2.1 removal of the corresponding T-cycles of K^* leaves no component of K^* with more than $(2/3)|V(K^*)|$ vertices. As each T-cycle has no more than $2(l''-l'-1)+1$ vertices of K^* one of which is t^*, we obtained a set C of $L(l') + L(l'') + 4(l''-l'-1)$ vertices of G whose removal leaves no component with more than $(2/3)n$ vertices. By (3.2) $|C| \leq \sqrt{12n}$. The components of $G-C$ can easily be combined to form sets A and B satisfying the theorem (see e.g. [6]). q.e.d.

Notice that we can prove also in a straightforward way a weighted version of Theorem 3.1 (see [6,1]) that is an improvement over the best known such theorem. Furthermore, given an embedding of G on the torus, one can find in a linear time a partition A, B, C satisfying the theorem by applying the techniques from [6].

4. A lower bound

To find the lowest constant that can replace $2\sqrt{3} \approx 3.4641$ from Theorem 3.1. we shall follow the approach from [1].

It was proved in [1] that no $\sqrt{n}$-separator theorem for the class of planar graphs with $\beta = 2/3$ and $\alpha < (\sqrt{4 \quad \sqrt{3}})/3 \approx 1.555 = \alpha_1$ exists. Obviously, α_1 is a lower bound on the best constant α_2 that can replace $\sqrt{12}$ from Theorem 3.1. It is natural to expect that a more precise analysis of the class of toroidal graphs will yield a better estimate on α_2 than by using directly the result from [1].

Intuitively, the approach of [1] (for planar graphs) is the following. We choose a sphere E_0 as a representative of surfaces of genus zero. An n-vertex graph H_n is defined whose vertices are "distributed uniformly" on E_0. Then we relate the problem of finding a separator of H_n, for n sufficiently large, to the problem of dividing E_0 into two disjoint regions whose areas s_1 and s_2 have ratio $s_1/s_2 = 1/2$. The same approach, however, can not be directly applied to toroidal graphs due to the much more complicated structure of the torus compared to that of the sphere. We are going to deal with the problem by implicitly introducing a special (non-Eucledean) metric on the torus.

We shall make use of the following geometric fact:

LEMMA 4.1. Among all simple closed curves in the plane surrounding a region with unit area the circle has a minimum length.

Next we define a sequence $G_1, G_2, \ldots$ of graphs such that $|V(G_N)| = N^2$ and the genus of G_N is 0 or 1 and prove that the minimum separator of G_N has size at least cN for some constant c to be specified below.

Introduce a coordinate system in the plane P whose axes have unit length and angle between them $\pi/3$.

Define an infinite graph G_∞ such that the vertices of G_∞ are all points in P with integer coordinates and for each vertex $w = (x, y)$ the set of its neighbors is

$$\{(x,y+1),(x,y-1),(x+1,y),(x-1,y),(x+1,y-1),(x-1,y+1)\}.$$

Let R_N be the rhombus in P bounded by the axes and the lines $\{(x,y): x = N\}$ and $\{(x,y): y = N\}$ and G_N^* be the (embedded) subgraph of G_∞ induced by the set of vertices of G_∞ in R_N (Figure 4.1).

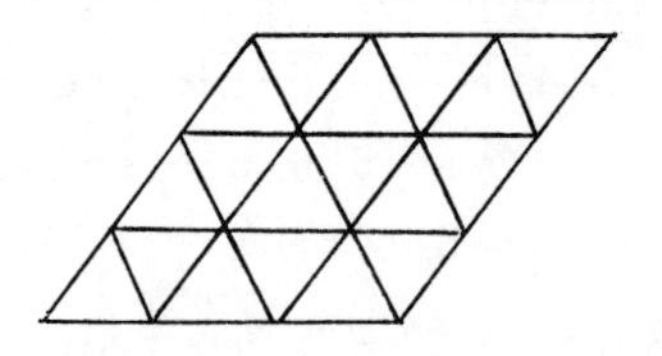

Figure 4.1. The graph G_N^*

DEFINITION. Let u and v be vertices of a graph G. By *merging* u and v we mean the operation of deleting v from G and adding edges (u,w) for all vertices $w \neq u$ that are adjacent to v and not adjacent to u.

Merge vertices (N,i) with $(0,i)$ and (i,N) with $(i,0)$ in G_N^*, for $i=0,\ldots,N$. The resulting graph which we denote by G_N can be embedded on the torus (with an embedding consistent with that of G_N^*).

DEFINITION. We say that the curve γ consisting of a finite number of simple closed nonintersecting curves *divides* a region R in the plane into subregions R_1 and R_2, if $R_1 \cup R_2 \cup \gamma = R$ and each continuous curve joining a point in R_1 with a point in R_2 intersects γ.

LEMMA 4.2. Let the curve γ divides R_N into two subregions R' and R'' with areas $s(R') \leq s(R'')$. Then the length of γ, which we denote by $l(\gamma)$, satisfies $l(\gamma) \geq 2\sqrt{\pi s(R')}$,

PROOF: Suppose that R' is a connected region. Then R' will consist of a region $D(R')$ homeomorphic to a disc and a finite number of holes inside $D(R')$. Let γ^* be the boundary of $D(R')$. Then by Lemma 4.1 we have

$$l(\gamma) \geq l(\gamma^*) \geq 2\sqrt{\pi s(D(R'))} \geq 2\sqrt{\pi s(R')} \tag{4.1}$$

If R' is not connected, let $R_1',\ldots,R_k'$ be all connected regions of R'. Applying (4.1) to each R_j', $j = 1,\ldots,k$, we obtain

$$l(\gamma) \geq \sum_{j=1}^{k} 2\sqrt{\pi s(R_j')} \geq 2\sqrt{\pi s(R')}.$$

THEOREM 4.1. The best constant that can replace $2\sqrt{3}$ from Theorem 3.1 is no smaller than $\sqrt{\pi 2}\,/\,\sqrt{3} \approx 1.9046$.

PROOF: Let N be fixed and C be a nonweighted minimal separator for G_N. Denote as usual by A and B ($|A| \leq |B| \leq 2N^2/3$) the corresponding partition of $V(G_N)\setminus C$. Denote by $R(A)$ the subregion of R_N containing all triangles t of the embedding of G_N such that at least one vertex of t belongs to A and let $R(B)$ contains all other triangles of the embedding. Denote by C^* the boundary of $R(A)$. It is clear that C^*

separates $R(A)$ and $R(B)$ and consists of vertices from C and edges between them. We need the following fact.

LEMMA 4.3. C^* consists of a finite number of simple closed non-intersecting curves.

PROOF: For the sake of simplicity let C^* denote the embedded subgraph of G_N corresponding to the curve C^*. The context will make it clear whether the subgraph or the curve is considered. We shall prove that all vertices of C^* have degrees two. Consider the following cases:

1. There exists a vertex v of degree 0 in C^*. Deleting v from C and adding it to A contradicts to the minimality of C.

2. There exists a vertex v of degree 1 in C^*. At least one of the vertices adjacent to v belongs to A. As no edge of G joins A and B then all vertices adjacent to v belong to A. Then v can be deleted from C and added to A.

3. There exists a vertex v of degree 3 in C^*. Then at least one edge e of C^* is incident either with two triangles from $R(A)$ or with two triangles from $R(B)$. In any such case e does not belong to the boundary of $R(A)$ which violates the definition of C^*.

4. There exists a vertex v of degree 4 in C^*. Using the fact that the triangles t_1 and t_2 adjacent to a edge from C^* have to belong to $R(A)$ and $R(B)$ respectively, it is easy to see that the only possibility in this case is the one illustrated on Figure 4.2.

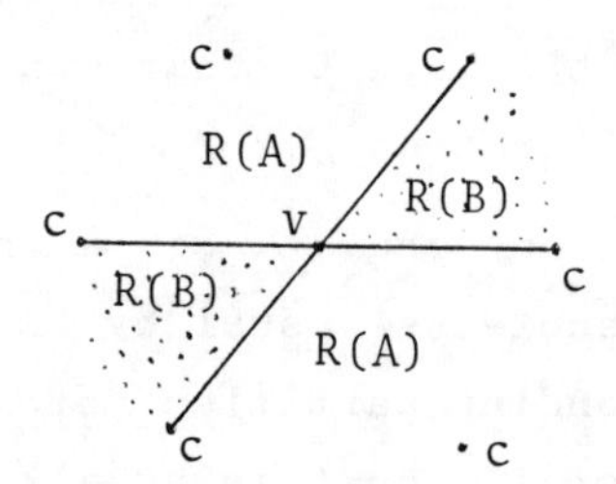

Figure 4.2. Illustration for Case 4.

In this situation deleting v from C and its adding to A again leads to a contradiction.

5. There is a vertex of degree 5 or 6. This case is impossible since there can not exist two triangles from $R(B)$ adjacent to an edge of C^*.

Thus the only remaining possibility is that all vertices of G^* have degree 2. q.e.d.

To complete the proof of the theorem assume that $|C| < 2N$. Consider the case where C^* is connected. As $|C^*| \leq |C|$, there exist integers x_0 and y_0 such that for any vertex $(x,y) \in C^*$, $x \neq x_0$ and $y \neq y_0$. Without a loss of generality assume that $x_0 = y_0 = N$. By cutting the embedding of G_N along lines $x_0 = N$ and $y_0 = N$ we obtain a planar embedding homeomorphic to the rhombus R_N. Then the curve C^* separates R_N into two regions containing $|A|$ and $N^2 - |A| - |C^*|$ vertices of G_N respectively.

Since each vertex from A is incident to six triangles and each triangle is incident to no more than three vertices of A, then $s(R(A)) \geq (\sqrt{3/2})|A|$. Thus we have

$$|C| \geq |V(C^*)| = l(C^*) \geq 2\sqrt{\pi s(R(A))} \geq \sqrt{2\sqrt{3}\pi |A|} \geq \sqrt{2\sqrt{3}\pi (N^2/3 - |C|)} \geq \sqrt{2\sqrt{3}\pi (N^2/3 - 2N)} = \sqrt{2\pi/\sqrt{3}}\, N - o(N).$$

The case where C^* consists of more than one cycle can be considered as in the proof of Lemma 4.2.

We are currently working on exploiting the upper and the lower bound techniques developed here to estimate the optimal constants in the separator size for graphs of arbitrarily large genus.

References:

[1] H. N. Djidjev, On the problem of partitioning planar graphs, SIAM J. Alg. Discr. Methods, vol. 3 (1982), 229-240.

[2] H. N. Djidjev, A separator theorems for graphs of fixed genus, Serdica, vol. 11 (1985), 319-329.

[3] H. N. Djidjev, Linear algorithms for graph separation problems, Proc. SWAT'88, Lecture Notes in Computer Science, vol. 318, North-Holland, Berlin, Heidelberg, New York, Tokyo, 1988, 216-221.

[4] H. Gazit, An improved algorithm for separating planar graphs, manuscript.

[5] J.R. Gilbert, J.P. Hutchinson, R.E. Tarjan, A separator theorem for graphs of bounded genus, Journal of Algorithms, No 5, 1984, pp. 391-398.

[6] R.J. Lipton, R.E. Tarjan, A separator theorem for planar graphs, SIAM J. Appl. Math., Vol. 36, No. 2, 1979, pp. 177-189.

ON SOME PROPERTIES OF (a,b)-TREES

Renzo Sprugnoli
Dipartimento di Matematica Pura e Applicata
Universitá di Padova (Italy)

Elena Barcucci, Alessandra Chiuderi, Renzo Pinzani
Dipartimento di Sistemi e Informatica
Universitá di Firenze (Italy)

ABSTRACT

In this paper some properties of (a,b)-trees are studied. Such trees are a generalization of B-trees and therefore their performances are dependent upon their height. Hence it is important to be able to enumerate the (a,b)-trees of a given height. To this purpose, a recurrence relation is defined which cannot be solved by the usual techniques. Therefore some limitations are established to the number of (a,b)-trees of a given height and an approximate formula is given which shows that the sequence of such numbers is a doubly exponential one. As a consequence of this fact, some properties, as for instance the probability that the height grows by the effects of an insertion, are studied using "profiles". The proofs are given of how profiles are generated and how from a profile one can get the number of corresponding (a,b)-trees.

INTRODUCTION

Let a and b be two integers, with $a \geq 2$ and $b \geq 2a-1$. An (a,b)-tree T is defined recursively by the following rules:

T1) all leaves are at the same level;

T2) every internal node has at most b subtrees;

T3) every internal node has at least a subtrees;

T4) the root (whenever it is not a leaf) has at least 2 subtrees

T5) if an internal node has k subtrees, then it contains (k-1) keys.

(a,b)-trees were introduced by Mehlhorn [5] as a generalization of B-trees and 3-2-trees. In fact, B-trees correspond to $(a,b)=(\lceil m/2 \rceil, m)$ and 3-2-trees to (2,3)-trees. We remark that an (a,b)-tree with n keys is built up by successive insertions and deletions, in such a way that it contains exactly n keys. Searching, inserting and deleting follow the same lines as B-trees (see, e.g. Comer [2]), and we do not repeat here the corresponding algorithms and related definitions, such as the concept of "splitting" a node. However, we note that the time complexity of these operations is proportional to the height of the tree, i.e. to the number of its levels.

In the present paper we give some properties of (a,b)-trees. In section 1 we count (a,b)-trees with a given height h; their number grows very rapidly and, indeed, they constitute a class of "doubly exponential sequences" as studied by Aho and Sloane [1]. For every a and b, the number of (a,b)-trees of height h is:

$$T_h(a,b) \simeq \alpha^{b^{h-1}}$$

where α is a constant depending on a, b and $T_0(a,b)$. We prove the existence of α in a different way from Aho and Sloane's; our method is specific to the class considered but gives a deeper result, that is, a sharp limitation for the value of α. In section 2 we study the "profile" of (a,b)-trees. The profile of 3-2-trees was introduced by Miller, Pippenger, Rosemberg and Snyder in [6]. The use of profiles is to collapse (a,b)-trees to a much simpler structure containing very few elements, as opposed to the class of all (a,b)-trees. Many properties of (a,b)-trees can be studied using profiles; for example, if we wish to use a computer to analyze a property of (a,b)-trees, we can only generate n keys (a,b)-trees for, say, $n \leq 10$. On the contrary, we can generate all the profiles up to n=50 or more, thus obtaining much accurate information. We show this method by studying the average number of splits in (a,b)-trees with n keys.

1. THE NUMBER OF (a,b)-TREES ACCORDING TO HEIGHT

It is very simple to find limitations for the height of (a,b)-trees with n keys. In fact, counting the number of keys at every level, we find that an (a,b)-tree of height h contains at most

$$(b - 1)\,(1 + b + \ldots + b^{h-1}) = b^h - 1$$

keys and at least

$$1 + 2(a - 1)\,(1 + a + \ldots + a^{h-2}) = 2a^{h-1} - 1$$

keys. Hence we have:

$$2a^{h-1} - 1 \leq n \leq b^h - 1$$

and taking logarithms

$$\lfloor \log_b(n+1) \rfloor \leq h \leq \left\lceil 1+\log_a\left(\frac{n+1}{2}\right) \right\rceil$$

This generalizes well-known bounds for B-trees and 3-2-trees. It is much more difficult to find the average height of (a,b)-trees with n nodes; however, Flajolet and Odlyzko [3] give a method to compute this value. Here we consider a different problem, i.e.

the problem of counting (a,b)-trees of height h.

The first difficulty is the fact that the root of an (a,b)-tree has not the same structure as the other nodes in the tree; in fact, it may contain less than (a-1) keys. So, let $T_h(a,b)$ be the number of (a,b)-trees of height h and $t_h(a,b)$ be the number of (a,b)-trees of height h having at least (a-1) keys in the root. We can easily give a recurrence relation for $t_h(a,b)$ (we shall omit the qualification (a,b) every time it can be understood):

$$\begin{cases} t_h = t_{h-1}^{b} + \dots + t_{h-1}^{a} = \dfrac{t_{h-1}^{b+1} - t_{h-1}^{a}}{t_{h-1} - 1} \\ t_1 = b - a + 1 \end{cases}$$

In fact, an (a,b)-tree of height h containing (a-1) keys in the root is composed by the root and a number k of subtrees of height (h-1), where $a \leq k \leq b$. Consequently, for $T_h(a,b)$ we have:

$$\begin{cases} T_h = t_{h-1}^{b} + \dots + t_{h-1}^{3} + t_{h-1}^{2} = \dfrac{t_{h-1}^{b+1} - t_{h-1}^{2}}{t_{h-1} - 1} \\ T_1 = b - 1 \end{cases}$$

We observe that for a=2, $T_h \equiv t_h$. We will now study the sequences $t_h(a,b)$, as a function of a and b. We consider (2,3)-trees (i.e., 3-2-trees) as a special case: it is the only case for which a=b-1, and this changes things completely, as we shall see.

Let us begin by proving:

Lemma 1.1: given two integers a,b with $a \geq 2$, $b \geq 2a-1$ and $b > 3$, there exists a real number x_o such that for every $x > x_o$ we have:

$$\left(x + \frac{1}{b}\right)^b \leq x^b + x^{b-1} + \dots + x^a \leq \left(x + \frac{1}{b-1}\right)^b$$

Proof: first we observe that the expression $x^b + x^{b-1} + \dots + x^a$ extends at least to the first three terms, because of our hypothesis on a and b. Since we have:

$$\left(x + \frac{1}{b}\right)^b = x^b + x^{b-1} + \frac{b-1}{2b} x^{b-2} + \dots$$

there exists a real number x_1 such that for every $x > x_1$

$$\left(x + \frac{1}{b}\right)^b \leq x^b + x^{b-1} + \dots + x^a$$

Analogously since

$$\left(x + \frac{1}{b-1}\right)^b = x^b + \frac{b}{b-1} x^{b-1} + \dots$$

there exists a number x_2 such that for every $x > x_2$ we have

$$x^b + x^{b-1} + \dots + x^a \leq \left(x + \frac{1}{b-1} \right)^b$$

Taking $x_o \geq \max(x_1,x_2)$, the conclusion of the lemma follows immediately. □

This lemma shows the doubly exponential nature of the numbers $t_h(a,b)$. To have deeper knowledge of these sequences, we must sharpen the conclusion of the preceding lemma. As we shall see, the left inequality is of paramount importance, and we can show:

Lemma 1.2: for every $x \geq 1$ we have (in the same hypothesis as Lemma 1.1):

$$\left(x + \frac{1}{b} \right)^b < x^b + x^{b-1} + \dots + x^a$$

Proof: because of our hypothesis on a and b, the sum on the right extends at least to the first three terms; hence, it is sufficient to show that for every $x \geq 1$ we have $(x+\frac{1}{b})^b < x^b+x^{b-1}+x^{b-2}$. For $x=1$, $(1+\frac{1}{b})^b < e < 3$ for every value of b and the inequality is obviously verified. So let us suppose $x>1$; we have:

$$x^b + x^{b-1} + x^{b-2} - \left(x + \frac{1}{b} \right)^b = x^b + x^{b-1} + x^{b-2} - x^b \left(\left(1 + \frac{1}{bx} \right)^{bx}\right)^{\frac{1}{x}} >$$

$$> x^b + x^{b-1} + x^{b-2} - x^b e^{\frac{1}{x}} = x^b + x^{b-1} + x^{b-2} - x^b \left(1 + \frac{1}{x} + \frac{1}{2x^2} + \frac{1}{6x^3} + \dots \right) =$$

$$= x^b + x^{b-1} + x^{b-2} - x^b - x^{b-1} - \frac{x^{b-2}}{2} - \frac{x^{b-3}}{6} - \dots =$$

$$= \frac{x^{b-2}}{2} - \frac{x^{b-3}}{6} \left(1 + \frac{1}{4x} + \frac{1}{20x^2} + \frac{1}{120x^3} + \dots \right) > \frac{x^{b-2}}{2} - \frac{x^{b-3}}{6} e^{\frac{1}{4x}} >$$

$$> \frac{x^{b-2}}{2} - \frac{ex^{b-3}}{6} > \frac{x^{b-2}}{2} - \frac{x^{b-2}}{2} = 0$$

and this proves completely the assertion of the lemma. □

The important point of Lemma 1.2 is that the inequality holds for every value of h, i.e. for the complete sequence $\{t_h(a,b)\}_{h \geq 1}$. So we can prove an important bound for $t_h(a,b)$ (note that the bound does not depend on a):

Lemma 1.3: in the hypothesis of Lemma 1.1, we have, for $h>1$:

$$t_h(a,b) \geq \left(t_1(a,b) + \frac{1}{b}\right)^{b^{h-1}} - \frac{1}{b} = \left(b - a + 1 + \frac{1}{b}\right)^{b^{h-1}} - \frac{1}{b}$$

Proof: from Lemmas 1.1 and 1.2 we have:

$$t_h + \frac{1}{b} \geq \left(t_{h-1} + \frac{1}{b}\right)^b$$

If we set $v_h = t_h + \frac{1}{b}$, we obtain $v_h \geq v_{h-1}^b$ and we can solve the recurrence $v_h = v_{h-1}^b$ with the initial condition $v_1 = b-a+1+\frac{1}{b}$. Taking logarithms we have a linear recurrence $u_h = bu_{h-1}$, where $u_h = \log v_h$ and $u_1 = \log(b-a+1+\frac{1}{b})$. The solution is obviously $u_h = b^{h-1}u_1$, and taking exponential:

$$v_h = v_1^{b^{h-1}}$$

From this the conclusion of our lemma follows immediately. □

Many times it is also possible to give an upper bound to the values of $t_h(a,b)$, that is:

$$t_h(a,b) \leq \left(t_1(a,b) + \frac{1}{b-1}\right)^{b^{h-1}} - \frac{1}{b-1}$$

This, however, is less important than the lower bound and we do not insist here on the point, although we shall consider it later for some observations on sequences with $a \simeq \frac{b}{2}$. For the moment we use the right inequality of Lemma 1.1 to argue that there exists a number α such that $t_h \leq (t_1+\alpha)^{b^{h-1}}$. In fact, we can find an index h_o (depending on x_2 of Lemma 1.1) such that

$$t_h \leq \left(t_{h_o} + \frac{1}{b-1}\right)^{b^{h-h_o}} \qquad \forall h \geq h_o;$$

expressing now $t_{h_o}+\frac{1}{b-1}$ as $(t_1+\alpha)^{b^{h_o-1}}$, which is always possible, we have the desired result. Unfortunately, from this we cannot conclude $\alpha=\frac{1}{b-1}$, which is false for (5,9)-trees, as an example.

Because of these bounds, we can consider the function:

$$f_h(\delta) = (t_1 + \delta)^{b^{h-1}} - \delta$$

since $f'_h(\delta) = b^{h-1}(t_1+\delta)^{b^{h-1}-1} - 1$ is always greater than 0 for $\delta>0$ and $h>1$, the function $f_h(\delta)$ is increasing, and we have:

$$f_h\left(\frac{1}{b}\right) \leq t_h \leq f_h(\alpha)$$

So, let us define δ_h as the only value for which $f_h(\delta_h)=t_h$. Since $\frac{1}{b}<\delta_h<\alpha$, we may hope that there exists a limiting value $\hat{\delta}=\lim_{h\to\infty} \delta_h$, so that we can write $t_h \simeq (t_1+\hat{\delta})^{b^{h-1}} - \hat{\delta}$. This is indeed the case, as we will now show.

Lemma 1.4: let a and b be as in Lemma 1.1; the sequence $\{\delta_h\}_{h>1}$ defined by the relation:

$$(t_1 + \delta_h)^{b^{h-1}} - \delta_h = t_h$$

is decreasing.

Proof: we begin by expressing δ_{h+1} as a function of δ_h. Let us write

$$t_{h+1} = t_h^{\,b} + t_h^{\,b-1} + \ldots + t_h^{\,a} = \left(t_h + \frac{1}{b}\right)^b - \left[\left(t_h + \frac{1}{b}\right)^b - \left(t_h^{\,b} + \ldots + t_h^{\,a}\right)\right]$$

where the expression within brackets is of two orders less than $(t_h+\frac{1}{b})^b$. Equating this to

$t_{h+1} = (t_1+\delta_{h+1})^{b^h} - \delta_{h+1}$, we have:

$$(t_1 + \delta_{h+1})^{b^h} = \left(t_h + \frac{1}{b}\right)^b - \left[\left(t_h + \frac{1}{b}\right)^b - (t_h^b + \dots + t_h^a)\right] + \delta_{h+1}$$

Using the expansion $(x+h)^{1/k} = x^{1/k}\,(1+\frac{h}{kx}+\dots)$, we obtain:

$$t_1 + \delta_{h+1} = \left(t_h + \frac{1}{b}\right)^{b/b^h}\left(1 - \frac{B + \delta_{h+1}}{b^h\left(t_h + \frac{1}{b}\right)^b} + \dots\right) = \left(t_h + \frac{1}{b}\right)^{1/b^{h-1}}\left(1 - O\left(\frac{1}{b^h t_h^2}\right)\right)$$

if B denotes the expression within brackets. Using the expansion a second time, we have:

$$\left(t_h + \frac{1}{b}\right)^{1/b^{h-1}} = t_h^{1/b^{h-1}}\left(1 + \frac{\frac{1}{b}}{b^{h-1}t_h} + \dots\right) = t_h^{1/b^{h-1}}\left(1 + O\left(\frac{1}{b^h t_h}\right)\right)$$

and hence we can ignore the preceding correction. In this way:

$$t_1 + \delta_{h+1} \simeq \left(t_h + \delta_h + \frac{1}{b} - \delta_h\right)^{1/b^{h-1}} = (t_h + \delta_h)^{1/b^{h-1}} + \frac{\left(\frac{1}{b} - \delta_h\right)(t_h + \delta_h)^{1/b^{h-1}}}{b^{h-1}(t_h + \delta_h)} + \dots =$$

$$= t_1 + \delta_h + \frac{(1 - b\delta_h)(t_1 + \delta_h)}{b^h(t_h + \delta_h)} + \dots$$

The relation connecting δ_{h+1} and δ_h is therefore:

$$\delta_{h+1} \simeq \delta_h + \frac{-b\delta_h^2 + (1 - bt_1)\,\delta_h + t_1}{b^h\,(t_h + \delta_h)}$$

It is now obvious that for $\delta_h > \frac{1}{b}$ the numerator is negative, and so $\delta_{h+1} < \delta_h$, as desired. □

Summing up all the considerations above, we may conclude with:

Theorem 1.1: For every integer a, b, $a \geq 2$, $b \leq 2a-1$, $b > 3$, there exists a constant $\delta > \frac{1}{b}$ (depending on a and b) such that:

$$t_h(a,b) \simeq \left(t_1(a,b) + \delta\right)^{b^{h-1}} - \delta$$

Proof: in Lemma 1.4 we have shown that the sequence $\{\delta_h\}_{h>1}$ is decreasing; since it is bound from below by $\frac{1}{b}$, it converges to $\delta = \lim_{h\to\infty} \delta_h$, and the conclusion of the theorem follows. □

Let us call *seed* of the sequence $\{t_h(a,b)\}_{h\geq 1}$ the number $t_1(a,b) + \delta$. In general, for doubly exponential sequences, only in a very few cases has the seed been proved to be a known constant (see Aho and Sloane [1] or Greene and Knuth [4]). Therefore, often the determination of the seed relies only on the sequence and we must be content with extracting the b^{h-1}-th root of t_h. Since the sequence converges very rapidly, we easily can obtain an approximation of the seed with a number of decimal digits. These, however, soon become insufficient to compute exactly the successive elements in the sequence, as a result of their rapid growth. In view of this fact, it may be interesting to have a method to compute a lower and an upper bound to the seed of a given sequence; these bounds, in fact, will give an idea of the error that would result by using an approximate value of the seed. Let us see how this can be done using the preceding results.

Let us write $\sigma(a,b)$ for the seed $t_1(a,b) + \delta(a,b)$ of a given sequence. By Lemma 1.4, the sequence $\{t_1 + \delta_h\}_{h>1}$ approaches σ from above. It is a simple matter to show that the sequence:

$$\left\{t_h^{1/b^{h-1}}\right\}_{h>1}$$

approaches σ from below. In fact, from the obvious relation $t_{h+1} > t_h^b$ it follows immediately

$$t_{h+1}^{1/b^h} > t_h^{1/b^{h-1}}$$

In practice, taking h=2, we may obtain useful limitations for σ. The lower bound $\sqrt[b]{t_2}$ is immediate. The upper bound requires some tedious computations. However we can use the following method to simplify things.

By Lemma 1.4, δ_2 is the solution of the equation $(t_1 + \delta)^b - \delta = t_2$; that is, it is a root of $f(\delta) = (t_1+\delta)^b - \delta - t_2$. To approximate the root we can start with $\delta' = t_2^{1/b} - t_1$, for which we have $f(\delta') < 0$, because of our previous observations, and use the Newton-Raphson method to obtain a new approximation:

$$\delta'' = \delta' - \frac{(t_1 + \delta')^b - \delta' - t_2}{b\,(t_1 + \delta')^{b-1} - 1}$$

Since $f(\delta)$ is convex in a neighbourhood of δ, we have $f(\delta'') > 0$, and $t_1 + \delta''$ is an upper bound for σ. Obviously we have:

$$t_1+\delta'' = t_1+\delta' - \frac{(t_1+\delta')^b - \delta' - t_2}{b(t_1+\delta')^{b-1}-1} = t_2^{1/b} - \frac{t_2 - t_2^{1/b} + t_1 - t_2}{b\,t_2^{(b-1)/b} - 1} =$$

$$= \frac{bt_2 - t_1}{bt_2^{(b-1)/b} - 1}$$

and hence:

$$t_2^{1/b} \leq \sigma \leq \frac{bt_2 - t_1}{bt_2^{(b-1)/b} - 1}$$

To show a typical example, let us consider (2,5)-trees. The sequence begins: 4, 1360, $4.656011*10^{15}$, $2.188114*10^{78}$, ... The lower bound for $\sigma(2,5)$ is $1360^{1/5} = 4.233580106$, and the corresponding upper bound is 4.2337256197. With these values we find:

$$3.51*10^{48961} < t_8(2,5) < 5.15*10^{48962}$$

Obviously better bounds can be found applying the same method to h=3.

In table 1 we give the value of $\sigma(a,b)$ for b=3,4,....,12 and the corresponding values of a. The case of (2,3)-trees does not fall in the discussion above. The situations, however, are very similar, and we can limit ourselves to sketching the derivation of existence of $\sigma(2,3)$.

We begin by observing that:

$$t_h + \frac{1}{3} < \left(t_{h-1} + \frac{1}{3}\right)^3 \quad \text{and} \quad t_h + \frac{1}{4} > \left(t_{h-1} + \frac{1}{4}\right)^3$$

as follows immediately from the definition of $t_h = t_{h-1}^3 + t_{h-1}^2$ and the development of $\left(t_{h-1} + \frac{1}{3}\right)^3$ and $\left(t_{h-1} + \frac{1}{4}\right)^3$. This shows that:

$$\left(2 + \frac{1}{4}\right)^{3^{h-1}} - \frac{1}{4} < t_h < \left(2 + \frac{1}{3}\right)^{3^{h-1}} - \frac{1}{3}$$

and these are different from the inequalities in Lemma 1.1, although valid for every $h \geq 1$. Defining the sequence of δ_h as in Lemma 1.4, we find that it is increasing, because the relation connecting δ_{h+1} to δ_h is:

$$\delta_{h+1} \simeq \delta_h - \frac{3\delta_h^2 + 5\delta_h - 2}{3^h (t_h + \delta_h)}$$

and the numerator is negative for $\delta_h < \frac{1}{3}$, as it should be. This proves the existence of a seed $2 + \delta < 2 + \frac{1}{3}$ (and also greater than $2 + \frac{1}{4}$). In fact we find $\sigma = 2.309926325......$.

b \ a	2	3	4	5	6
3	2.309926325				
4	3.290632938				
5	4.233704670	3.229319282			
6	5.189184653	4.193745142			
7	6.158310095	5.161720995	4.165488724		
8	7.136188673	6.138294853	5.141222450		
9	8.119579414	7.120926673	6.122772639	5.125336002	
10	9.106631466	8.107541476	7.108740588	6.110383161	
11	10.09624247	9.096885715	8.097705531	7.098785669	6.100264969
12	11.08771554	10.08818701	9.088772223	8.089518026	7.090500553

Table 1 - The seed of the first (a,b)-sequences.

b \ a	36	37	38	39	1/b
72	37.01408271				0.01408451
73	38.01388464	37.01388976			0.01388889
74	39.01369219	38.01369697			0.01369863
75	40.01350511	39.01350959	38.01351431		0.01351351
76	41.01332318	40.01332738	39.01333180		0.01333333
77	42.01314618	41.01315012	40.01315427	39.01315863	0.01315789
78	---------------	---------------	41.01298150	40.01298560	0.01298901

Table 2 - Extremal values of $\sigma(a,b)$.

As a concluding remark, we wish to observe that $\sigma(a,b) < t_1 + 1/(b-1)$ whenever $b \leq 2a$. When b is odd, a may assume the value $\lceil b/2 \rceil$ and for $b \geq 9$ we have $\sigma(\lceil b/2 \rceil, b) > t_1 + \frac{1}{b-1}$. In table 2, we give the values obtained using the maximum available precision on a TI-74. Although $\sigma(\lceil b/2 \rceil, b) \simeq t_1 + \frac{1}{b-1}$ and also $\sigma(\lfloor b/2 \rfloor, b) \simeq t_1 + \frac{1}{b-1}$, the former remains larger than the limiting value, as the latter is always smaller. For the moment, however, we do not give a formal proof of this fact.

2. PROFILES OF (a,b)-TREES

As we have previously seen, $t_h(a,b)$ values grow very quickly as far as h increases; for instance $t_8(2,5)$ is $O(10^{48962})$. So an experimental study of (a,b) trees is an onerous task even when supported by a computer.

In order to study some properties of (a,b)-trees we will use the concept of a "profile" (see Miller et al. [6]).

Let T be an (a,b)-tree of height h. We define the profile of T as the h-uple $(a_1, a_2, \ldots, a_h)$ such that a_i is the number of keys at the i-th level. It is useful to set $a_o = 1$.

We note that generally several (a,b)-trees of height h correspond to a given profile. Nevertheless we will see that some notable properties depend only on the number of keys at each level but not on the shape of the tree itself.

It is obvious that

$$n = a_1 + a_2 + \ldots. + a_h$$

where n is the total number of keys in T.

Furthermore, given a profile $K = (a_1, \ldots, a_h)$ it is possible to calculate the number f_i of nodes at the i-th level, in fact it is:

$$f_i = a_o + a_1 + \ldots + a_{i-1}$$

This relation may be proved true by induction. In fact, at the first level there is always one node, the root, thus $f_1 = 1$; at the second level there are as many nodes as the keys contained in the root plus one:

$$f_2 = 1 + a_1 = a_o + a_1$$

Suppose $f_i = a_o + a_1 + \ldots + a_{i-1}$. By definition, the number of sons of every node (which is not a leaf) of an (a,b)-tree is given by the number of keys in it contained plus one. Let $a_{i,j}$ be the number of keys contained in the j-th node at the i-th level, the total number of the sons of the nodes at the i-th level, i.e. f_{i+1}, is given by:

$$f_{i+1} = \sum_{j=1}^{f_i} (a_{i,j} + 1) = \sum_{j=1}^{f_i} a_{i,j} + f_i = a_i + f_i$$

and then, by the induction hypothesis:

$$f_{i+1} = a_o + a_1 + \ldots. + a_{i-1} + a_i$$

Obviously, not all of the $\binom{n-1}{h-1}$ h-uples $(a_1,a_2,...,a_h)$ such that $a_1+a_2+...+a_h=n$ are admissible profiles. In fact we can prove the following

<u>Theorem 2.1</u>: a h-uple $(a_1,...,a_h)$ is a profile of some (a,b)-trees if and only if:

1) $1 \leq a_1 \leq b-1$.
2) $(a - 1)(a_o + a_1 + ... + a_{i-1}) \leq a_i \leq (b - 1)(a_o + a_1 + ... + a_{i-1})$

<u>Proof</u>: from the definition of (a,b)-tree it follows that $1 \leq a_1 \leq b-1$. Furthermore, since each node at the i-th level contains at least (a-1) and at most (b-1) keys, we have:

$$(a - 1)(a_o + a_1 + ... + a_{i-1}) \leq a_i \leq (b - 1)(a_o + a_1 + ... + a_{i-1}) \qquad \square$$

This relation allows us to determine all the profiles $(a_1,...,a_h)$ such that $a_1+...+a_h=n$ for a given n. In fact, given a number n of keys, when i=h, we have:

$$(a - 1)(a_o + ... + a_{h-1}) \leq a_h \leq (b - 1)(a_o + ... + a_{h-1})$$

i.e.

$$(a-1)(a_o+...+a_h) \leq a_h + (a-1)a_h \quad \text{and} \quad (b-1)(a_o+...+a_h) \geq a_h + (b-1)a_h$$

and thus

$$(a-1)(n+1) \leq aa_h \qquad \text{and} \qquad (b-1)(n+1) \geq ba_h$$

i.e.

$$\left\lceil \frac{a-1}{a}(n+1) \right\rceil \leq a_h \leq \left\lfloor \frac{b-1}{b}(n+1) \right\rfloor \qquad (*)$$

Hence the number of keys at the last level can assume

$$\left\lfloor \frac{b-1}{b}(n+1) \right\rfloor - \left\lceil \frac{a-1}{a}(n+1) \right\rceil + 1$$

different values.

Corresponding to each value ν assumed by a_h it is possible to determine new bounds for a_{h-1}, using again (*):

$$\left\lceil \frac{a-1}{a}(n-\nu+1) \right\rceil \leq a_{h-1} \leq \left\lfloor \frac{b-1}{b}(n-\nu+1) \right\rfloor$$

Iterating this procedure we can determine all the admissible profiles which correspond to (a,b)-trees containing n keys (from now on we will simply use "profiles" instead of "admissible profiles").

We report here the 11 profiles corresponding to (2,4)-trees with 15 keys: (1,2,4,8), (3,4,8), (2,5,8), (1,6,8), (2,4,9), (1,5,9), (2,3,10), (1,4,10), (1,3,11), (1,2,12), (3,12); while in column 1 of table 3 we report all the profiles which correspond to (2,4)-trees of height 3 with 30 keys. We remark that the total number of (2,4)-trees of height 3 is 189,004,023, whereas the number of their profiles is only 423.

Given a profile $K=(a_1,...,a_h)$ we know the number of nodes and keys at each level of every (a,b)-tree corresponding to that profile. From this we can discover in how many ways it is possible to distribute the a_i keys into the f_i nodes at the i-th level.

Obviously, at the first level there is only the root which contains exactly a_1 keys.

Generally, for i=2,3,...,h, the number we are looking for is the number $C_{a,b}(f_i,a_i)$ of compositions of a_i indistinguished objects in f_i cells in such a way that each cell contains at least (a-1) and at most (b-1) objects, where:

Lemma 2.1:

$$C_{a,b}(m,n) = \sum_{k+(b-a+1)j=n-(a-1)m} (-1)^j \binom{m}{j} \binom{m+k-1}{k}$$

Proof: the generating function of $C_{a,b}(m,n)$ for each m is:

$$(t^{a-1}+ t^a+....+t^{b-1})^m = t^{(a-1)m} (1 + t + ... + t^{b-a})^m = t^{(a-1)m} (1 - t^{b-a+1})^m (1 - t)^{-m} =$$

$$= t^{(a-1)m} \sum_{j=0}^{m} \binom{m}{j} (-1)^j t^{(b-a+1)j} \sum_{k=0}^{m} \binom{m+k-1}{k} t^k$$

On the other hand this generating function corresponds to $\sum_{n=0}^{\infty} C_{a,b}(m,n)\ t^n$ where $C_{a,b}(m,n) \neq 0$ only when $m(a-1) \leq n \leq m(b-1)$.

It follows that:

$$C_{a,b}(m,n) = \sum_{k+(b-a+1)j=n-(a-1)m} (-1)^j \binom{m}{j} \binom{m+k-1}{k} \qquad \square$$

Finally we can determine the multiplicity M_k of profile K, that is, the number of distinct (a,b)-trees which admit K as a profile:

Theorem 2.2:

$$M_k = \prod_{i=2}^{h} C_{a,b}(f_i,a_i)$$

Proof: it follows immediately from the preceding lemma. $\square$

In column 3 of table 3 we report the multiplicity of the profiles corresponding to (2,4)-trees of height 3 with 30 keys.

Since the insertion of a key in an (a,b)-tree is carried out in the same way as it is done in a B-tree, a split will occur whenever a key is inserted into a node which already contains (b-1) keys. Inasmuch as the splits can reach the root of the tree and cause its height to rise, it is useful to know the probability that a split will occur when a key is inserted in a given (a,b)-tree.

To do this we shall again use the profiles, and we will prove that, given the profile of a certain (a,b)-tree, it is possible to determine the average probability that a split will occur and reach the upper levels, following the insertion of a key in one of the M_k (a,b)-trees corresponding to that profile.

Let us suppose we want to insert a key at the h-th level of a given (a,b)-tree having n keys and profile $K=(a_1,a_2,...,a_h)$.

The probability that this insertion may cause a split is given by the ratio of the number of times the insertion causes a split at the h-th level to the number of ways it is possible to insert the key.

As we stated at the beginning of this section, such a number is given by

$$a_o + a_1 + \dots + a_h = n + 1$$

while the numerator of such a ratio will be given by the number of nodes at the h-th level containing exactly (b-1) keys times b, the number of ways of inserting a key in each one of these nodes.

Analogously, the probability that a split occurs by the insertion of a key at th i-th level will be:

$$\frac{b\ (\text{number of nodes containing b-1 keys})}{a_o + \dots + a_i}$$

We want now to calculate the number of nodes containing exactly (b-1) keys at the i-th level, for i=2,3,..,h, in all the trees corresponding to a given profile $K=(a_1,\dots,a_h)$.

Let us denote by $C_{a,b}(m,n,p)$ the number of compositions of n objects into m cells with at least (a-1) and at most (b-1) objects in every cell and exactly p cells full.

It will be

$$C_{a,b}(m,n,p) = \begin{cases} 0 & \text{if } n<(b-1)p+(a-1)(m-p) \text{ or if } n>(b-1)p+(b-2)(m-p) \\ \binom{m}{p} C_{a,b-1}(m-p,n-(b-1)p) & \text{otherwise} \end{cases}$$

The total number of full nodes at the i-th level in all the trees corresponding to the profile $K=(a_1,\dots,a_h)$ is henceforth:

$$\sum_{p=1}^{f_i} p\ C_{a,b}(f_i,a_i,p)$$

Now we are able to express the probability of a split at the h-th level as:

$$p_h = \frac{b\sum_{p=1}^{f_h} p\ C_{a,b}(f_h,a_h,p)}{(n+1)\ C_{a,b}(f_h,a_h)} \qquad (**)$$

and in general:

$$p_i = \frac{b\sum_{p=1}^{f_i} p\ C_{a,b}(f_i,a_i,p)}{(a_o + \dots + a_i)\ C_{a,b}(f_i,a_i)}$$

At this point we can also calculate the probability that the splitting process will continue to the upper levels. In fact the probability that a split will reach the (h-1)-th level is $p_h p_{h-1}$, and so on for all the higher levels.

Hence the probability that the insertion of a key in one of the (a,b)-trees corresponding to $K=(a_1,\ldots,a_h)$ will cause a split that will reach the root and consequently increase the height of that tree is $p_h p_{h-1} \ldots p_1$.

profiles	# of nodes	multiplicity	probability
3 11 16	1 4 15	60	0.
3 10 17	1 4 14	5,460	0.029
3 9 18	1 4 13	80,080	0.064
3 8 19	1 4 12	308,484	0.091
2 9 19	1 3 12	16,236	0.
3 7 20	1 4 11	317,680	0.092
2 8 20	1 3 11	59,565	0.
3 6 21	1 4 10	83,500	0.077
2 7 21	1 3 10	50,100	0.
3 5 22	1 4 9	3,528	0.
2 6 22	1 3 9	6,174	0.
3 4 23	1 4 8	8	0.
2 5 23	1 3 8	48	0.
1 6 23	1 2 8	8	0.

Table 3 - (2,4)-trees of height 3 with 30 keys: profiles, number of nodes at each level, multiplicity of profiles, and probability of having a split that reaches the root.

In column 4 of table 3 we reported the probability that a split may increase the height of a (2,4)-tree of height 3 with 30 nodes.

Let us try to understand what this means in terms of profiles. The relation (**) tells us the probability of a split occurring when a key is inserted at the h-th level. If a split occurs at the h-th level with probability p_h, this means that, after an insertion of a key in one of the M_k trees corresponding to the profile $K=(a_1,\ldots,a_{h-1},a_h)$, we shall have, with probability $1-p_h$, the profile $K'=(a_1,\ldots,a_{h-1},a_h+1)$, whereas p_h is the probability of keeping a_h keys on the h-th level and adding the new key in the upper levels. Thus the probability of stopping the splitting process at the (h-1)-th level is $p_h(1-p_{h-1})$, i.e. the probability of split at the h-th level times the probability of having no splits at the (h-1)-th level. In terms of profiles this means that K will originate the new profile $K'=(a_1,\ldots,a_{h-1}+1,a_h)$ with probability $p_h - p_h p_{h-1}$.

Generally a given profile K will generate a new profile $K'=(a_1,\ldots,a_i+1,\ldots,a_h)$ with a probability:

$$P_i = S_{i+1} - S_i \qquad i=1,2,\ldots,h.$$

where $S_i = p_i p_{i+1} \ldots\ldots p_h$.

It will be useful to set $S_o=0$, so $P_o=S_1$ will denote the probability of generating a new root, i.e. that the splitting process will reach the root, or in other words that $K'=(1,a_1,...,a_h)$.

CONCLUSIONS

In this paper we pointed out some properties of (a,b)-trees, showing, among the other things, that the numbers of these trees of height h constitute a doubly exponential sequence. To overcome this difficulty, we used the concept of a "profile" in order to study the properties of such trees. The use of this concept has been shown to be a convenient tool for our purposes. It is our opinion that the use of this technique, in particular when studying (a,b)-trees, which are a generalization of other tree structures, might constitute a basis for a systematic study of the properties of every tree structures.

REFERENCES

[1] Aho A. V., Sloane N. J. A.: "Some Doubly Exponential Sequences", Fibonacci Quarterly 11, 4 (1973), 429-437

[2] Comer D.: "The Ubiquitous B-Tree", ACM Computing Surveys, 11, 2 (1979), 121-137

[3] Flajolet P., Odlyzko A.: "Exploring Binary Trees and Other Simple Trees", Journal of Computers and System Science, 25, 2 (1982), 171-213

[4] Greene D. H., Knuth D. E.: *Mathematics for the Analysis of Algorithms*, Birkhäuser, Boston, Mass., 1982

[5] Mehlhorn K.: *Data Structures and Algorithms* 1: *Sorting and Searching*, Springer-Verlag, Berlin, 1984

[6] Miller R. E., Pippenger N., Rosenberg A. L., Snyder L.: "Optimal 2-3-trees", SIAM Journal on Computing, 8, 1 (1979), 42-59

DISASSEMBLING TWO-DIMENSIONAL COMPOSITE PARTS VIA TRANSLATIONS+

by

Doron Nussbaum[1]
Tydac Technologies Inc.
1600 Carling Avenue, Ottawa
and
Jörg-R. Sack
School of Computer Science, Carleton University
Ottawa, Canada K1S 5B6

Abstract

This paper deals with the computational complexity of disassembling 2-dimensional composite parts (comprised of simple polygons) via collision-free translations. The first result of this paper is an $O(Mn + M \log M)$ algorithm for computing a sequence of translations (performed in a common direction) to disassemble composite parts. The algorithm improves on the $O(Mn \log Mn)$ bound previously established for this problem and is easily seen to be optimal. The algorithm solves the problem posed by Nurmi and by Toussaint.

The second result of this paper is an $\Omega(Mn + M \log M)$ lower bound proof for the problem of detecting whether a composite part can be disassembled, or contains interlocking subparts. Thus, detecting the existence of a disassembly sequence is as hard as computing one. As a consequence, the algorithm for computing a disassembly sequence is optimal also for the detecting problem.

+ Research supported by Natural Science and Engineering Research Council of Canada.

1 Research on this paper was carried out while the author was at Carleton University.

1. MOTIVATION AND PROBLEM DESCRIPTION

The rapid development in the fields of computer graphics, CAD-CAM systems, and robotics has attracted considerable attention to the problem of moving objects, e.g. line segments or polygons, in two and three dimensional space. The types of motion (such as single translation, k translations, sequences of translations and rotations,...) and the types of objects (e.g. line segments, convex polygons, polyhedra,...) determine a variety of motion problems.

Here we are interested in a motion planning problem which appears in the following "equivalent" formulations: (a) in the context of separability motions, where one wishes to separate each object from a collection of objects via collision-free motions, and (b) in the context of machine assemblies where a robot is to assemble/disassemble a composite machine part out of/into "simple" parts by moving the parts one by one.

In its most general form this problem has been shown to be P-space hard [R]. One computationally feasible variant of this problem has received considerable attention (see e.g. [COSW], [DS], [GY], [N], [Na], [OW], [NS], [ST], [T]). It assumes that all disassembly motions are translations which are performed in a common direction (typically specified by the user) and one translation motion per object. More formally, the problem is expressed as follows: Can a given composite part $P = \{P_1,P_2,\ldots,P_M\}$ composed of M n-vertex (sub)parts be disassembled, by translating one part at a time in a specified direction over an arbitrarily large distance and only once for each object, in such a way that no object (when translated) will collide with any of the objects not yet moved? If a disassembly of the composite part is possible then, to actually perform it, one must determine a sequence in which the parts are to be moved. We refer to these problems as disassembly detection and disassembly (sequence) determination. (We refrain from defining conceptually clear notions like e.g. collide, translate, etc.). Next, we state some terminology and review results which of immediately relevance to our work.

2. RELEVANT RESULTS

2.1 Terminology

We say that a composite part $P = \{P_1,P_2,\ldots,P_M\}$ is *simple* if each of the M n-vertex (sub)parts P_i, is a simple polygon and $P_i \cap P_j = \emptyset$ for i≠j. Any polygon P_i is represented as a list of its vertices as they appear on the boundary of P_i in clockwise order. Unless otherwise stated, all composite parts considered in this paper are assumed to be simple. Such a composite part is *convex* if each P_i is convex. Similarly, a composite disassembly is *rectilinear* if each P_i is rectilinear (also called orthogonal). A composite part P is *monotone in a direction d* if each of its parts P_i is monotone in d. A composite part P is *monotone* if there exists a direction d such that P is monotone in d. (The reader is referred to [PS] for basic definitions and techniques from the field of computational geometry.)

2.2 Existing and New Results

Guibas and Yao [GY] have shown that a disassembly sequence for a composite parts of M rectangles (convex n-gons) can be determined in $O(M \log M)$ time (respectively, $O(Mn + M \log M)$ time). Ottmann and Widmayer [OW] gave an algorithm for translating a set of line segments. Their algorithm provides an alternate solution for solving the problems studied in [GY] achieving the same asymptotic time bound, but reducing the two-pass algorithm by Guibas and Yao to one-pass.

While for convex composite parts (as well as line segments) a disassembly sequence exists for any direction specified, an arbitrary simple composite part may not admit disassembly, i.e. may contain a subset of the polygons that are interlocking.

For rectilinear composite parts, Chazelle, Ottmann, Soisalon-Soininen, and Wood [COSW] showed that the problem of detecting whether a disassembly exists can be solved in $O(Mn \log Mn)$ time, while computation of a disassembly sequence is done in $O(Mn \log^2 Mn)$ time. Nurmi [N] proved that for arbitrary simple composite parts both problems can be solved in $O(Mn \log Mn)$ time. Nurmi posed the question of whether an $O(Mn + M \log M)$ algorithm exists for solving these problems. Toussaint gave alternate solutions for these problems. Also in his solutions, the complexities for detection and computation differ; he conjectures that faster algorithms may exist for either problem.

In the following table we give a succinct summary of the existing results together with those results obtained in this paper. It is assumed that $P = \{P_1, P_2, \ldots, P_M\}$ is a composite part and each P_i has n vertices. The type of the assembly is determined by the type of the parts.

The first result proposed in this paper is an $O(Mn + M \log M)$ algorithm for disassembling arbitrary simple composite parts. The algorithm improves on the $O(Mn \log Mn)$ bound previously established for this problem [N] and is optimal. This solves the problem posed in [N] and [T]. In addition, it turns out that the algorithm is simpler than the one presented in [N].

The second result of this paper is an $\Omega(Mn + M \log M)$ lower bound for the problem of detecting whether a composite part can be disassembled or not. Thus, detecting the existence of a disassembly sequence is as hard as computing one. As a consequence, the algorithm for computing a sequence is optimal also for the detection problem. (The lower bound does not depend on the time it takes to read the data.)

Authors	Type of Parts	Computing a Disassembly Sequence	Detecting the Existence of a Disassembly Sequence
Guibas, Yao	rectangles	O(M log M)	exists always
Guibas, Yao; Ottmann, Widmayer	convex polygons	O(Mn + M log M)	exists always
Chazelle, Ottmann, Soisalon-Soininen, Wood	rectilinear polygons motion parallel to axes	O(Mn $\log^2$ Mn)	O(Mn log Mn)
Toussaint	simple polygons	O(min (M^3 n, M^2 n log Mn))	O(min(M^2n, Mn logMn)
Nurmi	simple polygons	O(Mn log Mn)	O(Mn log Mn)
here	**simple polygons**	**Θ(Mn + M log M)**	**Θ(Mn + M log M)**

Table 1 Existing and new results

3. ALGORITHMS FOR DISASSEMBLING 2-DIMENSIONAL COMPOSITE PARTS

To determine a disassembly sequence for a composite part $P = \{P_1, P_2, \ldots, P_M\}$ of M n-vertex polygons clearly $\Omega(Mn + M \log M)$ time is required. This is true since the input size is $\Omega(Mn)$ and sorting reduces to finding a disassembly sequence for a set of M distinct points on the x-axis. This does, however, not answer the question whether detecting the existence of an assembly sequence is easier than computing one. Indeed, some of the existing algorithms exhibit running times which are different for the two problem instances (see Table 2.1). In this section, we give lower bounds and matching upper bounds for both problems.

When simple composite parts are to be disassembled in the positive x-direction the visibility hulls of the individual parts are found. The *visibility hull* in the x-direction of a polygon P_f is obtained by augmenting P_f by all points lying on a horizontal line segment connecting two points in P. The visibility hulls (in the x-direction) are polygons monotone in the y-direction. If the direction of disassembly, d, is not the positive x-axis then the composite part is rotated until the direction is the positive x-direction.

We recall the following relations introduced earlier by Guibas, Yao [GY] and Ottmann, Widmayer [OW]: *dominance* (*dom*), *immediate dominance* (*idom*), its transitive closure $idom^+$ and *below*. Let A and B be two objects. The *idom* (immediate dominance) relation - A *idom* B if, and only if, there exists a point p_A on A and a point p_B on B such that the line segment from p_A to p_B, parallel to the direction of translation, does not intersect any other object. The transitive closure of idom is denoted by $idom^+$. When the direction of translation is the positive x-axis, then *idom* is denoted as *ixdom*, which refers to an immediate dominance in the x-direction; similarly $ixdom^+$ is defined. If two objects A and B cannot be related by using $ixdom^+$ then there exists a line L parallel to the x-axis such that A (B) is above L and B (A) is below L. In such a case, A (B) is *below* B (A). Objects A and B are related by the *dom* relation as A *dom* B if and only if (A $ixdom^+$ B) or (not (B $ixdom^+$ A) and (A *below* B)) denoted by $dom = ixdom^+ | below$. The relation *dom* is totally defined whereas $ixdom^+$ and *below* may only be partially defined.

For the following lemma, let $P = \{P_1, P_2, \ldots, P_M\}$ be a simple composite part of M n-vertex polygons and let d (equal to the x-axis) be the direction of disassembly. Next we will observe that the relation *dom* defines a linear order if for each pair P_i, P_j, for which i≠j, holds $\mathrm{VH}(P_i, d) \cap \mathrm{VH}(P_j, d) = \emptyset$. Lemma 1 and Theorem 1 are easy to prove by using the results established in [ST] and [T].

Lemma 1 (a) The relation *ixdom* is asymmetric if, and only if, the $\mathrm{VH}(P_i, d) \cap \mathrm{VH}(P_j, d) = \emptyset$, for all $1 \le i, j \le M$ and $i \ne j$.

(b) Let P_i, P_j and P_k be three polygons of P such that their visibility hulls in d are pairwise non-intersecting. P_k *ixdom* P_j and P_j *ixdom* P_i implies that either P_k *ixdom* P_i, or P_k and P_i are incomparable in *ixdom,* when P_k and P_i are examined in isolation.

Theorem 1 Let $P = \{P_1, P_2, \ldots, P_M\}$ be as above and let d (the direction of translation) be the positive x-axis. $\mathrm{VH}(P_i, d) \cap \mathrm{VH}(P_j, d) = \emptyset$, for all $1 \le i, j \le M$ and $i \ne j$ if and only if the relation *dom* defines a linear ordering.

Thus, to find a disassembly sequence for P, if one exists, we first need to ensure that P contains no pair of polygons whose visibility hulls intersect. Then, given its existence, a disassembly sequence for P is computed by finding a disassembly sequence for the monotone part comprised of the (pairwise non-intersecting) visibility hulls of $P_1, P_2, \ldots, P_M$. This is done by computing a linear extension of *dom* whose optimal computation is addressed next.

3.1 Monotone Composite Parts

Let $S = \{P_1, P_2, ..., P_M\}$ be a set of M simple n-vertex polygons monotone with respect to the positive y-axis. The task is to (1) determine whether S contains a pair of intersecting polygons, and (2) to find a linear extension of the relation *dom* defined on S. Next, we sketch the underlying idea of the algorithm.

Our algorithm uses the plane-sweep paradigm as employed in [OW]. A linear list called *LinExt* contains a linear extension of *dom*. A balanced tree maintains the set of all polygons intersected by the sweep-line such that an in-order traversal of the tree yields the intersection points of the polygons with the sweep-line, sorted by x-coordinate.

Whenever the scan line stops at an upper endpoint, for example vertex v of polygon A, A is inserted into the balanced binary search tree. During the insertion into the tree, v is examined with each of the nodes on the path (from the root to the position of A in the tree) whether it is to the left of, right of, or contained in the polygon stored at the node. If it is contained inside the polygon (stored at the node) then an intersection occurs and the algorithm terminates. If v is not contained in any of the polygons then the successor A'' and the predecessor A' of A in the tree are found. A is then tested for an intersection with A' and with A''. If no intersection occurs between these polygons then A is inserted into a linked list, called *LinExt*, before A'' (its successor) looking from left to right. Whenever a lower endpoint, for example u of B, is encountered the polygon B ceases to be *active*, i.e. whose lower endpoint had been encountered earlier. The predecessor B' and the successor B'' of B are found and B' and B'' are tested for an intersection. If no intersection occurs between them then B is deleted from the tree, but it remains in the linked list. If an intersection has occurred in any one of the above cases the algorithm reports that an intersection has occurred and terminates by concluding that the composite cannot be disassembled in the specified direction. (See Figure 1 for an example). We give the pseudo-code description of the algorithm.

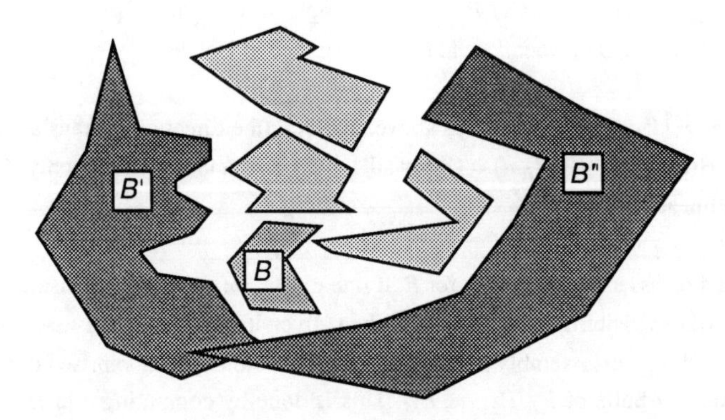

Figure 1 The intersection of the monotone polygons B' and B'' is detected when B ceases to be active.

Algorithm 1: Translating a set of simple monotone polygons.

{The direction of translation is the positive x-axis}

Input: A set $S = \{P_1, P_2, ..., P_M\}$ of M simple n-vertex polygons monotone with respect to the positive y-axis.

Output: A translation order of the polygons.

begin

Find the vertices of the polygons with the highest and the lowest y-value (denote these vertices as the endpoints of the polygons);

Initiate the balanced binary search tree and the doubly linked list;

For all endpoints of the polygons in descending order do:

if the endpoint is an upper endpoint of polygon P_f then

1. Insert P_f into the tree using the procedure Insert_polygon_Into_Tree(Root, P_f);
2. Find the predecessor P' and the successor P'' of P in the tree;
3. if P_f and P' intersect* then exit program; {NO TRANSLATION ORDER EXISTS ! }
4. if P_f and P'' intersect* then exit program; {NO TRANSLATION ORDER EXISTS ! }

if (P'' exists) then insert P_f as the new predecessor of P'' in the linked list;

else insert P_f as the new final element in the linked list;

if the endpoint is a lower endpoint of polygon P_f then begin

find the predecessor P' and the successor P'' of P_f in the tree;

5. if P' and P'' intersect* then exit program; {NO TRANSLATION ORDER EXISTS ! }

else delete P_f from the tree; {But not from the linked list}

end; {algorithm}

* If P' or P'', respectively does not exist then the polygons are assumed to be non-intersecting.

Procedure Insert_Polygon_Into_Tree (Root, P_f);

{This procedure inserts a polygon P into a binary balanced search tree. The relation between the polygon is *ixdom*$^+$.

Input: Root - the root of the tree.

P_f - a polygon.}

begin

While (P_f is not inserted into the tree) do begin

Find the two vertices, one each of the left and right chains (denoted as v_{Li} and v_{Rj} respectively), of the Root such that the y-value of the upper vertex of P_f (denoted as p) is less then the y-value of v_{Li} (v_{Rj}) and greater than the y-value of v_{Li-1} (v_{Rj+1}) (see Figure 2);

{the scan for the new vertices v_{Li} and v_{Rj} starts with the last v_{Lu} and v_{Rw}, of the Root, that have already been found}

If p is to the right of edge(v_{Li}, v_{Li-1}) and to the left of edge (v_{Rj}, v_{Rj+1}) then

exit program; {NO TRANSLATION ORDER EXISTS ! }

elseif p is to the left of edge(v_{Li}, v_{Li-1}) then

Insert_Polygon_Into_Tree (Left_Son_Of_Root, P_f);

else Insert_Polygon_Into_Tree (Right_Son_Of_Root, P_f); {p is to the right of edge (v_{Rj}, v_{Rj+1}) }

end;

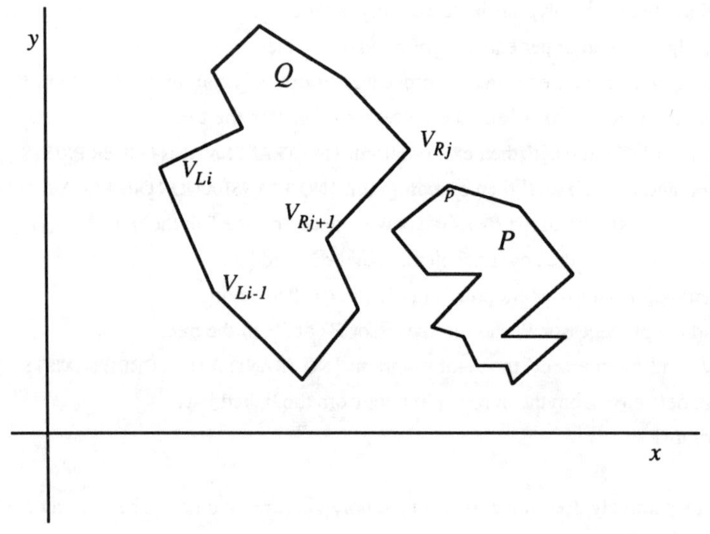

Figure 2 The vertices v_{Li} and v_{Rj} of Q are found when the vertex p of P_f is inserted.

Theorem 2 Algorithm 1 detects in $O(Mn + M \log M)$ time whether a set of M n-vertex polygons monotone in a common direction contains a pair of polygons which are intersecting.

Proof:

Complexity Analysis

For each polygon P_k of the composite part two pointers are maintained: (i) v_{Li} - that points to vertex i at the left chain of P_k; (ii) v_{Rj} - that points to vertex j at the right chain of P_k. (The left polygonal chain consists of the vertices between the bottom vertex and the upper vertex). Each time the algorithm inserts a new polygon P_f into the tree a search for the new vertices v_{Li} and v_{Rj} of the root is done. Pointer v_{Li} (v_{Rj}) points to vertex i (j) of the root such that the y-value of the upper vertex of P_f is less then the y-value of v_{Li} (v_{Rj}) and greater than the y-value of v_{Li-1} (v_{Rj+1}). The search starts from the previous v_{Lu}

and v_{Rw} that have already been found. If the new vertex v_{Li} (v_{Rj}) is the same as the previous vertex v_{Lu} (v_{Rw}) then O(1) time was involved in the computation of finding the new vertex v_{Li} (v_{Rj}). If the new vertex v_{Li} (v_{Rj}) differs from the previous vertex v_{Lu} (v_{Rw}) then there exists a chain of vertices $v_{Li+1}, \ldots, v_{Lu}$ ($v_{Rj-1}, \ldots, v_{Rw}$) that have been examined and discarded (as their *y*-coordinates are too large). Let C_i denote the number of vertices that have been discarded at the *i*-th search. Then the total cost of a search is less then C_i + O(1).

The time complexity of the *i*-th search is C_i + O(1), where $0 < C_i$. The algorithm inserts *M* polygons into the tree. Each time at most log *M* additional recursive searches are done and, thus, the maximum number of times that a search for vertices is done is *M* log *M*. Hence, the cost of the algorithm, during all its execution, is $\Sigma_{i=1..MlogM}(C_i + O(1)) = \Sigma_{i=1..MlogM}C_i + \Sigma_{i=1..MlogM}O(1)$. $\Sigma_{i=1..MlogM}C_i$ is the total number of edges that have been discarded. Each vertex is discarded at most once and, therefore, $\Sigma_{i=1..MlogM}C_i \leq Mn$. Thus, the total cost for all insertions (line 1) is $\Sigma_{i=1..MlogM}C_i + \Sigma_{i=1..MlogM}O(1) \leq Mn + M \log M$.

After the predecessor P' and the successor P'' of P_f have been identified in the tree structure (line 2), at a cost of O(log *M*), P_f is tested for intersection with P' and P'' at a cost of O(*n*) each (lines 3-4). For the deletion (line 5) one pair of polygons is tested for intersection. Since the total number of insertions and deletions is 2*M, the total cost of this step and thus of the entire algorithm is O(*Mn* + *M* log *M*).

Correctness

If there is no pair of intersecting polygons then the algorithm reports no intersection which is correct. Otherwise, we will show that the algorithm finds a highest intersection point (highest being defined as having maximum *y*-coordinate). Let *u* be a highest intersection point which occurs say between polygons P_i and P_j. Without loss of generality we may assume that the *y*-coordinate of the upper vertex of P_i is less than the *y*-coordinate of the upper vertex of P_j. Two cases arise: (a) the upper vertex of P_i is the intersection point *u*. (b) the intersection point is any other point of the boundary of P_i.

(a) When the algorithm encounters P_i at *u* all polygons encountered so far (i.e. appearing in *LinExt*) are either pairwise non-intersecting or have an intersection point below *u*. No polygons which has ceased to be active, intersects P_i. The binary search tree, maintains a sorted list of all active polygons (i.e. the in-order traversal of the tree yields the polygons in order of increasing *x*-values of points intersected by the sweep line). Thus the first if-statement of Procedure *Insert_Polygon_Into_Tree* finds that *u* is contained in P_j and the algorithm correctly reports that no translation order exists (which means that a pair of intersecting polygons has been found).

(b) Assume now that the intersection point *u* is a boundary point of P_i other than the highest vertex. W.l.o.g. assume that when P_i is inserted into the tree P_i and P_j are related as P_j *ixdom*$^+$ P_i. Two cases to be examined.

(i) P_j *ixdom* P_i Then, since P_j is successor of P_i, P_i and P_j are tested for intersection (line 4 of Algorithm 1), the intersection point is correctly found.

(ii) P_j *ixdom*$^+$ P_i but not(P_j *ixdom* P_i). P_i and P_j are active and u has not been found. Hence one or more polygons $P_{k'}$ exists such that P_j *ixdom* P_k' and P_k' *ixdom* P_i . Among these polygons let P_k be the polygon with lowest lower vertex. The y-coordinate of the lower vertex of P_k is greater than the y-coordinate of u and, therefore, P_k is deleted at which point polygon P_i becomes predecessor of P_j in the tree (before the algorithm reaches u). In line 5 of the algorithm, P_i and P_j are then tested for intersection and point u is correctly determined. q.e.d.

Theorem 3 Let $P = \{P_1, P_2, \ldots, P_M\}$ be a composite part (possibly non-simple) of M simple n-vertex polygons, monotone with respect to the y-direction. In $O(Mn + M \log M)$ time it can be determined whether a disassembly sequence for P exists, and, if so, a disassembly sequence for P can be computed.This is worst-case optimal.

Proof

By Theorem 1 and Theorem 2 the existence of a disassembly sequence can be determined in $O(Mn + M \log M)$ time. Thus we assume that a translation ordering for P exists and show that the algorithm correctly determines a disassembly sequence. The claim is that at any time during the execution of the algorithm the linked list *LinExt* maintained by the algorithm represents a linear extension of the dominance relation for all polygons encountered so far. This will imply in particular that upon termination of the algorithm a translation ordering for P is described in *LinExt* .

We establish the claim inductively assuming in addition that at any time an inorder traversal of the tree produces a subsequence of *LinExt*. Assume that this is true before vertex v of P_f is encountered. Two cases arise depending on whether v is a lower endpoint (case (a)) or an upper endpoint (case (b)).

(a) P_f is deleted then clearly the claim remains true.

(b) P_f is inserted into the tree. The inorder traversal of the tree gives the sweep line status which is the relation *ixdom* of all polygons in the tree, i.e. of all *active* polygons. If a successor P'' for P_f exists then P'' *ixdom* P_f and by transitivity of *dom*, P_f is also in the correct position (using *dom*) with respect to all polygons succeeding P'' in *LinExt* . If no successor P'' for P_f exists then P_f is appended to *LinExt* which is correct since for no P_i holds P_i *ixdom* P_f . If a predecessor P' for P_j exists then P_f *ixdom* P' and again P_f is in the correct position (using *dom*) with respect to all polygons preceding P' in *LinExt.* All other polygons P_j between P' and P'' are no longer active, thus P_f *below* P_j holds, implying P_f *dom* P_j. q.e.d.

3.2 Simple Composite Parts

As remarked earlier, we disassemble simple composite parts in the positive x-direction by first finding the visibility hulls of the individual parts. Each visibility hull can be determined in linear time by using any of the algorithms developed in [DSs], [EA], or [L]; thus the total time taken for this step is

$O(Mn)$. Then the visibility hulls are given as input to the Algorithm 1, the output of Algorithm 1 is a disassembly sequence for the simple composite part P.

Theorem 4 Let $P = \{P_1, P_2, ..., P_M\}$ be a (simple) composite part of M n-vertex polygons. In $O(Mn + M \log M)$ time it can be determined whether a disassembly sequence for P exists, and, if so, a disassembly sequence for P can be computed.

Clearly computing a translation ordering for a composite part is at least as hard as sorting and thus the following Corollary holds.

Corollary 1 For the problem of computing a disassembly sequence of simple composite part, the above algorithm is worst case optimal.

3.3 Lower Bound for Detecting the Existence of a Disassembly Sequence

It remains to see whether detecting the existence of a disassembly sequence is easier than computing one. For a variety of problems detecting a certain geometric property is easier than computing it; even sublinear solutions may exist, if the input is already stored in memory. These problems include computing lines of support for convex polygons, see e.g. [PS], detecting whether two convex polygons intersect [CD], etc.

The lower bound presented here for the disassembly detection problem will be independent on the time required to read in the polygons. We assume further that the visibility hull of each polygon in P has been previously computed and is stored in memory.

Chazelle and Dobkin [CD] have established a tight logarithmic lower bound for the Intersection Detection Problem for two convex polygons (stored in memory). Here, we first establish a linear lower bound on the corresponding problem for polygons monotone in a known direction. Given two polygons P and Q, monotone with respect to the same direction, detect whether P and Q intersect (where m and n are the number of vertices in P and Q respectively). A program that can solve this problem will be called ***IntersectionDetector***.

Lemma 2 At least $m + n - 4$ time is required to test whether two polygons monotone in a specified direction intersect, or not, where m, n are the number of vertices in the each of the polygons, respectively. The bound is independent on the time required to read the polygons into memory.

Proof We give an adversary-based argument. Let P and Q be two polygons monotone with respect to the y-axis. Let the vertices $q_2, q_3, \ldots, q_{n-1}$ of Q face the edge(p_m, p_1) of P. Let the vertices $p_3, p_4, \ldots, p_m$ of P face the edge(q_{n-1}, q_n) of Q, let the x-coordinate of p_1 (p_2) be less than the x-coordinate of q_1 (q_n) and let the y-coordinate of p_1 (p_2) be equal to the y-coordinate of q_1 (q_n) (see Figure 3).

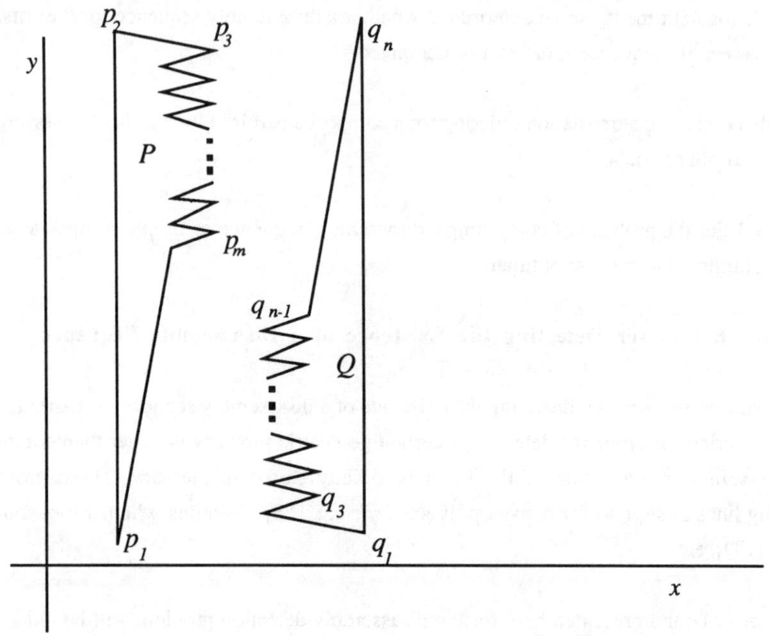

Figure 3 Construction of polygons P and Q for adversary argument.

Assume the contrary of Lemma 3.9, i.e. that there exists an IntersectionDetector that requires less then $m+n-4$ steps. To solve the problem, IntersectionDetector must be able to detect whether P and Q intersect. Whenever IntersectionDetector accesses vertex p_i (q_j) the adversary will give IntersectionDetector, for free, the information which edge(q_u, q_v) of Q (edge(p_r, p_s) of P) p_i (q_j) is facing and if p_i (q_j) intersects Q (P). The adversary maintains a list of all vertices that have not yet been examined by IntersectionDetector. If IntersectionDetector examines less than $m+n-4$ vertices then there exist at least one vertex among $p_3, p_4, \ldots, p_m$ and $q_2, q_3, \ldots, q_{n-1}$ that was not examined by IntersectionDetector. Assuming that vertex p_k, $2 < k \le m$, (q_r, $2 \le r < n$) was not examined. If IntersectionDetector outputs that the polygons intersect then IntersectionDetector fails because the polygons do not intersect. If IntersectionDetector outputs that the polygons do not intersect then just before the detector outputs the answer, that the polygons do not intersect, the adversary "moves" p_k (q_r), by increasing (decreasing) the x-coordinate of p_k (q_r), until p_k (q_r) intersect Q (P). "Moving" p_k (q_r) does not change the property of monotonicity of P (Q) with respect to the y-axis, as the x-values of the vertices are independent, and thus IntersectionDetector fails to give a correct answer. The adversary can find in $O(m+n)$ steps which vertex should be moved by traversing all the vertices. q.e.d.

It is easy to see that this construction generalizes to sets of monotone polygons. A similar argument can then be made for set of visiblity hulls instead of monotone polygons. Note however that it must be ensured that the input polygons are pairwise non-intersecting while the visibility hulls may, or may not, be intersecting.

Lemma 4 Detecting whether a simple composite part $P = \{P_1, P_2, \ldots, P_M\}$ of M n-vertex polygons can be disassembled in a given direction d requires at least $\Omega(Mn + M \log M)$ time.The bound is independent on the time to read the input.

Proof (a) "$\Omega(Mn)$" Follows from Lemma 3 and the previous discussion.

(b) "$\Omega(M \log M)$" The proof is based on a reduction from integer element uniqueness. The *Integer Element Uniqueness Problem* is defined as: given M integers N_i, $1 \le i \le M$, decide if any two are equal and has been shown to require $\Omega(M \log M)$ time [SP]. The construction needs to ensure that the input polygons are non-intersecting and that the visibility polygons intersect if, and only if, the M numbers are not all distinct. The transformation from the Integer Element Uniqueness to the Separability Detecting Problem is as follows. For each integer N_i, $1 \le i \le M$, of the Integer Element Uniqueness Problem a V-shaped simple 4-vertex polygon P_i, $1 \le i \le M$, is constructed (see Figure 4). Each polygon P_i, $1 \le i \le M$, is constructed such that its width is less than 1 (x-coordinate of $p_{i,4}$ - x-coordinate of $p_{i,2}$), also y-coordinate of $p_{i,3}$ - y-coordinate of $p_{i,1}$ < 1. Therefore, there is no pair of intersecting polygons.
Two polygons P_i and P_j are nested inside each other if, and only if, their visiblity hulls are intersecting. Two polygons P_i and P_j are nested if, and only if, $N_i = N_j$. Therefore, if the output of the intersection test is "YES" then no two integers are equal in the Integer Element Uniqueness Problem. If the output of the intersection test is is "NO" then there exists at least one pair of polygons that are nested and, therefore, there exist two equal integers in the Integer Element Uniqueness Problem that are equal.
Transforming the Integer Element Uniqueness Problem to the Separability Detecting Problem requires $O(M)$ time because each polygon consists of only four vertices. Let $f(M,n)$ denote the time required for testing the M polygons for separability.Converting the output of the intersection detection algorithm into a correct output for the Integer Element Uniqueness Algorithm requires $O(1)$ time.

Hence, the Separability Detecting Problem requires at least $\Omega(M \log M)$ - $O(M)$ - $O(1)$, which yields $f(M,n) \ge \Omega(M \log M)$. q.e.d.

Theorem 5 Detecting the existence of a disassembly sequence for a simple composite part is as hard as computing one. Both problems can be solved in $\Theta(Mn + M \log M)$ time.

Proof: Follows from Theorem 3.7 and Lemma 3.11. q.e.d.

REMARKS

This paper presents lower bounds and matching upper bounds for 2-dim. disassemblies via translations in a common direction. Related variants of the assembly problems allow e.g. multiple directions of translations for a composite 2-d part as discussed in [DS] and in [Na], or disassembly of a pair of polygons by multiple translations [PSS]; see also [T] for additional references. In [Nu] it is shown how to efficiently disassemble 3-dimensional composite parts and how to compute directions for disassembly.

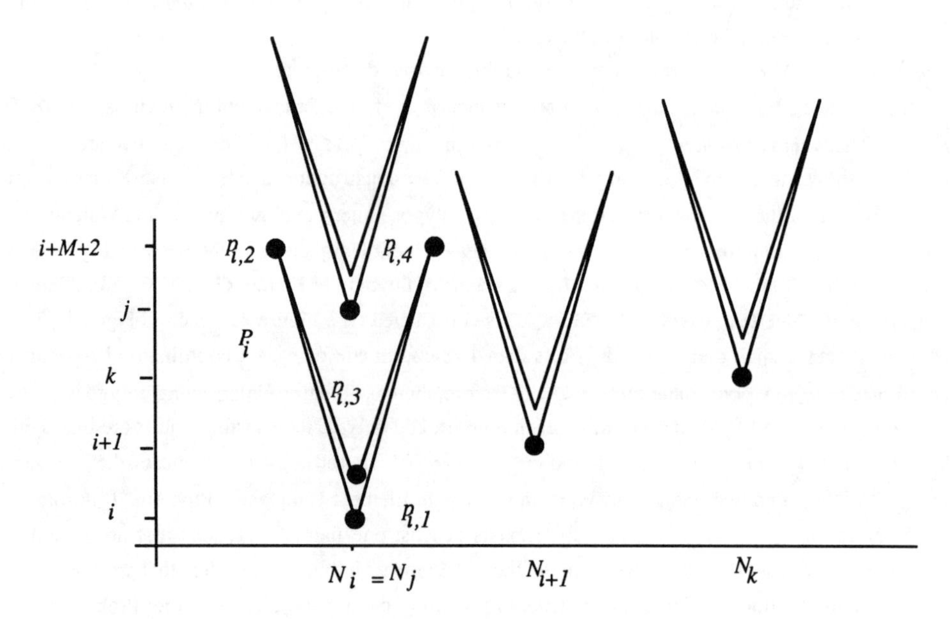

Figure 4 The reduction from element uniqueness.

REFERENCES

[CD] B. Chazelle and D. P. Dobkin, "Intersection of Convex Objects in Two and Three Dimensions", Journal of the ACM, Vol. 34, No. 1, January 1987, pp. 1-27.

[COSW] B. Chazelle, T. Ottmann, E. Soisalon-Soininen and D. Wood, "The Complexity and Decidability of Separation", Technical Report no. CS-83-34, University of Waterloo, Waterloo, Ontario, November 1983.

[D] R. Dawson, "On Removing a Ball Without Disturbing the Others", Mathematics Magazine, Vol. 57, No. 1, January 1984, pp. 27-30.

[DSa] J. Dean and J.-R. Sack, "Efficient Hidden-Line Elimination by Capturing Winding Information", Proceedings 23rd Allerton Conference on Communication, Control and Computing, Ill., Oct. 1985, pp. 496-505.

[DS] F. Dehne and J.-R. Sack, "Translation Separability of Sets of Polygons", The Visual Computer, No. 3, 1981, pp. 227-235.

[EA] H. ElGindy, D. Avis, "A Linear Algorithm for Computing the Visiblity Polygon from a Point", Journal of Algorithms, Vol. 2, No. 3, 1983, pp. 191-202.

[GY] L. J. Guibas and F. F. Yao, "On Translating a Set of Rectangles", in Advances in Computing Research Volume I: Computational Geometry, Ed. F. P. Preparata, JAI Press Inc., Greenwich, CO, 1983, pp. 61-77.

[L] D.T. Lee, "Visiblity of a Simple Polygon", Computer Vision, Graphics and Image Processing, Vol. 22, No. 2, 1983, pp. 207-221.

[N] O. Nurmi, "On Translating a Set of Objects in 2- and 3- Dimensional Space", Computer Vision, Graphics, and Image Processing, Vol. 36, 1986, pp. 42-52.

[NS] O. Nurmi and J.-R. Sack, "Separating a Polyhedron by One Translation from a Set of Obstacles", Proceedings Workshop on Graph Theory, Amsterdam 1988, Lecture Notes in Computer Science, Vol. 344, 1988, pp. 202-212.

[Na] B. K. Natarajan, " On Planning Assemblies", Proceedings of the Fourth Annual Symposium on Computational Geometry, Urbana-Champaign, Illinois, June 1988, pp. 299-308.

[Nu] D. Nussbaum, "Directional Separability in two and three Dimensional Space", School of Computer Science, Carleton University, 1988.

[OW] T. Ottmann and P. Widmayer, "On Translating a Set of Line Segments", Computer Vision, Graphics, and Image Processing Vol. 24, 1983, pp. 382-389.

[PS] F. P. Preparata and M. I. Shamos, *Computational Geometry An Introduction*, Springer-Verlag, Berlin, Heidelberg, New York, Tokyo, 1985.

[R] J. Reif, "Complexity of the Mover's Problem and Generalizations", Proceedings 20th Symposium on the Foundations of Computer Science, 1979, pp. 560-570.

[PSS] R. Pollack, M. Sharir and S. Sifrony, "Separating Two Simple Polygons by a Sequence of Translations", Technical Report No. 59/87, Eskenasy Institute of Computer Science, School of Mathematical Science, Tel-Aviv University, Tel-Aviv Israel, January 1987.

[ST] J.-R. Sack and G.T. Toussaint, "Separability of Pairs of Polygons Through Single Translations", Robotica, Vol. 5, 1987, pp. 55-63.

[T] G. T. Toussaint, "Movable Separability of Sets", Computational Geometry Ed. G. T. Toussaint, North Holland Amsterdam, New York, Oxford, 1985.

[Ta] R. E. Tarjan, "Depth First Search and Linear Graph Algorithms", SIAM Journal on Computing, Vol. 1, 1972, pp. 146-160.

WHICH TRIANGULATIONS APPROXIMATE THE COMPLETE GRAPH?

(Conference Abstract)[1]

GAUTAM DAS – University of Wisconsin

DEBORAH JOSEPH – University of Wisconsin

Abstract. Chew and Dobkin et. al. have shown that the *Delaunay triangulation* and its variants are sparse approximations of the complete graph, in that the shortest distance between two sites within the triangulation is bounded by a constant multiple of their Euclidean separation. In this paper, we show that other classical triangulation algorithms, such as the *greedy triangulation*, and more notably, the *minimum weight triangulation*, also approximate the complete graph in this sense. We also design an algorithm for constructing extremely sparse (nontriangular) planar graphs that approximate the complete graph.

We define a sufficiency condition and show that *any* Euclidean planar graph constructing algorithm which satisfies this condition always produces good approximations of the complete graph. This condition is quite general because it is satisfied by all the triangulation algorithms mentioned above, and probably by many other graph algorithms as well. We thus partially answer the question posed by the title.

From a theoretical standpoint, our results are interesting because we prove nontrivial properties of minimum weight triangulations, of which little is currently known. From a practical standpoint, the graph algorithms we consider are good alternatives to the Delaunay triangulation, particularly when designing a sparse network under severe constraints on the total edge length. Finally, our general approach may help in identifying or designing other algorithms for constructing sparse networks.

1. INTRODUCTION

The *complete graph* represents an ideal communication network between n sites. However, to conserve resources, sparse networks are often designed which approximate the complete graph in some sense. This problem has also been studied from a geometric context, where the graphs are Euclidean. In particular, recent research [C, DFS] has shown that the *Delaunay triangulation* and its variants are good approximations

---

[1] This work was supported in part by the National Science Foundation under grant DCR-8402375. The authors' address is: Computer Sciences Department, University of Wisconsin, 1210 West Dayton St., Madison, WI 53706, U.S.A.

of the complete graph, in that the shortest distance between any two sites within the triangulation is bounded by a constant multiple of their Euclidean separation.

In this paper, we show that other classical triangulation algorithms, such as the *greedy triangulation* [MZ], and more notably, the *minimum weight triangulation* [MZ], also produce good approximations of the complete graph in the above sense. We define a sufficiency condition, and show that *any* algorithm for constructing Euclidean planar graphs which satisfies this condition always produces good approximations of the complete graph. This condition is quite general; it is satisfied by all the three triangulation algorithms mentioned above, and by many other graph algorithms as well. We thus partially answer the question posed by the title. Finally, we apply this condition constructively to design an algorithm which generates extremely sparse (nontriangular) planar graphs. The length of the graph produced by this algorithm is proportional to the length of the *minimum spanning tree* between the sites.

From a theoretical standpoint, our results are interesting because we prove some nontrivial properties of minimum weight triangulations, which has always been an enigmatic construct in computational geometry. Other than the basic definition, very few properties are known about it, and its computational complexity is still unknown [L, K]. From a practical standpoint, the graph algorithms we consider are suitable alternatives to the Delaunay triangulation, particularly when designing a sparse network under severe constraints on the total edge length. For instance, an example in [MZ] shows that the ratio of the total edge length between a Delaunay and a greedy triangulation could be as large as $\Omega(n/\log n)$. Finally, our techniques may be useful in identifying or designing other algorithms that construct similar sparse graphs.

In this paper we will restrict our attention to Euclidean planar graphs, unless otherwise mentioned. Throughout we adopt the following definitions. If e is a line segment, $|e|$ denotes its length. If E is a set of line segments, $|E|$ denotes the total length of all segments, rather than the usual notion of cardinality. S is a collection of n points in the plane (called *sites*). Let $K(S)$ be the complete graph, containing all $n(n-1)/2$ straight edges that join the sites in S. A *triangulation* is a maximal nonintersecting subset of $K(S)$. A *graph algorithm* G takes input S, and produces an Euclidean planar graph $G(S)$. A *Delaunay triangulation*, $DT(S)$, is the straight edged Voronoi dual of S [DFS]. A *greedy triangulation*, $GT(S)$, is computed by a greedy algorithm which iteratively inserts the next shortest nonintersecting edge into a partially constructed triangulation [MZ]. A *minimum weight triangulation*, $MWT(S)$, is one with minimum total edge length [MZ]. $SP_{G(S)}(A, B)$ denotes the shortest path between sites A and B, composed of edges of $G(S)$. Let $measure(G(S)) = max_{A,B \in S} \frac{|SP_{G(s)}(A,B)|}{|AB|}$. Intuitively, this quantity measures how badly a graph approximates the complete graph.

The algorithm G produces *good approximations* if there exists a constant c such that, for all S, $measure(G(S)) \leq c$.

In [DFS] it was shown that, for all S, $measure(DT(S)) \leq \pi(1+\sqrt{5})/2$. Our methods are more general because we isolate a condition satisfied by many graph algorithms, including DT, GT, and MWT. (This is described in Section 3) First we show, in Section 2, that any graph algorithm satisfying this condition always produces good approximations of the complete graph. Finally in Section 4, we apply this condition in designing an algorithm for constructing general (nontriangular) graphs.

The reader should note that in this version of the paper, the various constants are roughly computed, and do not represent tight upper bounds. We feel that to establish tighter bounds, it is necessary to exploit properties unique to a particular graph algorithm. Our general approach thus helps in easily identifying algorithms that deserve further study.

2. GRAPHS WHICH APPROXIMATE THE COMPLETE GRAPH

We begin this section by defining the graph property that will be sufficient for our first result.

Definition. *Let α be an angle less than $\pi/2$, d a constant, and G a graph algorithm. The condition $P_{\alpha,d}(G)$ is defined as follows.*

$P_{\alpha,d}(G)$ = *true, if for every S, the following are true.*

i) The diamond property. For every edge e of $G(S)$, consider the two triangular regions defined on either side of e, such that e is the base of each triangle and the base angles are α (see Figure 1). Then at least one of the regions contains no other sites of S.

ii) The good polygon property. For every face of $G(S)$, let A and B be any two sites along its boundary such that the line segment AB lies wholly within the face (such segments are called chords). Then the shortest distance from A to B around the boundary of the face is smaller than $d|AB|$.

Intuitively, this property is useful because (i) ensures that an edge cannot act as a long obstacle between a pair of sites and (ii) ensures that the distance "around the boundary" of a graph face can not be too much longer than the distance "across" the face. In fact a formalization of this proves our first result. We will spend the remainder of this section showing that any graph algorithm satisfying this property always produces good approximations of the complete graph. Then in the next section we will show that the Delaunay triangulation, the greedy triangulation and the minimum weight triangulation satisfy the property. Finally, we will use this property to design an algorithm for constructing sparse nontriangular planar graphs.

For simplicity we assume throughout that π is a multiple of α.

Theorem 1. *For any angle $\alpha < \pi/2$ and any constant d, there exists a constant $c_{\alpha,d}$ such that, if $P_{\alpha,d}(G) = true$ for some graph algorithm G, then for all S, $measure(G(S)) \leq c_{\alpha,d}$.*

Proof. Our proof has some flavor of the techniques employed in [DFS]. The latter exploits specific properties of Delaunay triangulations, while ours is a more general approach and is thus more involved.

Let $G'(S)$ be the graph $G(S)$ augmented with all possible chords of faces. Clearly $G'(S)$ may not be planar. We introduce the notion of *pseudo paths*, which are paths in $G'(S)$. We also define *real paths*, which are paths in $G(S)$. The good polygon property ensures that if there exists a pseudo path of length l between sites A and B, then there exists a real path between them of length at most dl. Thus our efforts will be directed towards constructing a short pseudo path between any pair of sites. Henceforth, all paths will be pseudo, unless otherwise mentioned.

The diamond property ensures that every edge e of $G(S)$ is the base of at least one empty triangular region with base angles α. (If both regions are empty, select any one). Let $t(e)$ and $v(e)$ denote this triangle and its third vertex respectively. Note that this vertex need not be a site in S.

Let A and B be two sites such that AB is not in $G(S)$. The two half planes defined by extending AB will be referred to as *Top* and *Bottom* respectively. (Figure 2). We first construct a subgraph of $G'(S)$ within which a short pseudo path between A and B resides. Let $e_1, e_2, \ldots, e_m$ be the sequence of edges of $G(S)$ that intersect with AB. Let u_i and l_i refer to the upper and lower sites of e_i. (For simplicity we shall assume these edges do not share vertices). Clearly every adjacent pair of edges e_i and e_{i+1} in this sequence belong to the same face. Let UC_i be the upper convex chain from u_i to u_{i+1}, such that the boundary of the face from u_i to u_{i+1} is above it. The lower convex chain, LC_i, is similarly defined. Let $u_{i,1}l_{i,2}$ and $l_{i,1}u_{i,2}$ be the two common tangents between the two chains. Now consider the graph consisting of the edges of $G(S)$ that intersect AB, along with all upper and lower chains, and all common tangents. Clearly this graph (denoted as $G'(A, B)$) is a subgraph of $G'(S)$. We shall construct a path from A to B in this subgraph that is not too long with respect to AB. Sites in this subgraph may be classified into two groups, those that are end points of edges of $G(S)$ that intersect with AB (called *major sites*), and those that belong to upper and lower chains (called *minor sites*). Each group may be further classified into upper and lower sites in a natural manner.

Any simple path in this subgraph from A to B consists of alternating portions of the upper and lower boundary, connected by tangents. We now construct a *direct path* of this sort as follows. Begin with A on the path. Suppose u_i (a major site) is the last site of the partially constructed path. If $v(u_{i+1}l_{i+1})$ is inside *Bottom* (lies below AB), then the path is extended to u_{i+1} via the upper chain, UC_i. If $v(u_{i+1}l_{i+1})$ is inside *Top* (lies above AB), then the path is extended to l_{i+1} as follows. The path is first extended via the upper chain to $u_{i,1}$, then along a common tangent to $l_{i,2}$, and eventually along the lower chain to l_{i+1}. If l_i is the last major site, the extension is similar. We denote such direct paths as $DP(A,B)$. The following claim illustrates an important property of direct paths.

Claim 1: Suppose $DP(A,B)$ is *one sided*, that is, the path consists entirely of the upper boundary or entirely of the lower boundary of $G'(A,B)$. Then $\frac{|DP(A,B)|}{|AB|} \leq b_\alpha$, where $b_\alpha = \frac{2\pi}{\alpha \sin \alpha/4}$.

Proof. Without loss of generality, assume the upper boundary is the direct path. Let $R = \mathit{Bottom} \cup t(e_1) \cup \dots \cup t(e_m)$. Let $R(A,B)$ denote the boundary of R from A to B. Since $DP(A,B)$ is composed of convex chains, $|DP(A,B)| \leq |R(A,B)|$.

We now show an upper bound for $|R(A,B)|$. Partition the triangles $t(e_1), \dots, t(e_m)$ into $2\pi/\alpha$ groups, $G_0, G_{\alpha/2}, \dots, G_{\pi-\alpha/2}$, where G_θ contains those triangles whose left sides make angles in the range $[\theta, \theta + \alpha/2]$ with AB. Let $R_\theta = \mathit{Bottom} \cup$ *all triangles in* G_θ. But $R = R_0 \cup \dots \cup R_{\pi-\alpha/2}$, and hence

$$|R(A,B)| \leq |R_0(A,B)| + \dots + |R_{\pi-\alpha/2}(A,B)|.$$

Consider Figure 3, which illustrates the group G_θ. Clearly $R_\theta(A,B)$ is inside ACB, and is composed of three types of line segments. Segments that go up/down/horizontal make angles in the range $[\theta, \theta + \alpha/2]$ / $[\theta + \alpha, \theta + 3\alpha/2]$ / $[0, 0]$ with AB respectively. It is then easy to see that $|R_\theta(A,B)| \leq |AC| + |BC| \leq \frac{|AB|}{\sin \alpha/4}$, for any θ. Thus $|R(A,B)| \leq \frac{2\pi|AB|}{\alpha \sin \alpha/4}$, and the claim is proved. ∎

Since a direct path may not be one sided, we iterate the following modification exhaustively until we obtain a final path, $FP(A,B)$, which is almost one sided. Let u_i and u_j be two upper major sites on the direct path such that the major sites in between, $l_{i+1}, \dots, l_{j-1}$, are lower sites. Determine if a shortcut is necessary (this will be defined later). If not, retain the path, otherwise replace the portion of the path between $u_{i,1}$ and $u_{j-1,2}$ with the portion of the upper boundary of $G'(A,B)$ between them.

We now define when a shortcut is necessary. Recall that every edge e_k intersecting with AB is associated with an empty triangular region $t(e_k)$. We associate another empty region with each such edge, denoted as $t'(e_k)$, and defined as follows. Suppose $v(e_k)$ is below AB (Figure 4). Then $t'(e_k)$ is the closed empty region between AB, the common tangents $l_{k-1,1}u_{k-1,2}$ and $u_{k,1}l_{k,2}$, and the path along upper chains from $u_{k-1,2}$ via u_k to $u_{k,1}$. If $v(e_k)$ is above AB, then $t'(e_k)$ is symmetrically defined.

Partition the set of intersecting edges into two sets L and U. The edge e belongs to U (respectively L) if $v(e)$ belongs to *Bottom* (respectively *Top*). Define $R_U = Bottom \cup_{e \in U} [t(e) \cup t'(e)]$. R_L is symmetrically defined. It is easy to see that $|R_U(A,B)|$ and $|R_L(A,B)|$ are each smaller than $b_\alpha|AB|$.

Let θ be the angle between the tangents $u_{i,1}l_{i,2}$ and $l_{j-1,1}u_{j-1,2}$, when extended. It does not matter whether the extended diagonals meet in *Top* or *Bottom*. Two cases arise.

Case 1: $\theta > \pi/3$. The shortcut is not necessary.

Case 2: $\theta \leq \pi/3$. Intuitively, this means the diagonals are almost parallel. Consider Figure 5. The shaded region is bounded by the tangents $u_{i,1}l_{i,2}$ and $l_{j-1,1}u_{j-1,2}$, $R_U(A,B)$, and $R_L(A,B)$, and is empty of other sites. Let the upper (lower) boundary of this region be T_1 (T_2). We set up co-ordinate axes as shown in the figure. E (F) is the point on T_1 (T_2) with the minimum (maximum) y co-ordinate, and $h = y(E) - y(F)$. PEQ and VFW are parallel to the x axis. Let $w = max(|PQ|, |VW|)$. If $\frac{h}{w} \leq \frac{1}{2\tan\alpha/2}$, the shortcut is not necessary, otherwise the shortcut is taken.

The following claims lead to the final stages of the proof of the theorem. We again remind the reader that in this version of the paper, the calculations are rough, and the resulting value of $c_{\alpha,d}$ is not a tight upper bound.

Claim 2: If the shortcut is taken, then the length of the new portion of the path, which is the upper boundary of $G'(A,B)$ from $u_{i,1}$ to $u_{j-1,2}$, is smaller than $b_\alpha(|T_1| + |T_2|)$.

Proof. Let the lower convex hull of the upper boundary of $G'(A,B)$ from $u_{i,1}$ to $u_{j-1,2}$ be $u_{i,1} = z_1, z_2, \ldots, z_k = u_{j-1,2}$. Clearly $|z_1z_2| + |z_2z_3| + \ldots + |z_{k-1}z_k|$ is smaller than $|T_1|$. We also observe the following. If H and I are any two points above T_1, and J is any point below T_2, such that HJ and IJ intersect the shaded region, then the angle HJI is smaller than α, because of the shortcut condition. Due to this, for $1 \leq q \leq k$, $DP(z_q, z_{q+1})$ is one sided (where $DP(z_q, z_{q+1})$ is defined within $G'(z_q, z_{q+1})$), and the concatenation of all these direct paths is the upper boundary of $G'(A,B)$ from $u_{i,1}$ to $u_{j-1,2}$. Thus the length of the new portion is smaller than $b_\alpha(|T_1|)$, which is smaller than $b_\alpha(|T_1| + |T_2|)$. ∎

Claim 3: If the shortcut is not taken due to case 1, then the length of the old portion of the path between $u_{i,1}$ and $u_{j-1,2}$ is smaller than $3(|T_1| + |T_2|)$.

Proof. Clearly, the lower boundary of $G'(A,B)$ is smaller than $|T_2|$. We now compute the total length of the two tangents, $u_{i,1}l_{i,2}$, and $u_{j-1,2}l_{j-1,1}$. Let C be the meeting point of their extensions. Since $\theta > \pi/3$, if C is in *Top* (*Bottom*), their total length is smaller than $2|T_2|$ ($2|T_1|$). Thus their total length is always smaller than $2(|T_1| + |T_2|)$, and the claim is proved. ∎

Claim 4: If the shortcut is not taken due to case 2, then the length of the old portion of the path between u_i and u_j is smaller than $a_\alpha(|T_1| + |T_2|)$, where $a_\alpha = \frac{5}{\sin \alpha/2}$.

Proof. As before, the lower boundary of $G'(A,B)$ is smaller than $|T_2|$. We now compute the total length of the two diagonals. Since $\theta \leq \pi/3$, we see that

$$|u_{i,1}P| + |l_{i,2}V| + |u_{j-1,2}Q| + |l_{j-1,1}W| \leq 2(|T_1| + |T_2|)$$

irrespective of where C lies. Also,

$$|PV| + |QW| \leq 4h \leq \frac{2}{\tan \alpha/2}(|T_1| + |T_2|).$$

The claim follows after some manipulations. ∎

Thus, whichever case occurs, the length of $FP(A,B)$ from $u_{i,1}$ to $u_{j-1,2}$ is smaller than $[max(b_\alpha, 3, a_\alpha)](|T_1| + |T_2|) = b_\alpha(|T_1| + |T_2|)$. Summing over all such modifications, we conclude that $|FP(A,B)| \leq 2b_\alpha^2|AB|$. Thus $c_{\alpha,d} = \frac{8\pi^2 d}{\alpha^2 \sin^2 \alpha/4}$, and the theorem is proved. ∎

3. LAYOUT PROPERTIES OF CERTAIN TRIANGULATIONS

In this section we investigate the algorithms, Delaunay triangulations, greedy triangulations, and minimum weight triangulations, and show that all three satisfy the property of the previous section. Then in the next section, we consider more general graph algorithms.

Theorem 2. $P_{\pi/4,1}(DT) = true$.

Proof. Notice that condition (ii) of the property is trivially holds for any triangulation. As for condition (i), it is known that the circumscribing circle of any triangle in a Delaunay triangulation is empty of other sites. Thus every edge is the chord of two empty circles. If we select any one circle of any edge, it will wholly contain at least one of the triangular regions regions. Thus condition (i) holds. ∎

The next theorem describes a similar result for greedy triangulations. We first present a lemma.

Lemma 1. *Let A and B be any two sites such that AB is not an edge in $GT(S)$. Let e be the shortest edge of $GT(S)$ that intersects AB. Then $|e| < |AB|$.*

Proof. Suppose all edges of $GT(S)$ intersecting AB are longer. Consider the state of the algorithm just before the first of these edges is added to $GT(S)$. Clearly AB should be selected instead, because it is shorter, leading to a contradiction. ∎

Despite its simplicity this lemma is useful in the proof of the following theorem.

Theorem 3. $P_{\pi/8,1}(GT) = true.$

Proof. Consider Figure 6. AB is an edge of some $GT(S)$. CAB, DAB are triangles with base angles $\pi/8$. O is the center of AB. EAF and KBL are portions of circles centered at B and A respectively. We have to show that either CAB or DAB is empty of other sites.

Consider the left half of the figure. Let a site M be inside CAB. Draw a perpendicular MN onto AB. Let the triangles of $GT(S)$ that intersect with NM be $t_1, t_2, \ldots, t_k$, ordered from right to left along NM. Thus t_1 is the triangle immediately to the left of AB. Let t_i be the *first* triangle in the sequence such that one of its three sites, V, lies inside the region $GABIC$. Without loss of generality, assume V lies in the bottom half, $OBIC$. Consider the polygonal region $P = t_1 \cup t_2 \cup \ldots \cup t_i$. The portion of P's boundary from V to A (V to B) is the *upper chain* (*lower chain*).

We observe that the lower chain degenerates into a single edge, VB. If this was not so, then all edges of $GT(S)$ intersecting VB would be longer, contradicting Lemma 1. Two cases arise for the upper chain.

Case 1: The upper chain is a single edge, VA.

Case 2: The upper chain is not a single edge.
Let the sites along the upper chain be $V = u_0, u_1, \ldots, u_m, u_{m+1} = A$. Since $|VA| < |AB|$, by Lemma 1 at least one of these sites is inside EAG. We next observe that in fact u_m is inside EAG. If this was not so, let u_p be the last site along the upper chain to lie inside EAG, $1 \leq p < m$. Clearly all edges intersecting u_pA are longer, contradicting Lemma 1.

Now consider the right half of the figure, and assume that DAB also contains a site. A symmetric argument leads to the existence of a site, Q, which corresponds to V of the left half. Clearly Q is inside $HABJD$. Unlike the previous argument however, we shall surrender generality if we assume that Q lies in the bottom half of $HABJD$. Consequently four cases arise.

Case 3: Q lies in the bottom half, and the upper and lower chains are single edges, QA and QB.

Case 4: Q lies in the top half, and the upper and lower chains are single edges, QA and QB.

Case 5: Q lies in the bottom half, and the upper chain is not a single edge.

Case 6: Q lies in the top half, and the lower chain is not a single edge.

We are now ready to prove the theorem, by showing that any combination of the left half cases and right half cases leads to a contradiction. In this version of the paper, only two of the combinations are discussed, the others being similar.

Case 1 and Case 4: AB is the only edge intersecting VQ, and $|AB| > |VQ|$, which contradicts Lemma 1.

Case 2 and Case 6: All edges intersecting $u_m Q$ are longer, which contradicts Lemma 1.

A full analysis of all the combinations proves the theorem. ∎

The next theorem describes a similar result for minimum weight triangulations. Here the proof is more involved because we do not have a tool as powerful as Lemma 1. First we present some definitions. A *triangulation* of a n-site simple polygon P is the set of $n - 1$ boundary edges along with $n - 3$ nonintersecting diagonals that partition the interior into $n - 2$ triangles. A *minimum weight triangulation* of P, $MWT(P)$, is a triangulation with the minimum total edge length. Let $\mathbf{v}$ be some direction vector. A *plane sweep triangulation* of P is a triangulation constructed by the following plane sweep algorithm. The sweep line is oriented perpendicular to $\mathbf{v}$, and moves in the direction $\mathbf{v}$. On visiting a site, the algorithm constructs all possible nonintersecting diagonals between it and the previously encountered sites.

Theorem 4. $P_{\pi/8,1}(MWT) = true$.

Proof. The proof is very similar in structure to that of Theorem 3. Consider Figure 6 again, and let AB be an edge of some $MWT(S)$. We have to show that either CAB or DAB is empty of other sites.

Consider the left half of the figure, and let a site M be inside CAB. The same argument of Theorem 3 leads to the existence of a polygon P, which has a site V inside $OBIC$. Let the sites along the upper (lower) chain be $V = u_0, u_1, \ldots, u_m, u_{m+1} = A$ ($V = l_0, l_1, \ldots, l_p, l_{p+1} = B$). The following two claims lead us to conclude that the lower chain degenerates into a single edge, VB.

Claim 5: Either the upper or lower chain is a single edge.

Proof. Suppose this was false. Clearly $MWT(P)$ is a subset of $MWT(S)$. An example is in Figure 7. We now construct a different triangulation of P, called $T(P)$ (see Figure 8). First insert the diagonals VA and VB. Then construct the plane sweep triangulation of the polygon enclosed by the upper chain and VA (lower chain and

VB), with the sweep line moving in the direction **CG** (**CI**). We will now show that $|T(P)| < |MWT(P)|$, thus proving the claim.

Consider Figure 7. $MWT(P)$ has $m + p$ diagonals, and each connects an upper chain site with a lower chain site. Consider any site in $u_1, \ldots, u_m$ and $l_1, \ldots, l_p$. The rightmost diagonal terminating at the site is said to *belong* to the site. Thus in our example u_2l_2 belongs to u_2, while Al_3 belongs to l_3.

Consider Figure 8. $T(P)$ too has $m + p$ diagonals. We classify the sites $u_1, \ldots, u_m$ and $l_1, \ldots, l_p$ into three types, *peaks*, *troughs*, and *ordinary*. If both neighbors along the upper (lower) chain of some u_i (l_j) are encountered earlier by the sweep line, u_i (l_j) is a peak. If both neighbors are encountered later, it is a trough, otherwise it is an ordinary site. Thus in our example, u_1 is ordinary, l_3 is a peak, while l_2 is a trough. We observe that in each chain, there is one more of peaks than troughs, and if we ignore the other sites, the peaks and troughs alternate. Now consider all diagonals other than VA and VB. The polygon P is such that, during the plane sweeps, exactly 2/1/0 diagonals are added on encountering a trough/ordinary site/peak respectively. The diagonal inserted on encountering an ordinary site belongs to the site. The left of the two diagonals added on encountering a trough belongs to the trough, while the right diagonal is donated to the next peak along the chain. Thus in our example, u_1A belongs to u_1, Vl_2 belongs to l_2, while Bl_2 belongs to l_3. Note that the first peaks of each chain (in our case u_2 and l_1) do not possess any diagonals of $T(P)$.

We now consider $MWT(P)$ and $T(P)$ together and observe the following. First, $|VA| + |VB|$ is smaller than the sum of the lengths of the two diagonals of $MWT(P)$ that belong to the first peaks of each chain. In our example,

$$|VA| + |VB| < |u_2l_2| + |u_2l_1|.$$

Second, every site in $u_1, \ldots, u_m$ and $l_1, \ldots, l_p$ other than the first peaks has a diagonal of $MWT(P)$ and a diagonal of $T(P)$ of smaller length. In our example, for the site l_3, $|Bl_2| < |Al_3|$. Clearly $|T(P)| < |MWT(P)|$, and the claim is proved. ∎

Claim 6: In fact, the lower chain is a single edge, VB.

Proof. Recall that V lies inside $OBIC$ of Figure 6. Suppose the claim was false. By Claim 5, the upper chain is a single edge, VA, and $MWT(P)$ has p diagonals, connecting each of $l_1, \ldots, l_p$ with A. As in Claim 5, we construct a different triangulation, $T(P)$. First insert VB, then perform a plane sweep triangulation of the region enclosed by VB and the lower chain, with the sweep line moving in the direction **CI**. Now VB is shorter than the diagonal in $MWT(P)$ which connects the first peak with A. The proof is now similar to that in Claim 5. ∎

As in Theorem 3, two cases arise for the upper chain.

Case 1: The upper chain is a single edge, VA.

Case 2: The upper chain is not a single edge. In this case $MWT(P)$ has m diagonals, connecting each of $u_1, \ldots, u_m$ with B. We show that all of $u_1, \ldots, u_m$ lie within EAG. Suppose some were outside EAG. Then u_1B is the longest diagonal, because if u_iB was the longest diagonal for some u_i in $u_2, \ldots, u_m$, then within the quadrilateral $u_{i-1}u_iu_{i+1}B$, $|u_{i-1}u_{i+1}| < |u_iB|$, which is a contradiction. Now consider a different triangulation, $T(P)$, by first inserting VA, then triangulating the region enclosed by VA and the upper chain by a sweep line moving in the direction **CG**. Clearly u_1 is the first peak. Since $|VA| < |u_1B|$, the proof follows as in claim . We next observe that at any site u_i in $u_2, \ldots, u_m$, the angle $u_{i-1}u_iu_{i+1}$ internal to P is greater than *pi*, otherwise within the quadrilateral $u_{i-1}u_iu_{i+1}B$, $|u_{i-1}u_{i+1}| < |u_iB|$, which is a contradiction.

Now consider the right half of the figure, and assume that DAB also contains a site. The same argument of Theorem 3 leads to the existence of a site, Q, which corresponds to V of the left half. Similarly four cases for the right half arise. We are now ready to prove the theorem, by showing that any combination of the left half cases and right half cases leads to a contradiction. In this version of the paper, only the most complex combination is discussed.

Case 2 and Case 6: Consider Figure 9. $V, u_1, \ldots, u_m, A$ $(Q, l_1, \ldots, l_p, B)$ are the sites along the left (right) upper (lower) chain. Consider the polygonal region

$$R = Bu_mAQl_1 \ldots l_p.$$

We construct a different triangulation of R by inserting BQ, u_mQ, and l_iQ for each l_i in $l_2, \ldots, l_p$. Since $|BQ| < |AB|$, $|u_mQ| < |Al_1|$, and for each l_i in $l_2, \ldots, l_p$, $|l_iQ| < |l_iA|$, we have a contradiction.

A full analysis of all the combinations proves the theorem. ∎

4. A GENERAL PLANAR GRAPH ALGORITHM

Part of the motivation for designing Euclidean planar graphs is because their total edge length is small. Consider the set of all Euclidean planar graphs over some S. Although triangulations are planar, they are usually the longest graphs in this set. Thus it is motivating to look for extremely short (possibly nontriangular) planar approximations of the complete graph. However, a limiting factor is that the length of a *minimum spanning tree* ($MST(S)$) is a lower bound on the length of such graphs, for at the least these graphs should be connected.

We now introduce the following heuristic for constructing such graphs. We start with a triangulation of our choice, for example $GT(S)$. We then select some constant d which specifies the good polygon property of Theorem 1. Now the algorithm removes edges from the triangulation, while retaining both properties as required by Theorem 1. At any iteration, the triangulation has been reduced to some planar graph. All faces of the graph satisfy the good polygon property, and trivially all edges satisfy the diamond property. An edge may be removed if the new face formed by merging its two adjacent faces satisfies the good polygon property. The algorithm terminates if no such edge can be found.

Unfortunately, though this may perform well in practice, in theory we do not know exactly how short the graph is bound to become. To facilitate a more quantitative analysis, we have modified the heuristic into a more complicated algorithm, called A, which is as follows. The first step is to construct $MST(S)$, as well as the *convex hull* of S (denoted as $CH(S)$). This partitions the plane into empty polygons, and the sum of all edges is at most $3|MST(S)|$. The next step is to perform a greedy triangulation of each polygonal face. After that, a constant d is selected (whose value will be specified later) for the good polygon property. Finally, only the greedy triangulation edges are examined for possible removal, in *descending order* of edge length.

The following theorems and lemmas prove that $A(S)$ approximates the complete graph, and also lead to a quantitative upper bound on its length. Specifically, we prove that $|A(S)| = O(|MST(S)|)$.

Theorem 5. $P_{\pi/8,d}(A) = true$.

Proof. During edge removal, this property is maintained by the algorithm. Thus we only have to ensure that the edges of $MST(S)$, $CH(S)$, and the greedy triangulations of the polygonal faces satisfy the diamond property. It is known that if e is an edge of $MST(S)$, then the circle with e as diameter is empty of other sites. Thus all such edges satisfy the diamond property. If e is an edge of $CH(S)$, the exterior triangular region is empty, thus satisfying the diamond property.

Let $e = AB$ be a greedy triangulation edge. As in Theorem 3, consider the left half of Figure 6, and let a site M be inside CAB. This leads to the existence of a polygon P, which has a site V inside $OBIC$. It is easy to see that none of the diagonals of P belong to $CH(S)$. Similarly, none of them belong to $MST(S)$ either, otherwise the circle with the diagonal as diameter will contain M, which is a contradiction. Thus the diagonals are greedy triangulation edges, and so the proof of the theorem is identical to Theorem 3. ∎

The above theorem ensures that the output of the algorithm indeed approximates the complete graph. We now turn our attention to estimating the length of $A(S)$. Clearly the $MST(S)$ and $CH(S)$ edges add up to $O(|MST(S)|)$. We have to show that that the greedy triangulation edges remaining in $A(S)$ also add up to $O(|MST(S)|)$. Recall that the former edges are first added, which results in a planar graph with polygonal faces, hereafter referred to as *regions*. The greedy triangulation edges that are later added can be considered as *diagonals*, which split regions into *subregions*. Note that subregions are essentially faces of $A(S)$. We will consider each region in isolation, and it will be sufficient to prove that the length of all the diagonals is proportional to the length of its boundary. The following lemmas will be useful for the purpose. They describe local properties of $A(S)$, in particular how an edge relates to its neighboring subregions.

Lemma 2. *Let e be a diagonal of some region R. Then there exists a pair of sites A and B along the boundaries of its two adjacent subregions respectively, such that if e is removed, (i) AB will be a chord of the merged subregion, (ii) the shortest distance between A and B along the merged subregion boundary will be greater than $d|AB|$, and (iii) $|AB| \geq |e|$.*

Proof. The merged subregion clearly violates the good polygon property. Consider all pairs of sites where the violation occurs. Let A and B be the pair with the largest separation, $|AB|$ (Figure 10). To contradict the lemma, let us assume that $|AB| < |e|$. Clearly AB intersects e. Thus AB does not belong to the original greedy triangulation. But by Lemma 1, in the original greedy triangulation, there existed an edge f intersecting AB such that $|f| \leq |AB|$. Thus $|f| < |e|$. But since the candidates for removal were in decreasing length, while e was being examined, f was present in the graph. However, at that point f itself separated A and B into different subregions, so they could not have been a violating pair during e's examination. This contradiction proves the lemma. ∎

The next lemma describes how an edge relates to its next-to-adjacent subregions. Let $e = AB$ be a diagonal of some region R. Consider any side of e, and let the polygon P be the union of its adjacent subregion and next-to-adjacent subregions on that side (Figure 11). Let P's boundary without the edge AB be denoted as $P(A,B)$. It is composed of diagonals, as well as boundary edges of R. Let the set of diagonals along $P(A,B)$ be referred to as $diag(A,B)$.

Lemma 3. *Suppose algorithm A is designed with a (suitably large) constant $d > 1$ specifying the good polygon property. Then there exist constants $b > 1$ and $c > 1$ such that for any S, and any diagonal $e = AB$,*

$\frac{|P(A,B)|}{|AB|} \leq c$ implies $\frac{|AB|}{|diag(A,B)|} \geq b$.

Proof (sketch). The lemma in words is, if $P(A,B)$ is very close in length to AB, then the contribution to its length from diagonals is small. We shall give an intuitive sketch of the proof. Details are in [DJ].

Assume the contrary, that is even though $P(A,B)$ is short, most of it is formed by diagonals. Let $Q(A,B)$ refer to the boundary of the subregion adjacent to AB, apart from AB itself. $Q(A,B)$ is shorter than $P(A,B)$, and even more of its length is due to diagonals. Let the length of all its diagonals be l. By Lemma 2, every diagonal e on $Q(A,B)$ is associated with a portion of $P(A,B)$ of length at least $d|e|$. If each such portion is disjoint, we can conclude that $P(A,B)$ is at least as long as dl. By selecting a large enough d, we can arrive at a contradiction, because $P(A,B)$ is initially assumed to be small.

The complication is due to the fact that the above portions are *not* disjoint. However, we show in [DJ] that the overlap is not too much. ∎

We assume henceforth that the algorithm A is designed with the constant d as in Lemma 3. We are now ready for the final theorem.

Theorem 6. *Let R be a region. Then the length of all diagonals is of the order of the length of R's boundary.*

Proof (sketch). The details are in [DJ]. Figure 12 shows a region with all its subregions. Consider a subregion which has only one diagonal, PQ, along its boundary. We remove this subregion from R. Construct the *accounting tree* of edges as follows. Each node contains an edge, and PQ is at the root. The boundary edges are at the leaves, and the diagonals are at the internal nodes. The tree may be built conceptually in a top down manner by "exploring" subregions (Figure 13). First, the subregion adjacent to PQ is explored, and all its edges (other than PQ itself) are made children of PQ. Nodes containing diagonals are expanded in a natural manner, by exploring their adjacent subregions.

Our objective is to show that the total length of all edges is of the order of the total length of leaf edges. The internal nodes may be classified into two types, (a) and (b). Consider an internal node $e = AB$. If we start from e and explore two levels below into the tree, we shall effectively come across its adjacent and next-to-adjacent subregions. The leaf nodes at the first level along with all nodes at the second level compose $P(A, B)$, as in Lemma 3. If the condition before the implication in Lemma 3 is false, then e is a type (a) node, otherwise e is a type (b) node.

We omit further details and claim that such a condition throughout the tree is sufficient to prove the theorem. As an intuition, the theorem is easily seen to be true if the tree had only one of either type of nodes. ∎

5. CONCLUSIONS

We have identified a sufficiency condition such that, graph algorithms satisfying it produce good approximations of the complete graph. We have shown that this condition is quite general because it is satisfied by Delaunay triangulations, greedy triangulations, and minimum weight triangulations. We also use it to design an algorithm for constructing sparse nontriangular graphs. In network design this has some significance because these algorithms can sometimes be better alternatives to the Delaunay triangulation. Furthermore, we have increased our knowledge of minimum weight triangulations, of which little is currently known.

Since the upper bounds in our general results are not tight, our future research is aimed at improving them, possibly by discovering additional properties of the specific algorithms considered. An intriguing open problem is to design an algorithm which produces graphs with the *least measure*. It may be that Delaunay triangulations is the answer here. Finally, other sparse Euclidean graphs need to be investigated, including graphs that are not even planar.

6. ACKNOWLEDGEMENTS

We thank Meera Sitharam and Jorg Peters for several useful discussions.

7. BIBLIOGRAPHY

[C] Chew: There is a Planar Graph Almost as Good as the Complete Graph: ACM Symposium on Computational Geometry, 1986, 169-177.

[DJ] Das, Joseph: Planar Euclidean Graphs and their uses in Network Design: Technical Report, UW-Madison, in preparation.

[DFS] Dobkin, Friedman, Supowit: Delaunay Graphs are Almost as Good as Complete Graphs: IEEE Symposium on Foundations of Computer Science, 1987, 20-26.

[K] Klincsek: Minimal Triangulations of Polygonal Domains: Annals of Discrete Mathematics 9, 1980, 121-123.

[L] Lingas: On Approximation Behavior and Implementation of the Greedy Triangulation for Convex Planar Point Sets: ACM Symposium on Computational Geometry, 1986, 72-79.

[MZ] Manacher, Zobrist: Neither the Greedy nor the Delaunay Triangulation of a Planar Point Set Approximates the Optimal Triangulation: Information Processing Letters, July 1979, 31-34.

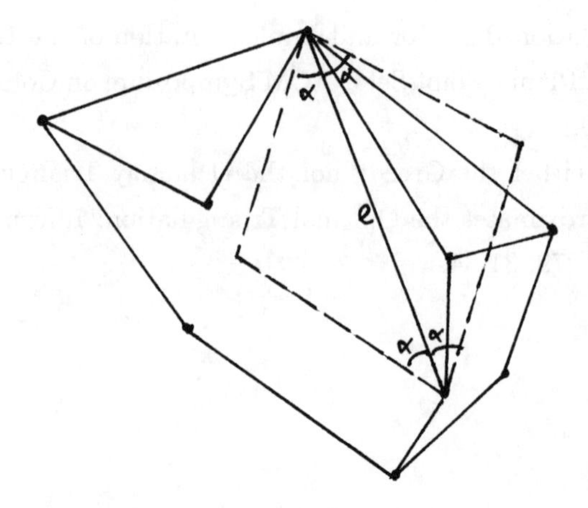

Figure 1

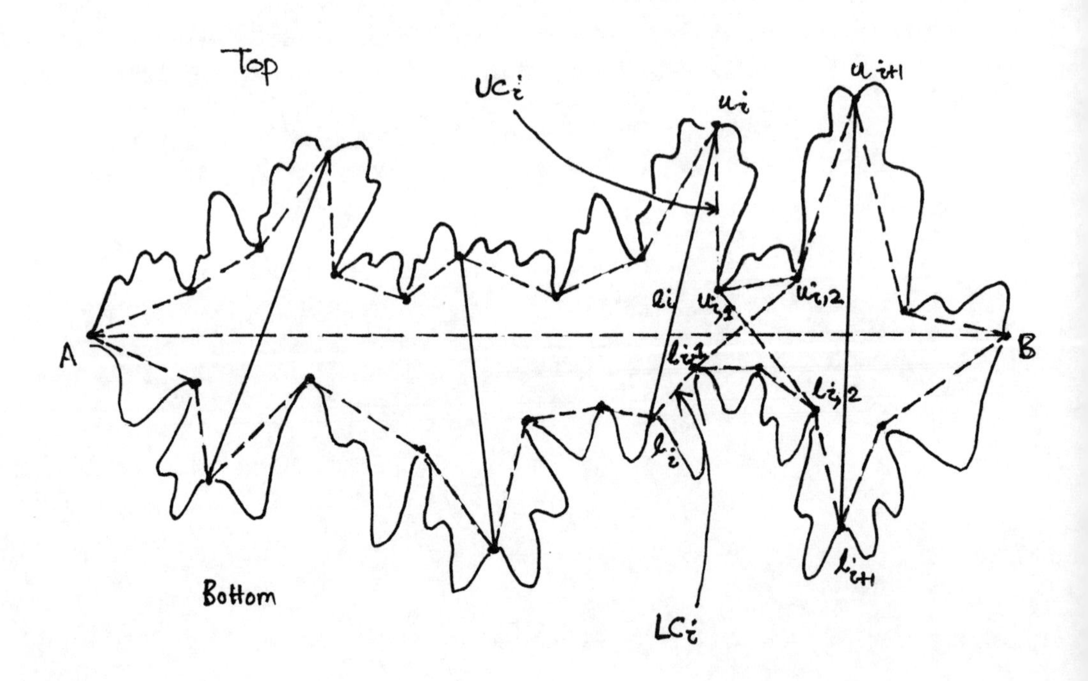

Figure 2

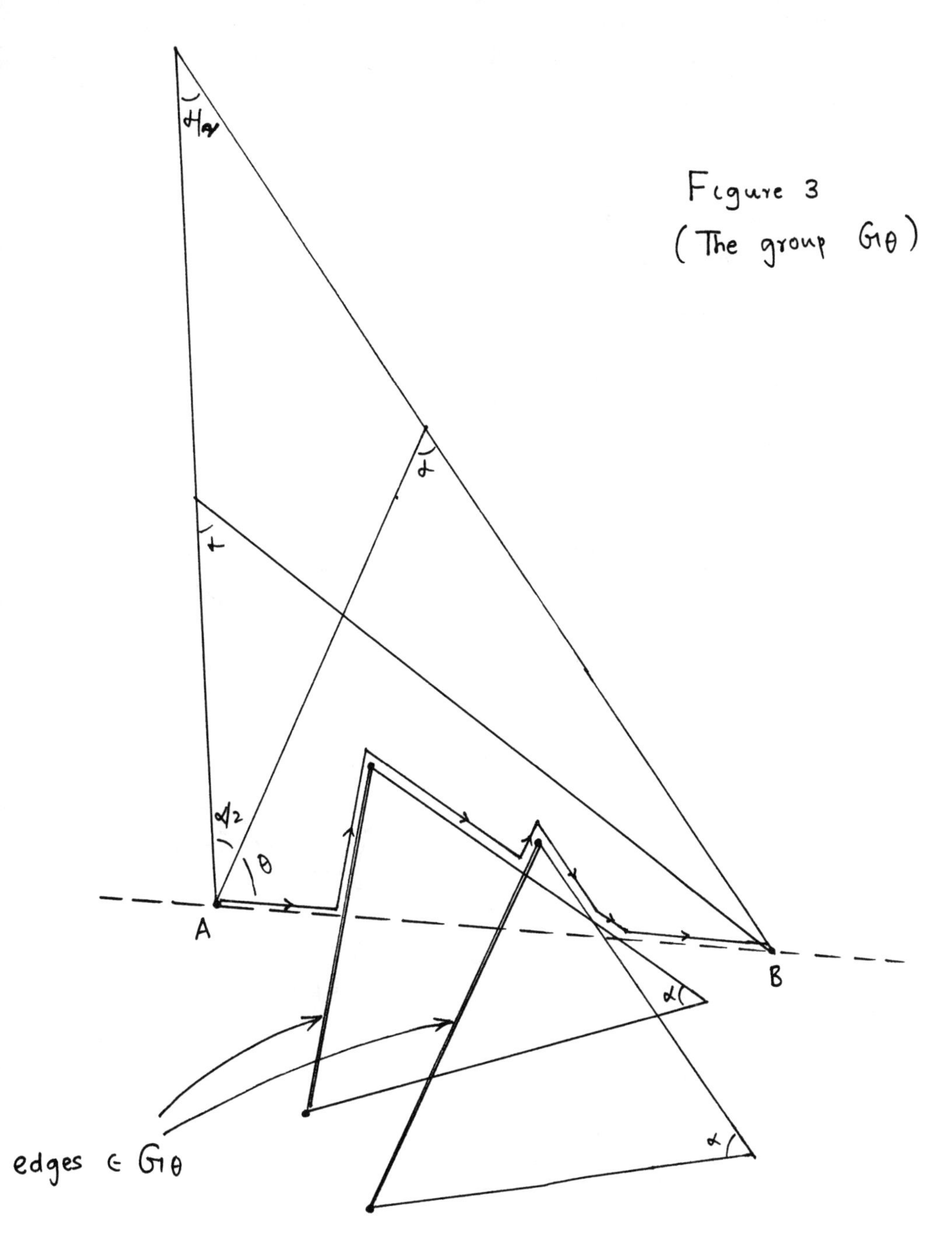
Figure 3
(The group G_θ)
A
B
edges $\in G_\theta$

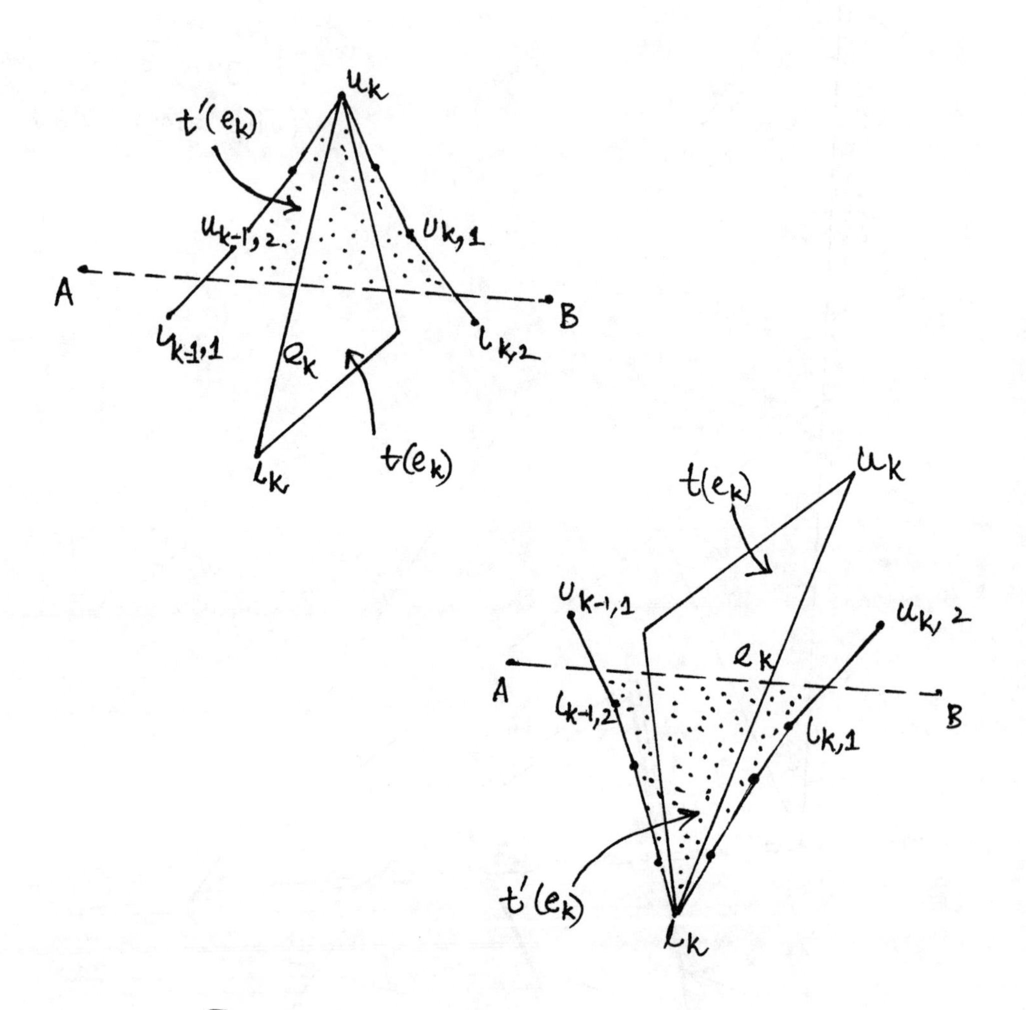

u_k
$t'(e_k)$
$u_{k-1,2}$
$u_{k,1}$
A
B
$l_{k-1,1}$
e_k
$l_{k,2}$
$t(e_k)$
l_k
u_k
$t(e_k)$
$u_{k-1,1}$
$u_{k,2}$
e_k
A
B
$l_{k-1,2}$
$l_{k,1}$
$t'(e_k)$
l_k

Figure 4

Figure 5

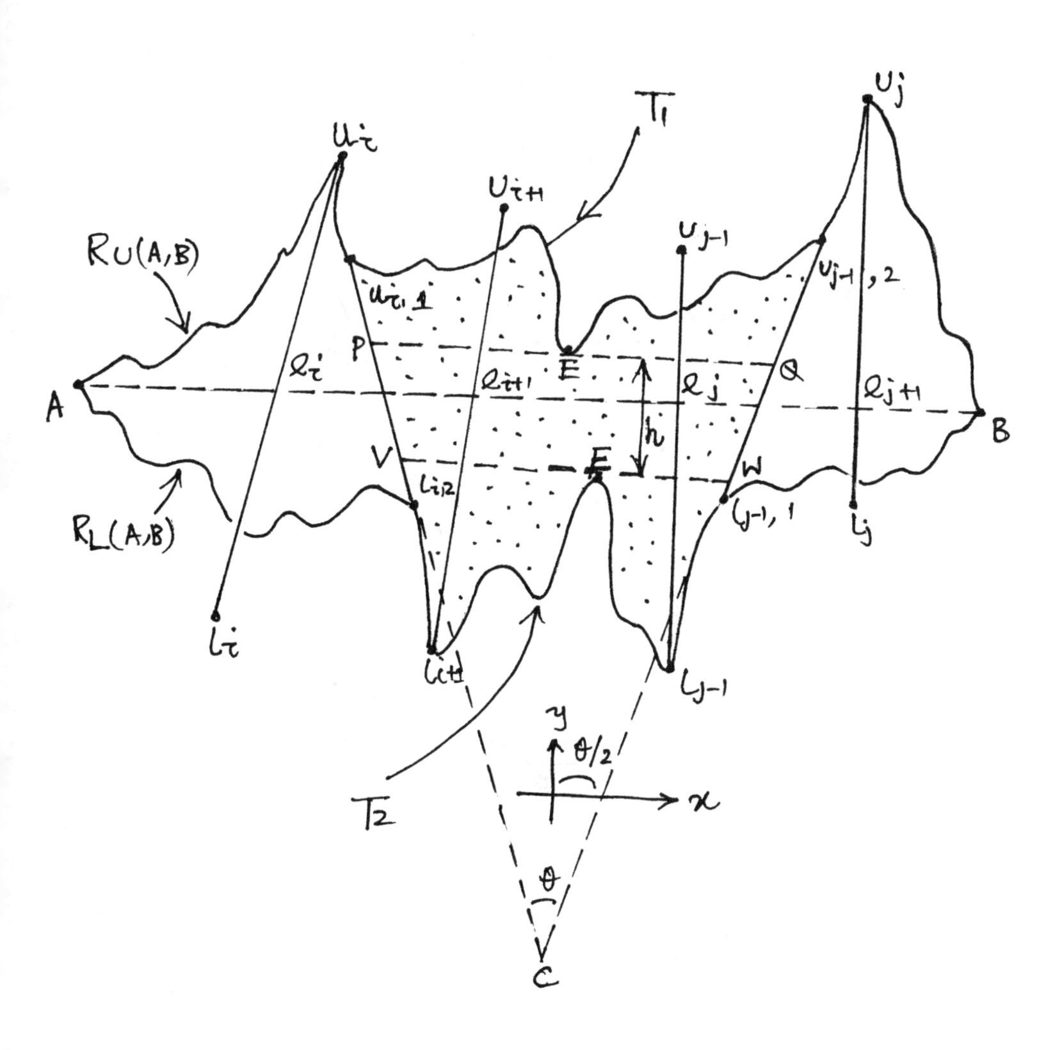

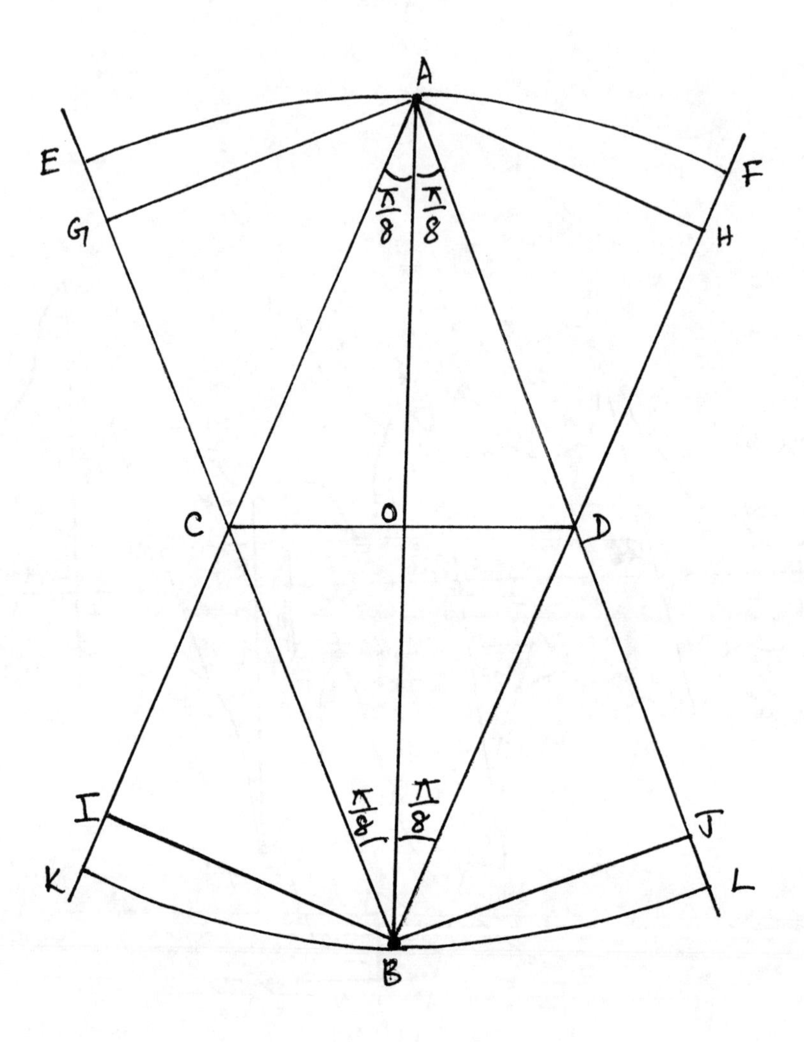

Figure 6

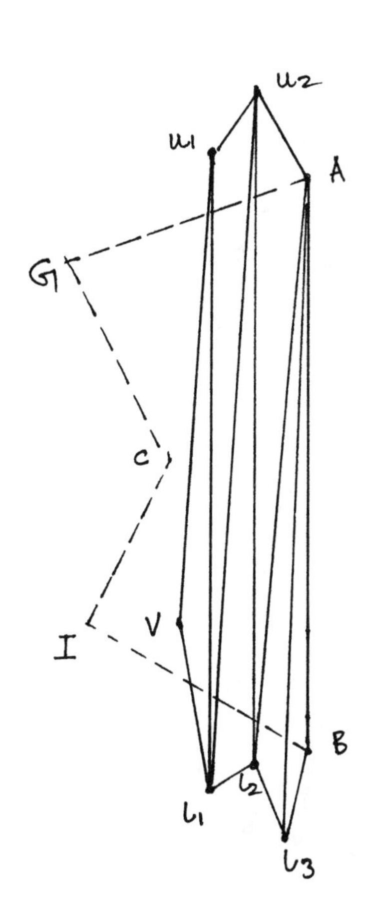

Figure 7

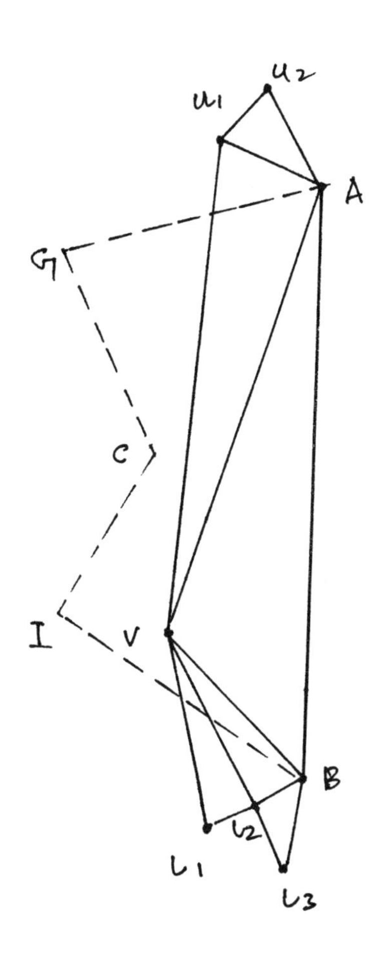

Figure 8

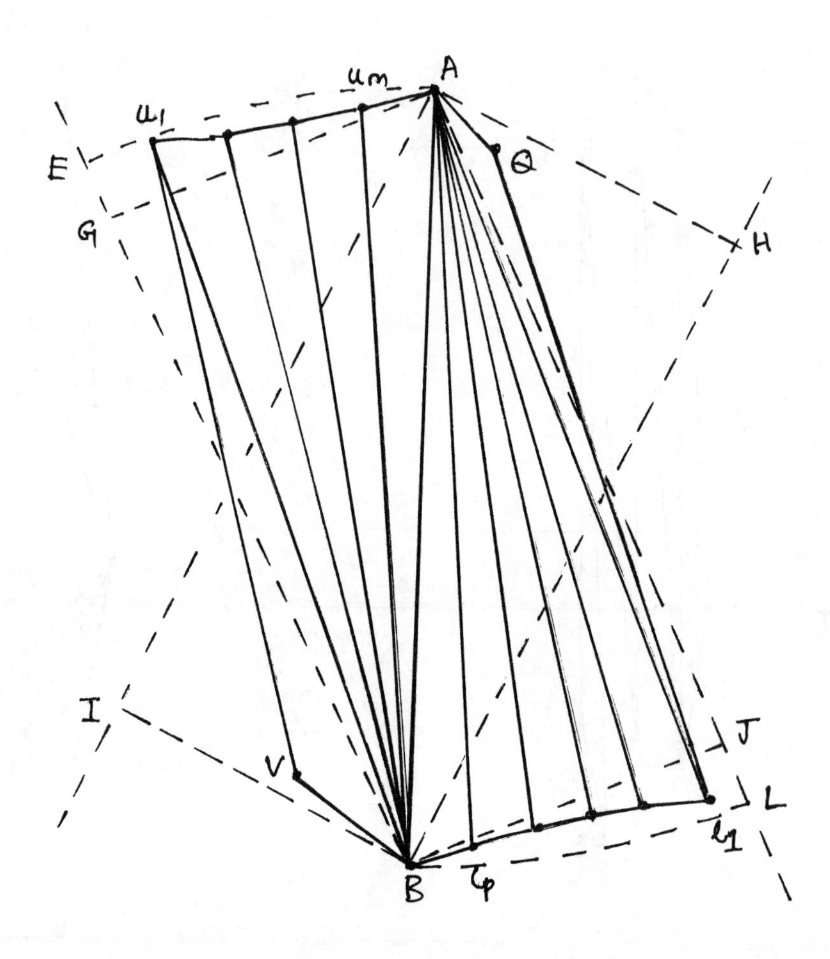

Figure 9

Figure 10

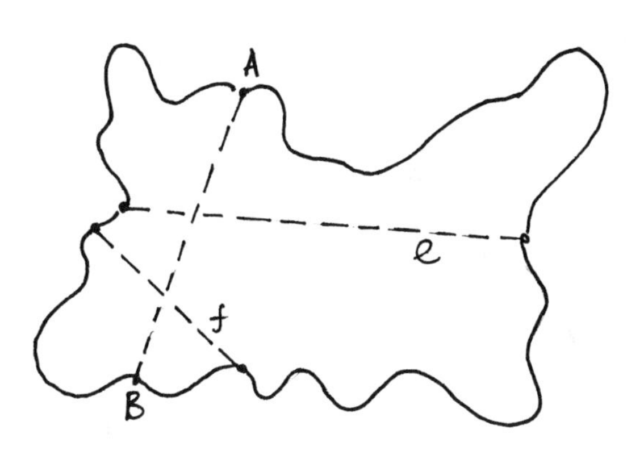

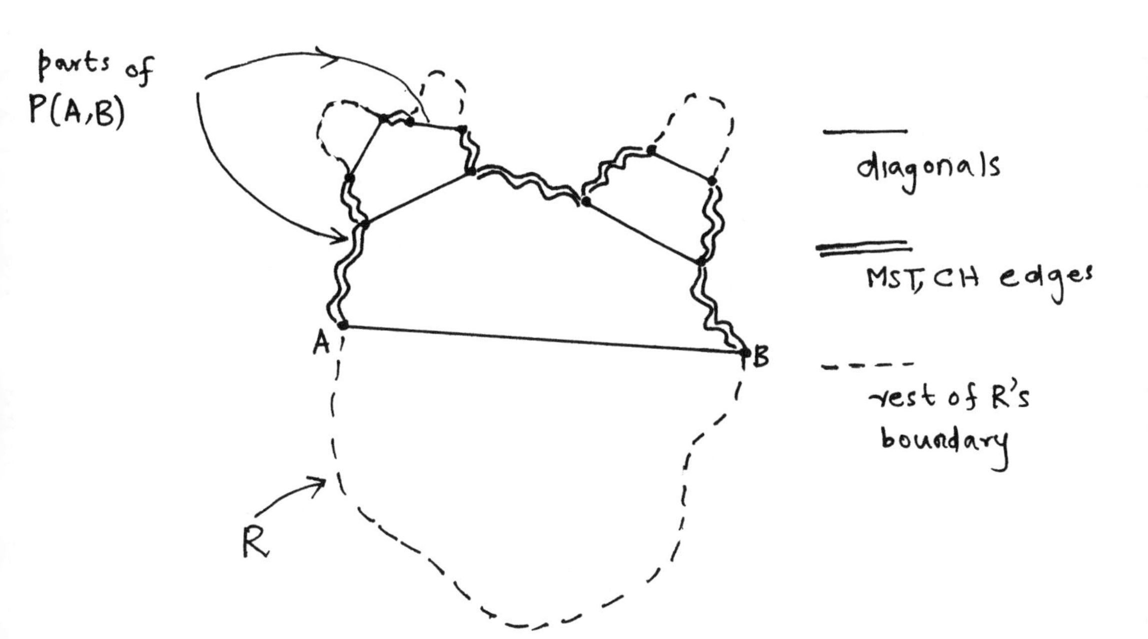

Figure 11

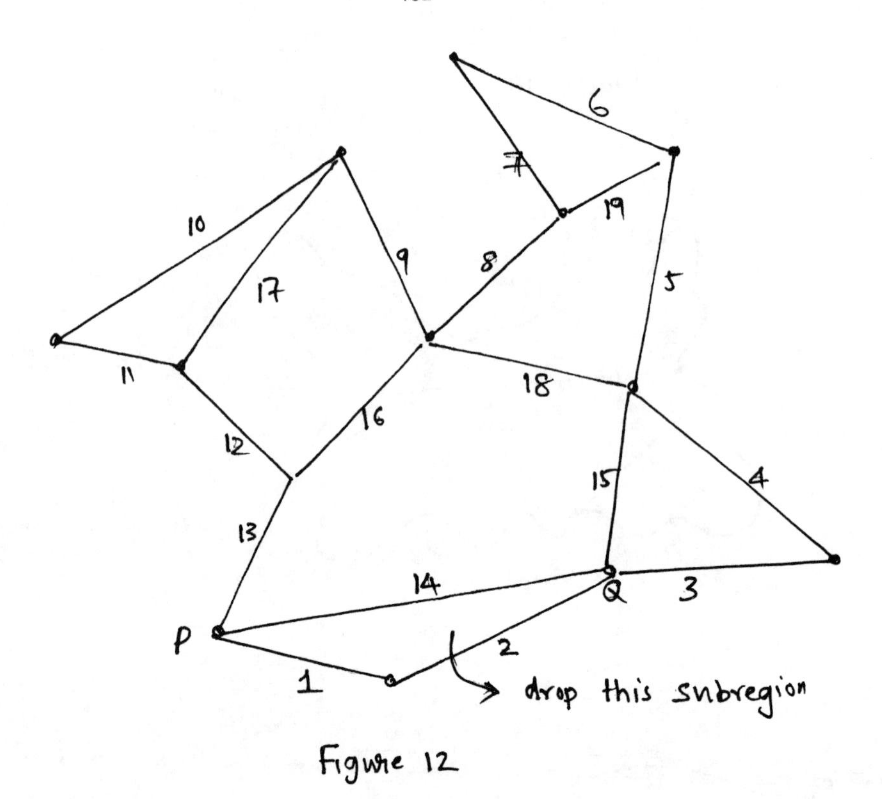

Figure 12

Accounting Tree

Figure 13

The approximability of problems complete for P

by

Maria Serna[+] and Paul Spirakis[*]

+ Polytechnic University of Catalonia, Barcelona, Spain

* Computer Technology Institute, Patras University, Greece

Abstract

We examine here the existence of approximations in NC of P-complete problems. We show that many P-complete problems (such as UNIT, PATH, circuit value etc.) cannot have an approximating solution in NC for any value of the absolute performance ratio R of the approximation, unless P=NC. On the other hand, we exhibit of a purely combinatorial problem (the High Connectivity subgraph problem) whose behaviour with respect to fast parallel approximations is of a threshold type. This dichotomy in the behaviour of approximations of P-complete problems is for the first time revealed and we show how the tools of log-space reductions can be used to make inferences about the best possible performance of approximations of problems that are hard to parallelize.

1. Introduction

Up to now, the best way to argue convincingly that a problem in P is not in the class NC (of the problems that admit fast parallel algorithms) is to prove that it is complete for P under log space or NC reductions, see e.g. [Cook, 85]. As [Anderson, Mayr, 86] suggest, for those problems we can lower our sights and attempt to find fast parallel algorithms that give approximate solutions. It seems that such approximations of P-complete problems do not fall in NC for all values of the approximation absolute performance ratio R (see [Garey, Johnson, 79]. In fact, we know of at least two P-complete problems, whose approximations exhibit a very different kind of behaviour. These are (1) the High Degree Subgraph problem, (HDS), (see [Anderson, Mayr, 86]) which cannot be approximated in NC by a performance ratio R<2 unless P=NC, but it can be approximated to within any R for R>2 by an algorithm in NC, and (2) the problem of computing the maximum depth at which the ones arrive at a monotone boolean circuit the "Circuit Depth

This research was done during the visit of the first author to Patras University, and is supported partially by a Spanish Research Scholarship and by the Ministry of Industry, Energy and Technology of Greece.

of Ones" CDOP, (see [Kirousis, Spirakis, 88] which cannot have an approximating solution in NC <u>for any R</u> in [1,+∞), unless P=NC. Let us note that another problem (one-dimensional bin packing) is more of the flavour of the High Degree Subgraph problem, since it is known that it is P-complete to produce the same packing as the First Fit (FFD) and Best Fit Decreasing (BFD) but there are algorithms in NC which find a packing asymptotically as good as the FFD and BFD packing (see [Warmuth, 85], [Anderson, Mayr, 86]).

In this paper we provide evidence that the dichotomy in the behaviour of the approximations of HDS and CDOP is followed by many other P-complete problems. Indeed, approximations to problems such as UNIT, PATH etc. (see e.g. [Jones, Laaser, 77] for the definitions of these problems) have a behaviour similar to CDOP: We show here that such problems cannot have an approximating solution in NC for any value of R in [1,∞], unless P=NC. On the other hand, we exhibit of a <u>purely combinatorial</u> problem (that of High Connectivity Subgraph, HCS) whose behaviour is similar to that of HDS. The High Connectivity Subgraph Problem is: Given a graph G=(V,E) and an integer k, does G contain an induced subgraph with vertex connectivity at least k? We first show that this is a P-complete problem for constant k and then examine its approximability. Such problems like the HCS or HDS are interesting because they not rely on some special mechanism such as an ordering or large integer weights to make them difficult.

We then proceed to define a special kind of log-space reductions that preserve approximability (in the sense of [Papadimitriou, Yannakakis, 88]. Since negative approximability results are usually based on the technique of the creation of a <u>gap</u> in the cost function of the optimization problem under study these reductions should preserve such gaps. We suspect that a new complexity class may be manifesting itself, within P. The characterization of this class through a syntactic definition is now under study. The very important problem of treating approximability in a unified way has been the object of intense research effort in the past (see e.g. [Ausiello, D'Atri, Protasi, 77], [Ausiello, D'Atri, Protasi, 80], [Ausiello, A. Marchetti-Spaccamela, Protasi, 80], [Paz, Moran, 81]), mainly for the class NP. Since the interest in parallel architectures and parallel algorithms has grown dramatically over the past few years, it seems to us that this problem will be revived in the context of P versus NC.

2. P-Complete problems that cannot be approximated in NC for any value of the performance ratio

In the sequel, $R=1/\varepsilon$.

2.1 Approximations to the PATH problem [Cook, 70]

The PATH problems is:

Given: a path system $P=(X,R,S,T)$ where X is a finite set, $S\subseteq X$, $T\subseteq X$, and $R\subseteq X\times X\times X$. An element x is called admissible iff $x\in T$ or y, $z\in X$ such that $(x,y,z)\in R$ and both y and z are admissible.

Question: Is there an admissible node in S?
The PATH has been shown P-complete by [Jones, Laaser, 77]. With T as above we define:

$$E(T) = \{x : (x,y,z)\in R \text{ and } y,z\in T\} \cup T$$

If we consider the subset family defined as

$$E^0(T) = \emptyset, \quad E^1(T) = T, \quad E^k(T) = E(E^{k-1}(T)), \quad k>1$$

then we clearly have

$$E^0(T) \subseteq E^1(T) \subseteq \ldots \subseteq E^k(T) \subseteq \ldots \subseteq X.$$

Since X is a finite set we can define

$$e(T) = \min \{k : E^k(T) = E^{k+1}(T)\}$$

Given the S, $T\subseteq X$ we define the distance between S and T as

$$\mathrm{dist}(S,T) = \max \{k : S\cap E^k(T) = \emptyset, \quad 0\leq k\leq e(T)\}.$$

The distance path problem (DPATH) is as follows:

Problem DPATH
Given a path system (X,R,S,T), compute dist (S,T).

For any $\varepsilon \in (0,1)$, the ε-approximation to DPATH, ε-DPATH is then the following problem:

Problem ε-DPATH
Given a path system (X,R,S,T) find an integer d such that

$$\mathrm{dist}(S,T) \geq d \geq \varepsilon\cdot\mathrm{dist}(S,T) .$$

The ε-DPATH problem is the approximation version of the optimization version of PATH.

Theorem 1 The ε-DPATH problem is P-complete under log-space reductions, for any $\varepsilon \in (0,1]$.

Proof To see that the problem is in P it suffices to consider that the operation E(T) and the test $S\cap E^k(T)=\emptyset$ can be done in polynomial time, and that $e(T)\leq|X|$.

We now consider the following log-space reduction from PATH:

Given a path system (X,R,S,T), let $L=\frac{|M|}{\varepsilon}$ where $M=|X|$ and construct the following path system (X',R',S',T')

$$X' = X \cup \{x_{-1}, x_0, x_1, \ldots, x_L\} \quad \text{(L+2 new symbols)}$$
$$R' = R \cup \{x_{k+1}, x_k, x_{k-1}), \quad 0 \leq k \leq L\}$$
$$S' = S$$
$$T' = T \cup \{x_{-1}, x_0\}$$

Then we have

(a) If there is an admissible node in S then

$$\text{dist }(S',T') = \text{dist }(S,T) < M$$

(b) If there is not an admissible node in S then

$$\text{dist }(S',T') = e(T') = L = \frac{M}{\varepsilon}$$

□

2.2 Approximations to the UNIT problem

The UNIT problem is:

Given: A propositional formula F in conjuctive normal form.

Question: Can the empty clause, , (indicating a contradiction) be deduced from F by unit resolution?

Unit was shown to be P-complete in [Jones, Laaser, 77].

Given F, let C = {w : w is a clause in F}.

We define

$$R(C) = \{u : u \text{ is the unit resolvent of two clauses in } C\} \cup C$$
$$R^k(C) = R(R^{k-1}(C)), \quad k>1$$
$$e(F) = \min \{k : R^k(C) = R^{k+1}(C)\}$$
$$D(F) = \max \{k : \square \notin R^k(C)\}$$

We consider the following problems:

DUNIT (Optimization version of UNIT)

Given F compute D(F)

ε-DUNIT (Approximation of UNIT)

Given F, $\varepsilon \in (0,1]$, find a d such that

$$D(F) \geq d \geq \varepsilon \cdot D(F).$$

Theorem 2 ε-DUNIT is P-complete under log-space reductions for any $\varepsilon \in (0,1]$

Proof To show ε-DUNIT in P, just note that R(C) can be computed in polynomial time and $e(F) \leq s(F) \leq M^c$ for some c (where $M=|F|$ and s(F) is the number of subclauses in F).

We now reduce UNIT to ε-DUNIT:

Let $L = \frac{s(F)}{\varepsilon}$

Let $F' = F \wedge (\bigvee_{i=1}^{L+1} x_i) \quad (\bigwedge_{i=1}^{L} \overline{x_i})$

where the x_i are new symbols and $\overline{x_i}$ is the complement of x_i.

Then the following holds:

$$F \in UNIT \Rightarrow D(F) = D(F') \ (\leq s(F))$$

$$F \notin UNIT \Rightarrow D(F') = L \ \left(= \left\lceil \frac{s(F)}{\varepsilon} \right\rceil\right) .$$

□

2.3 Approximations to the Circuit Value Problem

Consider (for simplicity) an alternating monotone circuit A together with its input value assignments, e. The depth d(g) of a gate g of A is the length of the longest path from some input to the output of the gate (gates are considered to be "nodes" in the directed acyclic graph representing the circuit). Let g_0 be the result gate of A and let $d(A)=d(g_0)$.

The depth of 1's, $d_1(A)$, is the maximum of the d(g) for all g such as their output value is 1 when the input values are e. Intuitively, $d_1(A)$ is the furthest point from the inputs that the ones can propagate to. For any particular $\varepsilon \in (0,1]$, the ε-Circuit Depth of Ones problem, ε-CDOP is

ε-CDOP: Given the alternating monotone circuit A find a d such that $d_1(A) \geq d \geq \varepsilon \cdot d_1(A)$

Theorem 3 [Kirousis, Spirakis, 88] ε-CDOP is P-complete for any $\varepsilon \in (0,1]$. To prove theorem 3, let $I = \left\lceil \frac{d(A)+1}{\varepsilon} \right\rceil$ and, given A, put it together with an alternating monotone circuit β with I levels and two gates per level. Fix the four inputs of β to 1 and connect the output y of A to the AND gates at level d(A). Let us call B the new circuit. Then

(a) if the result of A is 1 then $d_1(B) = \left\lceil \frac{d(A)+1}{\varepsilon} \right\rceil$

(b) if the result of A is zero then $d_1(B)=d(A)$.

2.4 Approximations to other P-complete problems, of the same flavour as ε-DPATH

The problems GEN and CFMEMBER have the same behaviour wrt approximations, as PATH, UNIT and CVP. One has to work in a similar way to construct optimization versions of these problems. E.g. for GEN.

Given: A set X, a binary operation * on X, a subset $S \subseteq X$ and an element $x \in X$.

Question: Is x contained in the smallest subset of X which contains S and is closed under * ?

Define

$$E(S) = \{x : x = y*z,\ y,\ z \in S\} \text{ union } S$$

and

$$E^k(S) = E(E^{k-1}(S)) \qquad k>1$$

Also

$$e(S) = \min \{k : E^k(S) = E^{k+1}(S)\}$$

Again

$$\text{dist } (S,x) = \max \{k : x \notin E^k(S),\ \ 0 \le k \le e(S)\}$$

The problem ε-DGEN is then:

ε-DGEN (Approx. to GEN)

Given X, *, $S \subseteq X$ and $x \in X$, find a d such that

$$\text{dist } (S,x) \ge d \ge \varepsilon \cdot \text{dist } (S,x)$$

Theorem 4: ε-DGEN is P-complete for any ε ε(0,1]. (Proof omitted).

3. P-complete problems that exhibit a threshold behaviour

3.1 The High Connectivity Subgraph problem

The High Degree Subgraph problem (HDS) (introduced in [Anderson, Mayr, 86] is: Given a graph G=(V,F) and an integer k, does G contain an induced subgraph with minimum degree at least k? [Anderson, Mayr, 86] show that HDS is P-complete and also that it can be approximated to within a factor of ε for any ε<1/2 in NC, but cannot be approximated by a factor ε for ε>1/2 in NC unless NC=P.

We consider here the High Connectivity Subgraph problem (HCS): Given a graph G=(V,E) and an integer k, does G contain an induced subgraph with vertex connectivity at least k?

Lemma 3.1 HCS is in P, for any G and any constant k

Proof (sketch) We need some definitions and results from graph theory

Definition 1 The vertex connectivity k(G) of an undirected graph G is the minimum number of vertices whose removal results in a disconnected graph or a trivial graph (of just one vertex)

Definition 2 A k-block of a graph G is a maximal k-connected subgraph.

Definition 3 [Matula, 78] A separation set S of G is a vertex set $S \subseteq V(G)$ such that G-S is disconnected. A minimum separating set $S \subset V(G)$ has $|S|=k(G)$.

Claim 3.1 [Matula, 78] Let $S \subseteq V$ be a minimum separating set of the noncomplete graph G with $A_1, A_2, \ldots, A_m$ ($m \geq 2$) the vertex sets of the components of G-S. Let $k \geq k(G)+1$. Then each k-block of G is a k-block of the subgraph induced by $A_i \cup S$ for precisely one value of i, and each k-block of $A_i \cup S$ for every i is a k-block of G.

The following procedure decides the HCS problem:

Test (G,k)

1. Test whether G is k-connected. If yes, return "true"
2. If G has less than k vertices return "false"
3. Find a minimum separating set S of G
4. Let $A_1, \ldots, A_m$ the vertex sets of the subgraphs induced by $A_i \cup S$ on G. Return $\bigvee_{i=1}^{m}$ Test ($\langle A_i \cup S \rangle$, k)

The above Test has a time complexity of $O(n^{k+1})$. □

Theorem 5 HCS is complete for P under log-space reductions, for any $k \geq 4$.

Proof (sketch) We reduce the monotone Circuit Value problem to HCS. The proof will be done for k=4. In the construction an input of CVP will correspond to a pair of nodes. Each gate will be performed by a subgraph in wich gate inputs and outputs correspond to node pairs. An OR-gate is the graph of figure 1.

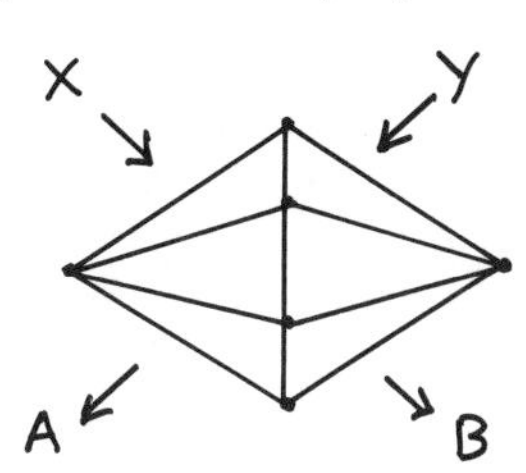

Fig. 1: OR-GATE

An and gate is the graph of Figure 2.

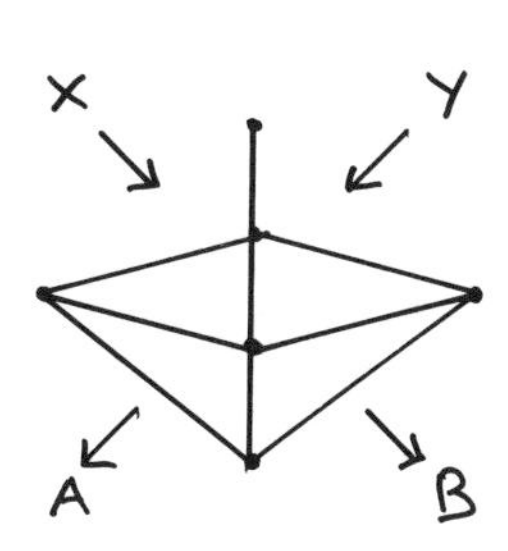

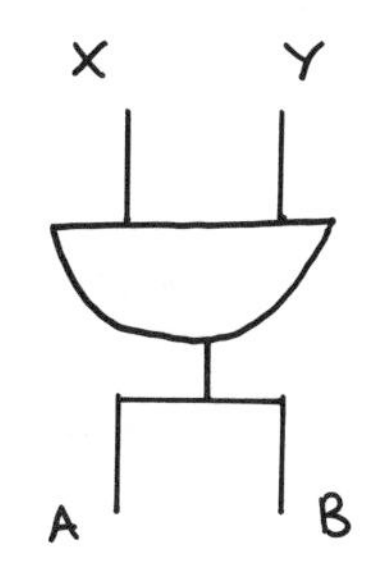

Fig. 2: AND-GATE

Each wire will be simulated by two nodes, these nodes will be connected to the corresponding input/output node pairs by all possible edges. Finally we connect all nodes corresponding to 1's to the node pair associated with the circuit output. For example, the circuit of figure 3.

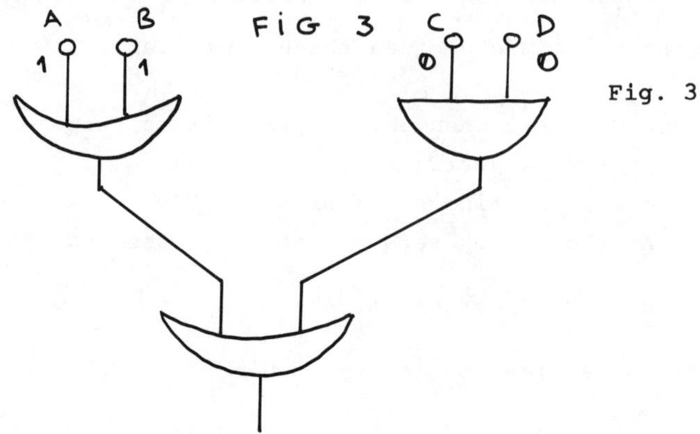

Fig. 3

Corresponds to the graph of figure 4.

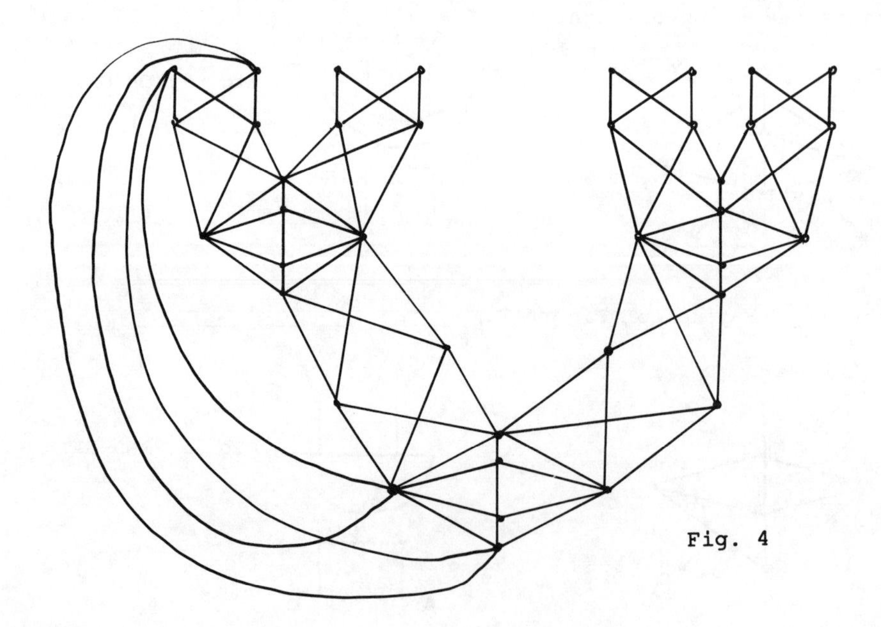

Fig. 4

Only the connections to input A from the output are shown. (Similar to B, C must be drawn).

Notice that the resulting graph contains at least a 3-connected component (each graph associated with a gate contains a 3-connected component). However, only if the result of the final gate is 1 (i.e. if there is a sequence of "conducting" gates up to the final gate) then the subgraph defined by the subset of the 1-inputs and the gates with output 1 which lead to the final gate, is 4-connected.

For k>4, the same basic construction will be used; from the graph G constructed as outlined above, we obtain a new graph G' by adding k-4 nodes that will be connected to all other nodes in G'. Then G' has a k-connected subgraph iff G has a 4-connected subgraph. □

Of course, it is straightforward that:

Lemma 3.2 HCS for k=2 is in NC

Proof sketch HCS for k=2 is equivalent to the question of whether G has a cycle. This can be tested in NC by computing the transitive closure of the adjacency matrix of G.

3.2 Approximations to HCS

An ε-approximation to HCS of G, k (where k is fixed) would be an algorithm that would return an answer which is either "G has no subgraph of vertex connectivity at least k" or "G has a subgraph of vertex connctivity at least $\varepsilon \cdot k$" (for some $\varepsilon \in (0,1]$).

Theorem 6 If P≠NC then there is no ε-approximation to HCS of G, k (k fixed) with $\varepsilon > 1/2$ in NC.

Proof sketch

We will provide a log-space reduction from a monotone circuit to a graph G which has a subgraph of vertex connectivity at least 2k if the output of the circuit is 1 and else no subgraph of G will be of vertex connectivity greather than k+1. The reduction is very similar to that of Theorem 5.

The 0's and 1's (inputs) correspond to k nodes. The OR gate is the gadget of Figure 5, where the boxes represent cliques of number of nodes equal to the number in the box, and circles represent as many isolated nodes as the number in the circle. In the figure an edge with a label on it actually means as many edges as the label indicated. Boxes are connected to nodes in the central circle so that all nodes in a box finally have degree 2k and all nodes in the circle have degree at most 4k-2.

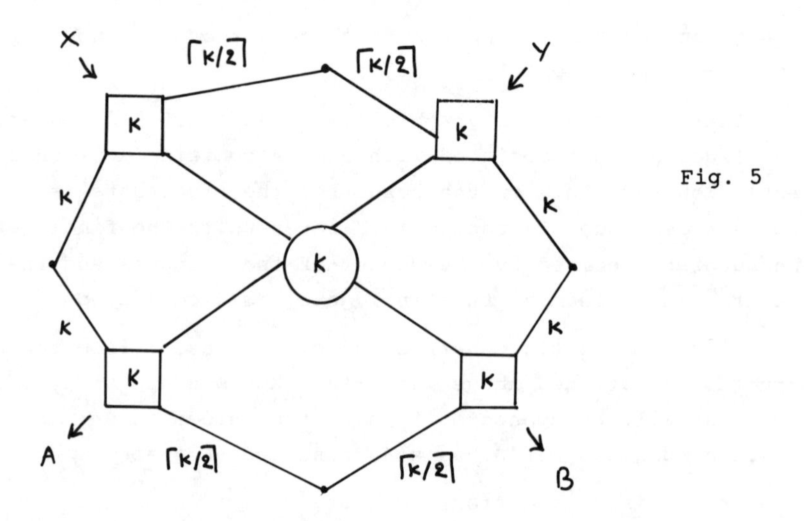

Fig. 5

The "and" gate is the gadget of Figure 6, it is obtained from the OR-gate by deleting the edges of the top node.

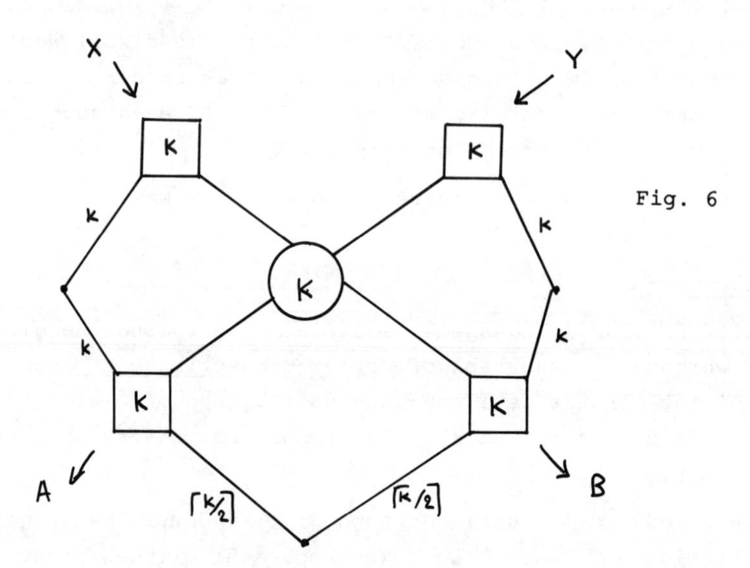

Fig. 6

In this case gate inputs and outputs correspond to sets of k nodes (for each side, the two distinguished nodes and k-2 nodes of the corresponding box). Each wire is simulated by k nodes connected by all possible edges to the two input/output sets of k nodes.

Again, the output of the final gate is connected to all the nodes associated with 1's. Note that if the output of the circuit is 1 then the subset of the 1's and the gates with output 1 which lead to the final gate define a 2k-connected subgraph. Else, the connectivity of any subgraph is at most k+1. □

The gap shown in the above construction, seems to be the best possible using this type of construction.

Let G be a graph and let f(n,m) be the connectivity of a subgraph that has to appear when G has n points and m edges. We conjecture that $f(n,m)=\theta(\frac{m}{n})$.

In this case using a technique similar to that of [Anderson, Mayr, 86] we can show that for any constant c<1/2 the optimization version of the HCS (for fixed k) can be solved by an algorithm in NC to a factor of c.

4. Linear reductions

Let (as in [Papadimitriou, Yannakakis, 88]) f be a log-space reduction from optimization (maximization) problem π to optimization (maximization) problem π'. We say that f is a linear reduction if there are constants $\alpha,\beta>0$ such that $\forall$ instance I of π:

(a) The optima of I and f(I), OPT(I) and OPT(f(I)) respectively, satisfy $OPT(f(I)) \leq \alpha \cdot OPT(I)$.

(b) For any solution of f(I) of cost c, we can find in log-space a solution of I of cost at least $OPT(I) + \beta(c-OPT(f(I)))$.

Clearly, linear reductions compose and also if π linearly reduces to π' and there is an NC-approximation algorithm for π' with worst-case error ε, then there is an NC-approximation algorithm for π with worst case error $\alpha\beta\varepsilon$.

For example, the reduction of GEN to CFMEMBER which is as follows: Given X,*,T and w construct the context free grammar G=(X,{a},P,w) where $P=\{x \rightarrow yz \text{ if } y*z=x\} \cup \{x \rightarrow \varepsilon \text{ if } x \in T\}$, is a linear reduction with $\alpha=\beta=1$.

References

[Jones, Laasser, 77] "Complete Problems for Deterministic Polynomial Time" by N. Jones and W. Laaser, Theor. Computer Science 3 (1977), 105-117.

[Papadimitriou, Yannakakis, 88] "Optimization, Approximation and Complexity classes", by C.H. Papadimitriou and M. Yannakakis, STOC 1988.

[Anderson, Mayr, 86] "Approximating P-Complete Problems", by R. Anderson and E. Mayr, TR Stanford Univ., 1986.

[Warmuth, 86] "Parallel Approximation Algorithms for one-Dimensional Bin Packing", by Warmuth, TR Univ. of California at Santa Cruz, 1986.

[Kirousis, Spirakis, 88] "Probabilistic log-space reductions and problems probabilistically hard for P", by L. Kirousis and P. Spirakis, SWAT 88.

[Ausiello, D'Atri, Protasi, 77] "On the Structure of Combinatorial Problems and Structure Preserving Reductions", by Ansiello, D'Atri and Protasi, Proc. 4th ICALP 45-57, 1977.

[Ausiello, D'Atri, Protasi, 80] "Structure Preserving Reductions among Convex Optimization Problems", by Ansiello, D'Atri and Protasi, J. Comp. Sys. Sc. 21, 136-153, 1980.

[Ausiello, Marchetti-Spaccamela, Protasi, 80] "Toward a Unified Approach for the Classification of NP-Complete Optimization Problems", by Ansiello, Marchetti-Spaccamela and Protasi, 12, 83-96, 1980.

[Garey, Johnson, 79] "Computers and Intractability - A Guide to the Theory of NP-Completeness", by M. Garey and D. Johnson, Freeman, 1979.

[Paz, Moran, 81] "Non Deterministic Polynomial Optimization Problems and their Approximation", by A. Paz and S. Moran, Theor. Computer Science, 15, 251-277, 1981.

[Matula, 78] "k-blocks and ultrablocks in graphs", by D. Matula, J. of Combinatorial Theory, 24, 1978, pp. 1-13.

A STRUCTURAL OVERVIEW OF NP OPTIMIZATION PROBLEMS †

DANILO BRUSCHI – Università degli Studi di Milano and University of Wisconsin

DEBORAH JOSEPH – University of Wisconsin

PAUL YOUNG – University of Washington and University of Wisconsin

1. INTRODUCTION

Although *NP* complete problems are typically stated as decision problems there are frequently underlying optimization problems that are of more interest. For example, in the *traveling salesperson* problem finding a minimum length tour is generally of more interest than knowing whether there is a tour of length less than k. However, early in the study of *NP* completeness it was recognized that computing the optimal solution of an *NP* optimization problem can be done in polynomial time only if there is a polynomial time solution for the corresponding decision problem. For this reason, and because nondeterministic functional computations are more difficult to formalize, the complexity theory for these problems has been built primarily on the study of the associated decision problems. Nevertheless, as the theory of *NP* complete sets and polynomial time reductions developed, it was also recognized that even the polynomial computable isomorphisms between known *NP* complete sets are not strong enough to preserve all of the underlying structure of the optimization versions of these problems. Furthermore, it was recognized early on (e.g., [GaJo 79], [GoSa 76], [Jo 74], and [PaMo 81]) that despite the fact that all known naturally occurring *NP* complete decision problems are polynomially isomorphic, ([BeHa 77],[JoYo 85]), some *NP* optimization problems have very good polynomially computable approximations while it is impossible for others to have good polynomially computable approximations unless $P = NP$.

---

† In this paper, we survey structural work on calculating optimal solutions for *NP* complete problems. This overview is not intended to be an exhaustive description of the area, nor even of all of the work in those papers which we discuss. Instead, it is an attempt to provide a general introduction to the area. As a consequence we have freely unified notations and simplified statements of results. We hope that this survey will provide a *guide* to some of the more interesting ideas and problems in the area, but readers wishing definitive statements of the results we discuss will need to consult journal versions of the papers we survey.

This work was supported in part by the Ministero della Pubblica Istruzione, through "Progetto 40%: Algoritmi e Strutture di Calcolo," the National Science Foundation under grant DCR-8402375, and by the Wisconsin Alumni Research Foundation under a Brittingham Visiting Professorship. The authors' 1988-89 address is: Computer Sciences Department, University of Wisconsin, 1210 West Dayton St., Madison, WI 53706, U.S.A.

Following the early work of Johnson, ([Jo 74]), on classifying approximation properties of *NP* complete optimization problems, in the late 1970's, Ausiello, D'Atri, and Protasi, and independently Paz and Moran moved beyond the study of approximation algorithms for specific *NP* complete problems and began the systematic study of the structure of *NP* optimization problems and their reductions. They addressed the question of what types of classifications can be made among *NP* optimization problems and the question of how the structural and combinatorial properties and reductions of *NP* problems relate to their ability to be polynomially approximated.

Their work was followed by work of Leggett and Moore, which considered the location of *NP* optimization problems within the polynomial time hierarchy. Among other results Leggett and Moore showed that *NP* optimization problems that arise from strongly complete *NP* decision problems are proper $\Delta_2^P(= P^{NP})$ problems. That is, these problems are not in $NP \cup coNP$ unless $NP = coNP$.

More recently, Papadimitriou and Yannakakis showed that certain languages associated with *NP* optimization problems are complete for complexity classes thought to lie just above *NP*. For example, they showed that the language $\{\langle G, k\rangle$: the maximum clique in G is of size $k\}$ is complete for $D^P(=$ all languages that can be expressed as the intersection of a language in *NP* and a language in *coNP*), while the language $\{G : G$ has a unique optimal traveling salesperson tour$\}$ is complete for Δ_2^P.

Following the work of Papadimitriou and Yannakakis, Krentel and Wagner began the task of developing a complexity theory for *NP* optimization problems based on traditional notions of complexity classes. Krentel, drawing from natural problems such as *clique* and *traveling salesperson,* defined a complexity class of *optimization functions* which he called *OptP*. To handle the problem that polynomial many-one reductions are not strong enough to preserve structural properties of optimization problems, Krentel defined the notion of a *polynomial time metric reduction*, which is a polynomial time truth-table reduction between functions that is allowed only one oracle computation. Using these reductions, Krentel showed that the optimization function for the *traveling salesperson* problem is complete for *OptP*, while the optimization function for the *clique* problem is complete for a weaker complexity class. He also showed that the optimization function for the *bin packing* problem is contained in a complexity class which is still weaker than the class related to the *clique* problem. Thus, by defining natural *functional* complexity classes for *NP* optimization problems, Krentel was able to use completeness for these classes as a way of distinguishing between optimization problems of varying difficulties.

Wagner's results have followed a similar vein using set classes rather than functional classes, but, in addition, he has developed close connections between the Boolean hierarchy over *NP* and sets related to *NP* optimization problems. Noting that all such *NP* optimization problems lie in classes at or below the functional analog of P^{NP}, and in fact lie in the *extended Boolean hierarchy* over *NP*, Wagner has classified problems by locating them in this Boolean hierarchy. His work carefully details both the types of functions that one might try to optimize for *NP* decision problems, (that is, the various ways that an optimization problem might be created from a decision problem), and the classes of the *Boolean hierarchy* for which the resulting problems are complete.

At the same time that the work on classification of optimization problems was taking place, additional research was being done on polynomial approximations. Papadimitriou and Yannakakis defined, using Fagin's logical characterization of *NP*, the classes *MAX NP* and *MAX SNP*, which characterize many *NP* optimization problems that are polynomially approximable. Similarly, Johnson, Papadimitriou, and Yannakakis defined the class *PLS* of functions that map instances of optimization problems to feasible solutions which witness local optima, and Krentel showed that the "local optima" function for *weighted satisfiability* is complete for *PLS*. On a more applied side, recent work in [HoKe 82], [JAMS 89], and [KGV 83] has provided unified techniques for polynomially approximating a wide variety of *NP* complete problems.

In the sections which follow we attempt to give an overview of the diverse work on *NP* optimization. In doing so, we trace the development of concepts and definitions that originated with Ausiello, D'Atri and Protasi, Moran and Paz, Leggett and Moore through the more recent work of Johnson, Papadimitriou, Yannakakis, Krentel and Wagner. Where appropriate we raise open questions.

2. NP OPTIMIZATION PROBLEMS

In general, to specify an *NP* problem B, one identifies a polynomial time testable predicate P and a polynomial p such that $P(x,y) \Rightarrow |y| \leq p(|x|)$, and defines B by

$$B(x) \Longleftrightarrow (\exists y)P(x,y).$$

Those elements y such that $P(x,y)$ are called *feasible solutions for* x, and the relation $|y| \leq p(|x|)$ guarantees that the sizes of feasible solutions are bounded by a polynomial in the size of x.

Without loss of generality, one can usually think of y as an accepting computation of a nondeterministic Turing machine on input x, although for our purposes it is probably better to think more directly of y as a potential solution to a question about x. For example, y might be a *clique* in a graph x, or a complete *tour* of the graph. It is also natural to have a polynomially computable *measure* or *valuation* function $m(x,y)$ of the goodness of y as a solution to the question about x, and to be interested only in feasible solutions which are sufficiently good. For example, in the case of *clique*[1] if $m(x,y)$ is the number of vertices in the clique y, then the clique problem, *CLIQUE* is specified by the special form,

$$CLIQUE(x,k) \Longleftrightarrow (\exists y)[P(x,y)\ \&\ m(x,y) \geq k],$$

where $P(x,y)$ asserts that y is a clique of the graph x. Similarly, the *traveling salesperson* problem, is specified by

$$TSP(x,k) \Longleftrightarrow (\exists y)[P(x,y)\ \&\ m(x,y) \leq k],$$

where $P(x,y)$ asserts that y is a tour of the graph x and $m(x,y)$ is a function giving the total length of the tour y.

In the setting of this paper an *NP* complete problem C that is specified in either of these forms is called an *NP optimization* problem, and we sometimes refer to the class of all such problems as the class *NPO*. In practice, for *NP* optimization problems one is often more interested in calculating some variation of the function

$$opt_C(x) =_{def} Max\{m(x,y) : P_C(x,y)\}, \quad \text{or}$$
$$opt_C(x) =_{def} Min\{m(x,y) : P_C(x,y)\}.$$

Proving that the underlying problem is *NP* complete shows that the problem of computing opt_C is probably hard, but it does not show *how* hard computing opt_C must be, nor does it answer the question of whether we can efficiently find approximate solutions for opt_C.

Of course, by simply taking $m(x,y)$ to be the characteristic function, $\chi_{P_C}(x,y)$, of P_C we can make the problem of solving the underlying *NP* complete problem fairly directly equivalent to calculating the function $opt_{P_C}(x)$. However, for more interesting cases, for example where $m(x,y)$ measures the size of a clique, or perhaps where $m(x,y)$ measures the weight of a tour of a graph, the reader will see that solving the underlying *NP* decision problem can be trivially reduced to calculating $opt_{P_C}(x)$, but it is not evident that the problem of calculating $opt_{P_C}(x)$ is so trivially reducible to solving the underlying *NP* decision problem.

---

[1] An informal description of some of the decision problems and the functional variants of the optimization problems considered in this paper is given in the Appendix. For a description of the remaining decision problems, we refer the reader to [GaJo 79].

Much work on optimization problems has focused, not on directly calculating the function opt_C, or even the more natural problem of finding a feasible solution y which optimizes the value of $m(x, y)$, but instead on using the already developed machinery for classifying set recognition problems to classify set recognition problems related to the optimization problem.

For example, in [LeMo 81], Leggett and Moore considered the difficulty, for an *NP* complete optimization problem C, of the related set

$$OPT(C) =_{def} \{\langle x, k\rangle : \; k = opt_C(x)\} \tag{$*$}$$

which is the problem, given x and k, of deciding whether k is the optimum value of any feasible solution of x. It is easily seen that for any *NP* complete optimization problem C, $OPT(C)$ is ***always*** in the class $\Delta_2^P =_{def} P^{NP} \subseteq \Sigma_2^P \cap \Pi_2^P$ of the polynomial time hierarchy. Leggett and Moore called a set, S, a ***proper*** Δ_2^P set if it is in Δ_2^P and if

$$S \in NP \cup coNP \implies NP = coNP.$$

With this definition, they were able to show that for many natural *NP* optimization problems C, the corresponding optimization problem $OPT(C)$ is a proper Δ_2^P set and is therefore very unlikely to fall in either *NP* or in *coNP*.

The basic tool Leggett and Moore used is the notion of a Σ-preserving reduction, $\leq_\Sigma^P$, which is a polynomial time reduction that preserves membership in the Σ_k^P levels of the polynomial time hierarchy. Examples of Σ-preserving reductions include polynomial time many-one reducibility, $\leq_m^P$, and polynomial time conjunctive truth-table reducibility, $\leq_c^P$. Their basic theorem is that if C is any *NP* optimization problem for which there exists some *NP* complete problem B satisfying $B \leq_\Sigma^P OPT(C)$, then $OPT(C)$ must be a proper Δ_2^P set.

This result can easily be applied to all *NP* complete optimization problems which are ***polynomially bounded***; i.e., to problems for which there is some polynomial p such that for each instance x of C, $opt_C(x) \leq p(|x|)$. It is not hard to see that for any polynomially bounded *NP* complete optimization problem C, $C \leq_c^P OPT(C)$, and thus $OPT(C)$ is a proper Δ_2^P set. It then follows trivially from this that the optimization problems for all ***strongly*** *NP* complete sets[2] are proper Δ_2^P sets. Since many natural

[2] The ***strongly NP complete*** problems as defined by Gary and Johnson ([GaJo 79]) are just those problems of the form $(\exists y)P(x, y)$, which remain *NP* complete when they are restricted to $(\exists y)[P(x, y) \& m(x, y) \leq p(|x|)]$ for some polynomial p. I.e., they are just those which remain *NP* complete when their feasible solutions are restricted to include just those which keep the problem ***polynomially bounded*** as defined by Leggett and Moore.

NP complete optimization problems are strongly *NP* complete, or even polynomially bounded, it turns out that many natural *NP* complete sets automatically give rise to optimization problems which are proper Δ_2^P sets.[3]

Examples of *NP* complete sets for which the general Leggett and Moore technique applies include the polynomially bounded problems CLIQUE, CHROMATIC NUMBER, and SAT, as well as KNAPSACK which is neither polynomially bounded nor strongly NP-complete.

Although the problems Leggett and Moore investigated are proper Δ_2^P sets, Leggett and Moore do not use the full power of Δ_2^P. Specifically, to be sure that k is the size of an optimal solution for an instance x one need only verify that there is some solution of x of size at least k and that there are no solutions of x of size greater than k. From this it is clear that all of the Leggett and Moore sets $OPT(C)$ can be defined as the intersection of a set in *NP* and a set in *coNP*.

In [PaYa 84] Papadimitriou and Yannakakis formally defined D^P to be the class of sets which can be written as the intersection of a set in *NP* and a set in *coNP*, and they showed that MAX CLIQUE $=_{def}$ OPT(CLIQUE) is $\leq_m^P$ complete for D^P. Other problems they showed to be complete for D^P include SAT-UNSAT and problems like CLIQUE FACETS, which is the problem, given a graph G and an inequality, of deciding whether the inequality is a facet of the clique polytope of G?

In [PaWo 85] and in [CaMe 87] the additional polynomially bounded *NP* complete optimization problems MINIMAL UNSATISFIABILITY and (for all k) MINIMAL-k-UNCOLORABILITY were shown to be $\leq_m^P$ complete for D^P.

Prompted by the Papadimitriou and Yannakakis definition of D^P, a number of people began investigating the *Boolean hierarchy* over *NP*, ([CGHHSWW 88], [CGHH-SWW 89]), and over other complexity classes as well, ([BBJSY 89]). Boolean hierarchies are typically formed by taking at the base level a class of sets which is closed under union and intersection but not complement, and then forming the $k+1^{st}$ level of the

---

[3] Leggett and Moore called polynomially bounded problems *essentially unary,* but the terminology *polynomially bounded* has since become more standard. *Traveling salesperson* is a natural example of a problem which is strongly *NP* complete but whose unrestricted version is not polynomially bounded.

Leggett and Moore go on to show that if one starts, not with optimization problems for *NP*, but rather with optimization problems in the Σ_k^P level of the polynomial hierarchy, then corresponding results to those for Δ_2^P will show that various optimization sets are proper Δ_{k+1}^P sets. These results are very similar to the results discussed here for Δ_2^P, but are beyond the scope of this brief survey.

Boolean hierarchy by iterated closure under k operations involving union, intersection, and complement. If done properly, D^P is just $NP[2]$, the second level of the Boolean hierarchy over NP.

Following the development of the Boolean hierarchy over NP and the proof by Papadimitriou and Yannakakis that MAX CLIQUE is complete for D^P, Wagner was interested in investigating when different variations of optimization problems are complete for the kth level of the Boolean hierarchy over NP. To do this, he defined a number of interesting variations of optimization problems. For our purposes, the most interesting of these is the following generalization of (*):

$$OPT_k(C) =_{def} \{\langle x, a_1, \ldots, a_k\rangle : opt_C(x) \in \{a_1, \ldots, a_k\}\ \}. \qquad (**)$$

Note that $OPT_1(C) = OPT(C)$. Wagner observed that if C is any NP optimization problem, then the corresponding problem $OPT_k(C)$ is in the $2k^{th}$ level, $NP[2k]$, of the Boolean hierarchy over NP.[4] He defined a valuation function, m to be *polynomially invertible* if the set, $Sol(x, k)$, of feasible solutions of size k,

$$Sol(x, k) =_{def} \{y : P_C(x, y)\ \&\ m(x, y) = k\},$$

is completely enumerable (from x and k) in polynomial time. He then observed that for every NP optimization problem, C, with an invertible valuation function, the optimization set $OPT_k(C)$ is in $coNP$, and hence these sets are very *unlikely* to be complete for $NP[2k]$. Wagner also showed that the one specific problem that he considered with an invertible valuation function, MAX SAT ASSIGN, is *complete* for $coNP$.

Wagner next considered nine common NP optimization problems for which the valuation function is not polynomially invertible, including seven which are polynomially bounded in the sense of Leggett and Moore's definition and two which are neither polynomially bounded nor polynomially invertible. (Of these latter two, TRAVELING SALESPERSON is strongly NP complete, but the other, SUM OF SUBSETS is *not* strongly NP complete.) Wagner gave detailed reductions showing that for each of these nine optimization problems, C, the corresponding optimization problem $OPT_k(C)$ is complete for $NP[2k]$.

---

[4] At first glance, one might expect it to be easier to answer whether $opt_{P_C}(x) \in \{a_1, a_2\}$ than to answer whether $opt_{P_C}(x) = a_1$, but note that it seems harder to answer that $opt_{P_C}(x)$ is *neither* a_1 *nor* a_2 than to answer that $opt_{P_C}(x)$ is *not* a_1, and this is what makes deciding $OPT_2(C)$ (presumably) harder than deciding $OPT_1(C)$.

Recall that Leggett and Moore's basic theorem gives very *general* conditions on NP complete optimization problems which force their associated optimization sets $OPT(C)$ to be proper Δ_2^P sets. Given the many examples of optimization problems that are complete for D^P, and given Wagner's nine examples of optimization problems that are complete for $NP[2k]$, it is natural to ask whether there are similar *general* conditions which force the NP complete optimization problems $OPT_k(C)$ to be *complete* for $NP[2k]$. For example, it is tempting to conjecture that for every NP complete optimization problem which is polynomially bounded, the set $OPT_k(C)$ is complete for $NP[2k]$. This seems an interesting problem even for the special case $k = 1$, where $NP[2] = D^P$.

Question 1. *Is every NP complete optimization problem which also is polynomially bounded, or even strongly NP complete,* $\leq_m^P$ *complete for* D^P*?*

In addition to the variations of optimization problems considered by Wagner, Papadimitriou investigated another type of optimization problem, optimization problems which have a *unique* solution. In [Pa 84] he proved that the problem of determining whether the optimum solution for a *traveling salesperson* problem is *unique*, the problem of determining whether the optimum solution for an *integer program* is *unique*, and the problem of determining whether the optimum solution for the *knapsack* problem is *unique* are each $\leq_m^P$ complete for Δ_2^P. This leads us to ask:

Question 2. *What type of conditions, such as those of Leggett and Moore, must one place on an NP complete optimization problem C in order to guarantee that* $UNIQUEOPT(C) =_{def} \{x : \exists! y[m(x,y) = opt_C(x)]\}$ *is complete for* Δ_2^P*?*

So far we have indirectly discussed how difficult it is to compute the *optimization function*

$$opt_C(x) =_{def} Max\{m(x,y) : P_C(x,y)\}$$

by instead examining the difficulty of the associated set

$$OPT(C) =_{def} \{\langle x, k\rangle : k = opt_C(x)\}.$$

Motivated partly by a desire to better understand which NP complete optimization problems can be approximated in polynomial time, Krentel, ([Kr 88]), initiated a direct study of the *function* class:[5]

---

[5] At this point notation becomes overbearing and one quickly realizes that the characters "P" and "Opt" are being greatly over used. In the literature the situation is commonly worse. Wagner and Wechsung, ([WaWe 86]) carefully provide unique names for all sets and functions of interest to the study of NP optimization. But our belief is that their naming does not solve

$$\textbf{OptP} =_{def} \{opt_C : C \text{ is an } NP \text{ optimization problem}\}.$$

To study differences within this class, he also introduced, for any "smooth" function, z, the subclasses:

$$\textbf{OptP}[z(n)] =_{def} \{f : f \in \textbf{OptP} \ \& \ |f(x)| \leq z(|x|)\}.$$

To study these functional classes, Krentel needed some suitable notion of a polynomial time reduction from one *function*, f, to another function, g. In the context of his work, the notion of a ***one-truth-table reduction*** turned out to be appropriate.[6]

Definition. *Let f and g be functions. We define $f \leq_{1tt}^{P} g$ if there exist polynomially computable functions t_1 and t_2 such that $f(x) = t_2(x, g(t_1(x)))$.*

These definitions enabled Krentel to separate various optimization problems by directly separating the *functional* versions of the problems. For example, he showed that for the *NP* complete optimization problems MAX WEIGHTED CLAUSES, MIN TRAVELING SALESPERSON, MAX SAT ASSIGN, 0-1 INTEGER PROGRAMMING, and MAX KNAPSACK the corresponding functions opt_C are all $\leq_{1tt}^{P}$ complete for *OptP*. He also showed that for the following *NP* complete optimization problems MAX SAT, MAX CLIQUE, MIN CHROMATIC-NUMBER, and LENGTH OF LONGEST CYCLE the corresponding functions opt_C are $\leq_{1tt}^{P}$ complete for $\textbf{OptP}[log(n)]$. Most, but not all, of the proofs of these results come by straightforward variations of standard reductions used to prove that the underlying *NP* complete sets are complete problems for *NP*.

Krentel also observed that most (but not all) of the $\leq_{1tt}^{P}$ reductions used in the preceding proofs can be taken to be *linear reductions*, i.e., reductions in which the outermost component t_2 of the reduction separates nicely into *linear components*, t_3 and t_4. I.e.,

$$f(x) =_{def} t_2(x, g(t_1(x))) = t_3(x) * g(t_1(x)) + t_4(x).$$

This enabled Krentel to directly relate completeness for the functional classes to completeness for the more standard set classes. For example:

---

the problem of providing *intuitive* names for the common functions and classes. (Surely this is an open problem which should be solved before someone introduces yet another class!)

[6] Krentel actually called these *metric* reductions, but since the reductions are basically polynomial time with one free *oracle computation* of the function g, they correspond directly to the classical notion of a polynomial time one-truth-table reduction.

Theorem.

- *If f is complete for* **OptP** *under linear reductions, then* $L_{2,f} =_{def} \{\langle x, k_1, k_2\rangle : f(x) = k_1 \pmod{k_2}\}$ *is* $\leq_m^P$ *complete for* Δ_2^P.
- *If f is complete for* **OptP**[2] *under linear reductions, then* $L_{3,f} =_{def} \{\langle x, k\rangle : f(x) = k\}$ *is* $\leq_m^P$ *complete for* D^P.
- *If f is complete for* **OptP**[1] *under linear reductions, then* $L_{4,f} =_{def} \{\langle x, k\rangle : f(x) \geq k\}$ *is* $\leq_m^P$ *complete for* NP.

For the same functions z used to define **OptP**$[z(n)]$, Krentel also establishes a close relationship between

$$\mathbf{FP}^{SAT}[z(n)] =_{def} \{f : \ f(x) \text{ is computable in polynomial time given } z(|x|) \text{ queries to an oracle for } SAT\}$$

and **OptP**$[z(n)]$. In particular, he proved, for suitable "smooth" f and g :

Theorem. *Suppose that for all n, $f(n) < g(n)$ and that $f(n) < (1 - \epsilon) * log(n)$ for some constant $\epsilon > 0$. Then*

$$\mathbf{FP}^{SAT}[f(n)] = \mathbf{FP}^{SAT}[g(n)] \quad \Longrightarrow \quad P = NP.$$

This result implies that it is unlikely that certain **OptP** complete functions can be approximated well by a polynomially computable function. For example, notice that if $opt_C(x)$ is an **OptP** function and if in polynomial time we can approximate it within an additive quantity $a(|x|)$, then we can use binary search to compute $opt_C(x)$ with $log(a(|x|))$ queries to a SAT oracle. This gives particularly intriguing information about MIN BIN PACKING. On the face of it, based on the size of its output, the optimization function opt_{BIN} for MIN BIN PACKING belongs to **OptP**$[log(n)]$, and hence to $\mathbf{FP}^{SAT}[log(n)]$. But Karmakar and Karp, ([KaKa 82]) have given a polynomial time algorithm which approximates the optimal solution to MIN BIN PACKING to within an additive constant of at most $O(log^2(n))$. As pointed out in [Kr 88], this implies that $opt_{BIN} \in \mathbf{FP}^{SAT}[O(loglog(n))]$ and that problems such as opt_{CLIQUE} cannot be reduced to opt_{BIN} under linear polynomial time reductions unless $P = NP$.

Krentel goes on to prove the following more general result about lower bounds for polynomial approximations of **OptP** complete functions:

Theorem. *Suppose $P \neq NP$. Suppose also that opt_C is* **OptP**$[f(n)]$ *complete, where $f \in O(log(n))$ is smooth. Then there exists an $\epsilon > 0$ such that any polynomial time approximation algorithm A for opt_C must have $|A(x) - opt_C(x)| \geq (1/2)^{2^{f(|x|^\epsilon)}}$ infinitely often.*

As we proceed, we shall see that to obtain such a nice result on the difficulty of *approximating* optimization problems, it is not surprising that Krentel was forced to use some form of *linear*, as opposed to just $\leq_{1tt}^P$, reductions.

Given that Karmakar and Karp's result places opt_{BIN} in $\mathbf{FP}^{SAT}[O(loglog(n))]$, it is natural to conjecture that opt_{BIN} is *complete* for $\mathbf{FP}^{SAT}[O(loglog(n))]$; but the best that is known is that opt_{BIN} is hard for $\mathbf{FP}^{SAT}[1]$. In light of these considerations Krentel raised the following question:

Question 3.

- *Which function classes $\mathbf{FP}^{SAT}[z(n)]$ contains opt_{BIN}?*
- *Are there natural complete problems for subclasses of FP^{SAT} other than $FP^{SAT}[O(log(n))]$ and $FP^{SAT}[n^{O(1)}]$?*

3. APPROXIMATING SOLUTIONS TO OPTIMIZATION PROBLEMS

In spite of the fact that finding optimal solutions of *NP* complete optimization problems is at least *NP* hard, it was recognized early on that some, but not all, such functions have "approximate" solutions which can be calculated in polynomial time. This phenomenon stimulated two lines of research on optimization problems. In one direction it stimulated research to find faster and better approximation algorithms, and in another direction it led to attempts to give structural properties common to all polynomially approximable optimization problems.

Among the techniques that have wide applicability, *neighborhood* or *local search* is perhaps the most used. An interesting problem is thus the extent to which this technique can be applied. For example: Is it possible to apply a local search algorithm to obtain approximate values for each function contained in **OptP** ? This question was first investigated by Johnson, Papadimitriou, and Yannakakis ([JPY 85]), and subsequently by Krentel ([Kr 89]). Their main results are briefly discussed in Section 3.1.

General questions related to the study of combinatorial or structural properties common to polynomially approximable optimization problems have been investigated by many people and we discuss a variety of these results in Section 3.2. Section 3.3 contains a brief discussion of various restricted reducibilities which have been used in structural studies of *NP* optimization problems.

3.1. The Difficulty of Local Search

Johnson, Papadimitriou and Yannakakis have noted that in practice local search is one of the most successful techniques used to compute polynomial time approximations for optimization functions. The applicability of this technique is related to the existence of a *neighborhood structure* which specifies for each feasible solution y a neighboring set of feasible solutions whose measure is "close" to the measure of y.

Given an optimization problem with a neighborhood structure, a local search algorithm operates as follows: Starting from a randomly chosen feasible solution, until no better neighboring solution exists it repeatedly replaces the current solution by a neighboring solution with the best measure. Once this has been done the algorithm has identified a "local optimum." Typically, th algorithm is repeated a certain number of times with different initial solutions and the locally optimal solution that is found is chosen.[7]

In [JPY 85], Johnson, Papadimitriou, and Yannakakis began an investigation of the class of problems which can be approximated through local search techniques. First they introduced a complexity class of functions called *PLS*. Specifically *PLS*, for *Polynomial Local Search*, is the class of functions that map instances of optimization problems with a given neighborhood structure to local optima. The *PLS* function is computed from an optimization problem and a neighborhood structure which must satisfy the following requirements:

- Given an instance of the problem, we must be able to produce in polynomial time some solution.
- Given an instance and a solution, we must be able in polynomial time to compute the measure of the solution.
- Given an instance and a solution, we must be able in polynomial time to determine whether that solution is locally optimal and if not to generate a neighboring solution of improved measure.

---

[7] Obviously, there are different variations of this technique. The algorithm that we have described is probably the most simple. "Simulated annealing" is another popular method, ([KGV 83]). But see [JAMS 89] for a critical evaluation. In the remainder of this paper, when we refer to a local search technique we implicitly intend any reasonable (polynomial time) local search technique.

The class *PLS* turns out to lie somewhere between *FP* and *FNP* (i.e. the functional analogues of *P* and *NP*). Johnson, Papadimitriou, and Yannakakis note that if *PLS* contains some *NP*-hard function, then $NP = coNP$; thus it seems unlikely that $PLS = FNP$. On the other hand $PLS = P$ implies the existence of a general method for finding local optima, and to date no such method is known. So the questions $FP =? PLS =? FNP$ have no obvious answers.

To gain insight into this new complexity class Johnson, Papadimitriou, and Yannakakis defined a reduction between the optimization problems which underlie *PLS* functions. Intuitively, these reductions not only must map instances of a problem A to instances of a problem B, but must also map local optima of A to local optima of B. More precisely, we say that a problem A in *PLS* is reducible to another problem B if there are polynomially computable functions f and g such that:

- f maps instances of A to instances of B,
- g maps $\langle$*solutions, instances*$\rangle$ pairs for instances in the range of f back to solutions of A,
- for all instances x of A, if s is a local optimum for the instance $f(x)$ of B, then $g(s, f(x))$ is a local optimum for x.

With this notion of reducibility, Johnson, Papadimitriou, and Yannakakis introduce the corresponding notion of *PLS* completeness and gave several examples of functional problems which are *PLS* complete. (The most natural of these is the problem, FLIP, which given a circuit produces a binary *input* whose *output* cannot be increased by flipping a single bit of the input. Feasible solutions which differ in only bit are then in the same *neighborhood,* so that in polynomial time one can generate and evaluate all feasible solutions which lie in the neighborhood of a given candidate solution.) These authors conjectured that for each *PLS* complete problem the problem of verifying local-optimality is itself *LOGSPACE*-complete for P. However, this conjecture is very unlikely to be true since Krentel, ([Kr 89]), has proven that the problem MAX WEIGHTED CLAUSES is *PLS* complete, but that the corresponding verification problem is in *LOGSPACE.*

3.2. Polynomially Approximable NP Complete Optimization Problems

One of the first people to study the problem of finding approximate solutions to *NP* optimization problems was Johnson, who in his seminal paper, ([Jo 74]), provided a series of results on the approximability of problems such as MAX CLIQUE, MAX SAT, and MIN CHROMATIC NUMBER and gave what are now the standard definitions of polynomial time approximability.

Definition. *Let C be an NP maximization problem. Let f be any polynomially computable function mapping instances x of C to feasible solutions of x. Let $approx(x) =_{def} m(x, f(x))$. We say that approx is a polynomial time approximation algorithm for $opt_C(x)$ if the ratio $\frac{opt_C(x)}{approx(x)}$ is bounded by some constant ϵ greater than one.*[8]

Note that the ratio $\frac{opt_C(x)}{approx(x)}$ always lies between one and (+) infinity. Thus to say that a *maximization* problem C has a polynomial time approximation algorithm merely says that the solution $approx(x)$ always comes within a multiplicative factor of $opt_C(x)$. Notice also that this notion of approximability differs from that used by Krentel to state his results about approximability. In fact, ϵ is a multiplicative factor here while Krentel's results require an additive factor.[9]

In order to preserve the fact that the ratio $\frac{opt_C(x)}{approx(x)}$ is in the range $[1, +\infty)$, for *minimization* problems Johnson reverses the ratio and requires that the ratio $\frac{approx(x)}{opt_C(x)}$ is bounded by some constant ϵ greater than one.

Of course, it is even better if we can approximate the optimal solution to within any desired accuracy. In this case, the function f which produces the approximating feasible solution will also have to be a function of some ϵ which gives the desired accuracy:

---

[8] Where necessary, we shall assume without further discussion that $opt_C(x)$ is bounded away from zero.

[9] Another difference is that Krentel's approximation algorithm A need not be computed through an intermediate function f which actually finds feasible solutions $f(x)$ such that $A(x) = m(x, f(x))$. In practice of course, if one is building approximation algorithms, one wants to find, not just an approximation to the *value* of the optimal solution, but also a *feasible solution* whose value gives a good approximation to the value of the optimal solution. Paz and Moran, ([PaMo 81]), call approximation algorithms which work by actually finding feasible solutions *constructive* approximation algorithms, using the unmodified term *approximation algorithms* for those which may merely find vaules which approximate the value of the optimal solution. They also prove a number of results relating the two concepts. However their terminology is not currently standard, and most authors require in their definitions, as we do, that all approximation algorithms be constructive in the sense that they work by actually finding feasible solutions. Thus in the remainder of this survey, we assume that approximation algorithms work by finding feasible solutions. Nevertheless many results, particularly those involving lower bounds, will hold for approximation algorithms which are not constructive in the sense of Paz and Moran, and the reader interested in such results will have to carefully consult the relevant literature.

Definition. *Let C be an NP optimization problem.*

- *We say that opt_C has a **polynomial time approximation scheme** if there exists some function $approx(x, \epsilon)$ such that for every ϵ greater than zero $approx(x, \epsilon)$ is polynomially computable as a function of $|x|$ and for every instance x of C, $\frac{opt_C(x)}{approx(x,\epsilon)} \leq 1+\epsilon$ if C is a maximization problem, (or such that $\frac{approx(x,\epsilon)}{opt_C(x)} \leq 1+\epsilon$ if C is a minimization problem).*
- *We say that opt_C has a **full polynomial time approximation scheme** if the time complexity of $approx(x, \epsilon)$ can be bounded by a polynomial in both $|x|$ and $1/\epsilon$.*

It is well-known that there are approximation algorithms for the optimization versions of *NP* complete problems such as TRAVELING SALESPERSON (with triangle inequality), VERTEX COVER, and SAT. It is also known that unless $P = NP$ these problems do not have full polynomial time approximation schemes.[10] This leaves as an important open question whether such problems have polynomial time approximation schemes.

An early study characterizing which *NP* complete optimization problems are approximable and which are fully approximable was given by Paz and Moran, ([PaMo 81]). They began by observing that in many optimization problems, once one *fixes* k, the set

$$\{x : opt_C(x) \leq k\} \qquad (***)$$

is recognizable in polynomial time. For example, for any fixed k it is possible to decide in polynomial time whether a given graph has a clique of size at least k. Paz and Moran called problems for which (***) is always solvable in polynomial time *simple,* and they called them *p-simple* if there is a *uniform* polynomial q such that (***) is always recognizable in time $q(|x|, k)$. Using the notions of simplicity together with multiplicative and polynomial bounds on the optimal solutions, they then went on to completely characterize those *NP* complete optimization problems which are polynomially approximable and those which have fully polynomial time approximation schemes:

---

[10] The formulation of these quite general results comes from the work of many researchers. Some of the key early contributions were made in [Jo 74], [GoSa 76], [IbKi 75], [Sa 76], [Ni 75], [GaJo 79]. However the reader should refer to [GaJo 79] or to [WaWe 86] for an exhaustive list.

Theorem. *An NP complete maximization problem, C, is polynomially approximable if and only if*

(1) C is simple, and

(2) there is a constant Q_0 such that for each instance x and each integer $h > 0$,

$$0 \leq opt_C(x)/h - b(x,h) \leq Q_0,$$

where b is a function mapping $\langle instance, nonzero\ integer \rangle$ pairs to integers whose time complexity is bounded by a polynomial in $|x|$ and $opt_C(x)/h$.

Theorem. *An NP complete maximization problem, C, has a fully polynomial time approximation scheme if and only if*

(1) C is p-simple, and

(2) there is a polynomial $q(n)$, such that for each instance x and for each integer $h > 0$:

$$0 \leq opt_C(x)/h - b(x,h) \leq q(|x|),$$

where b is a function mapping $\langle instance, nonzero\ integer \rangle$ pairs to integers whose time complexity is bounded by a polynomial in $|x|$ and $opt_C(x)/h$.

Exactly analogous theorems hold for *NP* complete *minimization* problems, but with the inequalities appropriately reversed.

While Paz and Moran's results give necessary and sufficient conditions for optimization problems to possess approximate solutions, the characterizations are not completely satisfying since the characterizations themselves are stated in terms of approximating solutions. Clearly what one would like are *intrinsic* characterizations of optimization problems which are themselves sufficient to guarantee the existence of approximating solutions.

One such *intrinsic* characterization was recently given by Papadimitriou and Yannakakis, ([PaYa 88]), who introduced yet another pair of complexity classes for optimization problems: *MAX NP* and *MAX SNP*. These classes give characterizations which are sufficient to guarantee the existence of polynomial time approximation algorithms.

To formalize the definitions of *MAX SNP* and *MAX NP*, Papadimitriou and Yannakakis used Fagin's logical characterization of *NP*, ([Fa 74]). Recall that Fagin showed that every predicate in NP can be expressed in the form $\exists S\phi(G,S)$ where G and S are structures and ϕ is a first order formula. For instance we can express SAT as:

$$\exists T\,\forall c\,\exists y\,[P(c,y)\,\&\,y\in T\ \vee\ N(c,y)\,\&\,y\notin T],$$

where $P(c,y)$ is *true* if the variable y occurs positively in the clause c and $N(c,y)$ is *true* if the variable y occurs negatively in the clause c. In general the first order predicate ϕ can always be expressed in the form $\forall\overline{x}\,\exists\overline{y}\,\psi(\overline{x},\overline{y},G,S)$, where ψ is quantifier free. Furthermore, in many cases with a little work ϕ can be expressed in the restricted form $\forall\overline{x}\,\psi(\overline{x},G,S)$.

The class *MAX NP* is then defined to be the following class of maximization problems:

$$\text{If}\ \ \exists S\,\forall\overline{x}\,\exists\overline{y}\,\psi(\overline{x},\overline{y},G,S)\in NP,\ \ \text{then}\ \ Max_S|\{\overline{x}:\exists\overline{y}\,\psi(\overline{x},\overline{y},G,S)\}|\in MAX\ NP.$$

Similarly, the class *MAX SNP* [11] is defined to be the following class of maximization problems:

$$\text{If}\ \ \exists S\,\forall\overline{x}\,\psi(\overline{x},G,S)\in NP,\ \ \text{then}\ \ Max_S|\{\overline{x}:\psi(\overline{x},G,S)\}|\in MAX\ SNP.$$

Obviously, MAX SAT is in *MAX NP*, and Papadimitriou and Yannakakis showed that MAX 3SAT is in *MAX SNP*. To investigate these classes, they defined the notion of an *L-reduction* from an optimization problem A to an optimization problem B (which are both assumed to be maximization problems) to be a polynomial transformation f such that there are two constants $\alpha,\beta>0$, and for each instance x of A:

- the optima of x and $f(x)$, $opt_A(x)$ and $opt_B(f(x))$ respectively, satisfy $opt_B(f(x))\ \le\ \alpha * opt_A(x)$ and
- for any feasible solution y of $f(x)$, we can find in polynomial time a feasible solution y' of x satisfying $opt_A(x)-m(x,y')\ \le\ \beta*[opt_B(f(x))-m(f(x),y)]$.

Using this notion they prove that every problem in *MAX SNP* has a polynomial approximation algorithm, and they state that the same is true for *MAX NP* . They also give a dozen examples of problems which are complete for *MAX SNP* with respect to L-reductions, of which the most natural are perhaps MAX 3SAT, MAX 2SAT, and MAX CUT.

Their work naturally suggests the following general problems:

---

[11] The name *SNP* comes from the fact that the formulas $\exists S\,\forall\overline{x}\,\psi(\overline{x},G,S)$, without the first order existential quantifier, are called "strict" Σ_1^1 formulas.

Question 4. *MAX NP and MAX SNP contain problems which are clearly NP optimization problems, but it is surely not true that every problem which we think of as an NP optimization problem is in MAX NP. Can one characterize in some more complexity theoretic fashion (other than the definition) those NP optimization problems that are in MAX NP? Can one give intrinsic complexity theoretic characterizations of other classes of NP complete optimization problems which always have some form of polynomial approximations?*

Question 5. *As an alternative to the second order characterization of Fagin, Hodgson and Kent, ([HoKe 82]), have given a first order characterization of NP by bounding the domain of existential and universal quantifiers. Can this characterization be used either to give an alternative definition of MAX NP and MAX SNP, or to define other classes of NP complete optimization problems which always have some form of polynomial approximations? If so, is this first-order approach more natural, and how do its "max classes" compare with those of Papadimitriou and Yannakakis?*

3.3. Reductions for NP Optimization Problems

The fact that all naturally occurring examples of *NP* complete problems are polynomial time isomorphic but not all are approximable tells us that the standard reductions of complexity theory are inadequate for studying relations among optimization problems. For example, we earlier saw that Krentel needed to introduce the notion of a "linear" reduction to study such relations, and we have just seen that Papadimitriou and Yannakakis needed to introduce "L-reductions" to study the relations between classes of optimization problems with similar approximation properties.

Earlier studies of relations among optimization problems also introduced similar restrictions on the reductions. For example, Paz and Moran, ([PM 81]), called the reductions used in their studies of approximations of *NP* optimization problems *polynomial time measure preserving reductions* and *polynomial time ratio preserving reductions.* (The *measure preserving* reductions are merely reductions among *NP* optimization problems which always carry instances of the reduced problem which have maximal solutions of size k to instances of another problem which have maximal solutions of size exactly $Q_0 * k$ where Q_0 is some multiplicative constant. The *ratio preserving* reductions are similar, but the size of the optimal solution is only preserved to within a multiplicative *interval.*)

Paz and Moran used these reductions to show that a variety of *NP* complete problems can be interreduced via reductions that preserve approximability properties. Thus they were able to transfer known approximability results (or nonapproximability results) from one *NP* complete optimization problem to another. Among other examples, they used this technique in reducing MAX SAT to MAX CLIQUE, MIN CLIQUE COVER to MIN CHROMATIC NUMBER, and MIN CHROMATIC NUMBER to MIN CLIQUE COVER.

In this context, it is interesting to note that Paz and Moran were the first to use such restricted reductions to obtain *complete* versions of optimization problems. For example they showed that, with respect to polynomial time measure preserving reductions, MAX WEIGHTED VARIABLE is complete for the class of *NP* optimization problems. Thus, if MAX WEIGHTED VARIABLE is approximable, so is *every NP* optimization problem.

More recently, Crescenzi and Panconesi ([CrPa 88]), building on earlier work by Orponen and Mannila ([OrMa 87]), have developed a theory of complete problems not only for NPO, the class of *NP* optimization problems, but more importantly, for APX, the class of problems which admit polynomial approximations, for $PTAS$, the class of problems which admit polynomial time approximation schemes, and for $FPAS$, the class of problems which are fully polynomially approximable. Obviously,

$$FPAS \subseteq PTAS \subseteq APX \subseteq NPO.$$

For each of these classes except $FPAS$, extending a definition due to Orponen and Mannila, Crescenzi and Panconesi defined an appropriate natural form of reduction[12] and then use these reductions to define *complete* sets for $PTAS$, APX, and NPO. The reductions are not only reasonably natural, but most important, just as any complete problem for *NP* falling into *P* implies that $NP = P$, if any of the complete problems for any of these three largest classes falls into the immediately smaller class, then the whole class collapses to the immediately smaller class. *Thus proving that a problem is complete for a class is strong evidence that it cannot be approximated by a stronger form of polynomial approximation.*

[12] The first two of the Crescenzi and Panconesi reductions are credited to Orponen and Mannila, ([OrMa 87]), and are very similar to the L-reductions of [PaYa 88].

Orponen and Mannila show that TRAVELING SALESPERSON and $0-1$ INTEGER PROGRAMMING are NPO complete. Following this, using techniques similar to those used by Paz and Moran for showing that MAX WEIGHTED VARIABLE is complete for NPO under measure preserving reductions, Crescenzi and Panconesi showed that BOUNDED MAX WEIGHTED VARIABLE is complete for APX, and that LINEAR BOUNDED MAX WEIGHTED VARIABLE (which they call LINEAR BSAT) is $PTAS$ complete under the reductions they use. They also show that NPO contains various problems which are not complete for NPO and which are not in APX unless $P = NP$.

Question 6, ([CrPa 88]). *Are there natural incomplete problems for NPO or natural sets which are complete for $PTAS$?*

Question 7. *The reductions used by Orponen and Mannila and those used by Paz and Moran are different. What is the relationship between the reductions and between the complete sets for NPO under these reductions? Almost every paper we have discussed introduces one or more new restricted reductions to study approximation problems. Although many ideas for the restrictions appear similar, it is not clear what the actual relations are among the different restricted reductions used by different authors. It would be interesting to know what the relationship between these various restrictions really is and whether all are necessary to obtain the results which we have surveyed. Is there some clear notion of what a "correct" reduction should be? Of the results surveyed here, the work of Crescenzi and Panconesi provides the strongest justification for the "correctness" of the Orponen-Mannila type reducibilities, (and, because of the similarities of the reductions, thus also for the L-reductions of Papadimitriou and Yannakakis). These reducibilities seem to give appropriately fine distinctions without begin overly cumbersome.*

While discussing restricted forms of reducibilities which preserve approximability properties, we should mention the early work of Ausiello, D'Atri and Protasi. In ([ADP 80]) they obtained information about the approximability of NP optimization problems by giving a very detailed examination of the combinatorial and internal structures both of the problems being reduced and of the reductions. In this paper they introduced the notion of the *structure* of an optimization problem, the notion of a *convex* optimization problem, and the notion of a ***structure preserving reduction.*** Informally, the structure of an instance x of an optimization problem is a list which contains the number of feasible solutions for x associated with each admissible value of the measure m. An optimization problem is said to be *convex* if for each instance x there is at least one feasible solution y such that $m(x, y) = h$ for every h such that $worst(x) \leq h \leq opt_C(x)$.

Using much more restricted reductions than the restricted reductions we have already discussed, Ausiello, D'Atri and Protasi specify reductions by pairs of polynomially computable functions $\langle f_1, f_2 \rangle$ such that an optimization problem A is reducible to an optimization problem B if each instance x, of A, is mapped by f_1 to an instance of B, f_2 carries the *measures* of feasible solutions of x to the *measures* of feasible solutions of $f(x)$. A polynomially structure preserving reduction is such a reduction which maps an optimization problem A to an optimization problem B in such a way that each instance of A is mapped to an instance of B with *exactly* the same structure. They showed that parsimonious reductions which satisfy very strong linearity conditions are always structural preserving.[13]

Ausiello, *et. al.*, useed these reducibilities to examine relationships among a variety of *NP* complete optimization problems. While from our current perspective the conditions on their structure preserving reductions may seem too restrictive to still be of general interest, Ausiello *et. al.* were among the first to give conditions under which combinatorial problems have similar approximability properties:[14]

Theorem. *Let A and B be two convex NP optimization problems which are both maximization or minimization problems. If there are reductions $\langle f_1, f_2 \rangle$ from A to B, and $\langle g_1, g_2 \rangle$ from B to A such that:*

- *both are structure preserving,*
- *both are strictly monotone,*
- *both are essentially linear:*

$$f_2(x,k) = a(x) + k \ , \quad g_2(y,h) = b(y) + h \ , \quad \text{and} \quad a(x) \geq -b(f_1(x)),$$

---

[13] P. Orponen has pointed out to us that, rather interestingly, motivated partially by Simon's results on "parsimonious" reductions, Lynch and Lipton, ([LyLi 78]), also investigated "structure preserving reductions" which are very similar in spirit to those introduced by Ausiello, *et.al.* However, unlike Ausiello, *et.al.*, Lynch and Lipton do not discuss the application of structure preserving reductions to approximating solutions for optimization problems.

[14] [ADP 80] gives a more general theorem than state here, but we have adopted the following version because it gives a good intuition while avoiding heavier notation.

The measure used here to evaluate the quality of an approximation algorithm is different from the standard one (namely the ratio $\frac{opt_C(x)}{approx(x)}$, or its reciprocal) introduced by Johnson. In particular, the measure used by Ausiello, D'Atri and Protasi is $\frac{opt_C(x)-approx(x)}{opt_C(x)-worst(x)}$, where worst$(x)$ is the value of the worst feasible solution for x. This measure is not only intuitively very appealing, but it has the nice property that it is symmetric with respect to maximization and minimization; a property that does not hold for the standard measure. Since the function *worst* can usually be taken to be zero for maximization problems, these measures are usually equivalent for maximization problems. But they are *not* equivalent for minimalization problems. Although this measure seems never to have been subsequently used, [ADP 80; p 145] contains a useful discussion of why the authors believe this measure should be preferred for minimization problems.

then B has a polynomial approximation algorithm if and only if A has a polynomial approximation algorithm.

Ausiello, D'Atri and Protasi then went on to group a variety of well-known *NP* complete optimization problems into distinct equivalence classes using this theorem, and they prove that some of the classes are distinct under these reductions.

4. BIBLIOGRAPHY

[ADP 80] G. Ausiello, A. D'Atri and M. Protasi, "Structure preserving reductions among convex optimization problems," *J Comput and System Sci*, **21** (1980), 136–153; (first appeared in *Proc 4th ICALP*, Lecture Notes in Computer Science, **52** (1977), 45-60.)

[BeHa 77] L. Berman and J. Hartmanis, "On isomorphism and density of *NP* and other complete sets," *SIAM J Comput,* **1** (1977), 305-322.

[BBJSY 89] A. Bertoni, D. Bruschi, D. Joseph, M. Sitharam and P. Young, "Generalized Boolean hierarchies and Boolean hierarchies over *RP*," *Univ Wisconsin CS Dept Tech Report*, **#809**, 1–40; (short abstract to appear in *FCT '89 Proceedings.*)

[CGHHSWW 89] J. Cai, T. Gundermann, J. Hartmanis, L. Hemachandra, V. Sewelson, K. Wagner, and G. Wechsung, "The Boolean hierarchy II: applications," *SIAM J Comput,* **7** (1989), 95–111.

[CrPa 88] P. Crescenzi and A. Panconesi, "Completeness in approximation classes," to appear in *FCT '89 Proceedings,* (1989), 16 pages.

[Fa 74] R. Fagin, "Generalized first-order spectra, and polynomial-time recognizable sets," *Complexity and Computations*, R. Karp (ed.), AMS, (1974).

[GaJo 79] M. Garey and D. Johnson, "Computers and Intractability: A Guide to the Theory of *NP*-Completeness," Freeman, 1979.

[GoSa 76] T. Gonzales and S. Sahni, "P-complete approximation problems," *J ACM*, **23** (1976), 555–565.

[HoKe 82] B. Hodgson and C. Kent, "An arithmetical characterization of *NP* ," *Theor Comput Sci*, **21** (1982), 255–267.

[HoSh 86] D. Hochbaum and D. Shmoys, "A unified approach to approximation algorithms for bottleneck problems," *J ACM*, **33** (1986), 533–550.

[IbKi 75] O. Ibarra and C. Kim, "Fast approximation algorithms for the knapsack and sum of subsets problems," *J ACM*, **22** (1975), 463–468.

[JAMS 89] D. Johnson, C. Aragon, L. McGeoch, and C. Schevon, "Optimization by simulated annealing: an experimental evaluation," *Operations Research*, to appear.

[Jo 74] D. Johnson, "Approximation algorithms for combinatorial problems," *J Comput and System Sci*, **9** (1974), 256–278.

[JPY 88] D. Johnson, C. Papadimitriou and M. Yannakakis, "How easy is local search?," *J Comput and System Sci*, **37** (1988), 79–100; (first appeared in *Proc 26th IEEE Symp Foundations of Comput Sci*, (1985), 39–42.)

[JoYo 85] D. Joseph and P. Young, "Some remarks on witness functions for polynomial reducibilities in *NP*," *Theor Comput Sci,* **39** (1985), 225-237.

[KaKa 82] N. Karmakar and R. Karp, "An efficient approximation scheme for the one-dimensional bin-packing problem," *IEEE 23rd Symp Foundations of Comput Sci*, (1982), 312–320.

[KGV 83] S. Kirkpatrick, C. Gelat, and M. Vecchi, "Optimization by simulated annealing," *Science*, **220** (1983), 671-680.

[Kr 88] M. Krentel, "The complexity of optimization problem," *J Comput and System Sci*, **36** (1988), 490–509; (first appeared in *Proc 18th ACM Symp Theory on Comput*, (1986), 69–76.)

[Kr 89] M. Krentel, "On finding locally optimal solutions," *Proc Structure in Complexity Conference*, (1989), 132-137.

[LeMo 81] E. Leggett and J. Moore, "Optimization problems and the polynomial time hierarchy," *Theor Comput Sci*, **15** (1981), 279–289.

[LyLi 78] N. Lynch and R. Lipton "On structure preserving reductions," *Siam J Comput,* **7** (1978), 119-126.

[OrMa 87] P. Orponen and H. Mannila, "On approximation preserving reductions: complete problems and robust measures," Tech Report, University of Helsinki, 1987.

[Ni 75] R. Nigmatullin, "Complexity of the approximate solution of combinatorial problems," *Dokl Akad Nauk*, SSSR **224**, (1975) 289–292 (in Russian), English translation in *Soviet Math Dokl*, **16**, 1199–1203.

[Pa 84] C. Papadimitriou, "On the complexity of unique solutions," *J ACM*, **31** (1984), 392–400.

[PaMo 81] A. Paz and S. Moran, "Non deterministic polynomial optimization problems and their approximation," *Theor Comput Sci*, **15** (1981), 251–277; (first appeared in *Proc 4th ICALP*, Lecture Notes in Computer Science, **52** (1977), 45-60.)

[PaWo 85] C. Papadimitriou and D. Wolfe, "The complexity of facets resolved," *J Comput and System Sci*, **37** (1988), 2–13; (first appeared in *Proc IEEE 26th Symp on Foundations of Comput Sci*, (1985), 74–78).

[PaYa 84] C. Papadimitriou and M. Yannakakis, "The complexity of facets (and some facets of complexity)," *J Computer and System Sciences*, **28** (1984), 244–259, (first appeared in *Proc 14th ACM Symp Theory of Comput*, (1982), 255–260.)

[PaYa 88] C. Papadimitriou and M. Yannakakis, "Optimization, approximation, and complexity classes," *Proc 20th ACM Symp Theory Comput*, (1988), 229–234.

[Sa 76] S. Sahni, "Algorithms for scheduling independent tasks," *J ACM*, **23** (1976), 116–127.

[Wa 86] K. Wagner, "More complicated questions about maxima and minima, and some closure properties of *NP*," *Theor Comput Sci*, **51** (1987), 53–80; (first appeared in *Proc 13th ICALP,* Lecture Notes Computer Science, **226** (1986), 434–443.)

[Wa 88] K. Wagner, "Bounded query computations," *Proc Structure in Complexity Conference,* (1988), 260–277.

[WaWe 85] K. Wagner and G. Wechsung, "Computational complexity," D. Reidel Publishing, 1986.

APPENDIX

Decision Problems

MINIMAL-k-UNCOLORABILITY

Instance: a graph G.

Question: Is G uncolorable with k colors, but by removing any node from G the resulting graph is k colorable?

MINIMAL UNSATISFIABILITY

Instance: a Boolean formula ϕ in conjunctive normal form with at most three literals per clause and at most two occurrences of each literal.

Question: Is ϕ unsatisfiable, but removing any clause renders it satisfiable?

SAT-UNSAT

Instance: two Boolean formulas ϕ and ψ.

Question: Is ϕ satisfiable and ψ is unsatisfiable?

SUM OF SUBSETS

Instance: a finite set S of positive integers and a positive integer b.

Question: Is there a set $A \subseteq S$ such that the sum of the elements is equal to b?

Functional Version of Optimization Problems

BOUNDED MAX WEIGHTED VARIABLES

(Same as MAX WEIGHTED VARIABLES except the input includes a constant W satisfying $\sum w_i \leq 2 * W$, the weight W is attached to any *unsatisfying* assignment, and the output is the maximum weight over *all* assignments, whether satisfying or unsatisfying.)

INTEGER PROGRAMMING

Instance: integer matrix A and integer vectors B and C.

Output: the maximum value of $C^T x$ over all vectors x of integers subject to the linear constraint $Ax \leq B$.

FLIP

Instance: a circuit with n inputs and n outputs.

Output: an input whose output (when viewed as a binary integer) cannot be reduced by flipping any single bit of the input.

LINEAR BOUNDED MAX WEIGHTED VARIABLES

(Same as BOUNDED MAX WEIGHTED VARIABLES except that the constant W satisfies $\sum w_i \leq (1 + \frac{1}{n-1}) * W$.)

LONGEST CYCLE

Instance: graph G.

Output: the length of the longest cycle in G.

MAX CLIQUE

Instance: graph G.

Output: the size of the largest clique in G.

MAX CUT

Instance: graph G with integer weights on the edges.

Output: the maximum k obtainable by partitioning the graph G into two subgraphs G_1 and G_2 and then summing the weights of the edges that have one endpoint in G_1 and the other endpoint in G_2.

MAX KNAPSACK

Instance: integers $x_1, \ldots, x_n, N$.

Output: the largest value of $\sum_{i \in S} x_i$ for $S \subseteq 1, \ldots, n$, which is less than N.

MAX SAT

Instance: Boolean formula in conjunctive normal form.

Output: the maximum number of simultaneously satisfiable clauses.

MAX SAT ASSIGN

Instance: Boolean formula $\phi(x_1, \ldots, x_n)$.

Output: the lexicographic maximum assignment which satisfies ϕ, or 0 if ϕ is not satisfiable.

MAX WEIGHTED CLAUSES

Instance: Boolean formula in conjunctive normal form with weights on the clauses.

Output: the maximum weight of any satisfying assignment, where the weight of an assignment is the sum of weights on the true clauses.

MAX WEIGHTED VARIABLES

Instance: Boolean formula in conjunctive normal form with nonnegative weights, w_i, on the variables, x_i.

Output: the maximum weight of any satisfying assignment, where the weight of an assignment is the sum of weights on the true variables; output negative if there is no satisfying assignment.

MIN BIN PACKING

Instance: finite set of items U, an integer "size" for each $u \in U$ and a positive integer bin capacity B.

Output: the minimum k such that there exists a partition of U into disjoint sets $U_1, \ldots, U_k$ and the sum of the size of the items in each U_i is less than or equal to B.

MIN CLIQUE COVER

Instance: a graph G.

Output: the minimum number of disjoint cliques in G whose union is G.

MIN TRAVELING SALESPERSON

Instance: graph G with integers weights on the edges.

Output: the length of the shortest traveling salesperson tour in G.

MIN VERTEX COVER

Instance: a graph G.

Output: the size of the minimum cover of G.

SORTING WITHIN DISTANCE BOUND ON A MESH-CONNECTED PROCESSOR ARRAY

Bogdan S. Chlebus

Instytut Informatyki, Uniwersytet Warszawski
PKiN, p. 850, 00-901 Warszawa, Poland

Abstract An algorithm is developed which sorts random sequences of keys on the $n \times n$ square mesh in the expected time $2n$. The algorithm is shown to be optimal, that is, the matching $\Omega(2n)$ lower bound on the expected-time of algorithms sorting randomly ordered inputs is proved.

1 Introduction

There are a number of results concerning the worst-case complexity of sorting on the $n \times n$ mesh of processors known in the literature. The first linear-time algorithm of complexity $6n$ was given by Thompson and Kung [10]. Schnorr and Shamir [9] and Ma, Sen and Scherson [8] developed $3n$ and $4n$ algorithms, respectively, and proved the matching lower bounds. The $3n$ lower bound was also proved by Kunde [7]. The apparent discrepancy between two distinct but optimal bounds $3n$ and $4n$ follows from certain additional technical constraints regarding indexing of processors and feasibility of concurrent routings in different directions. More on this can be found in the paper by Chlebus and Kukawka [2] giving an overview of the existing results and a classification of various models and problems concerning sorting on the grid of processors. The lower bounds for various indexings of processors were studied by Han and Igarashi [4]. The lower bound of $2.25n$ valid for all indexings of processors was proved by Han and Igarashi [4] and Kunde [6]. Quite recently this bound was improved to 2.27 by Han, Igarashi and Truszczynski [5].

In this paper we investigate the problem of sorting if the input is assumed to be a random sequence of keys. A certain indexing of processors is introduced and an algorithm is developed which sorts in this ordering in the expected time $2n$. We also show that any sorting algorithm must have at least $2n$ expected behaviour if it is evaluated on random inputs. Actually we exhibit sharp bounds on the involved probabilities, proving them to be exponentially close to 1.

2 Algorithm Nested Snake

The mesh of processors which is our model of a parallel computer is a $n \times n$ grid with processors placed at its vertices and connected to their neighbours along the links of the grid. The network operates in a synchronous way. In the starting position each processor has exactly one key and the whole sequence of keys is assumed to be in random order.

When referring to indexing of vertices of a rectangular grid, we call it a *row-major* one if the elements in row i precede all the elements in row j for all $i < j$. In the *snake-like* row-major indexing the rows are moreover ordered from left to right and from right to left alternatingly in consecutive rows. The *column-major snake-like* indexing is defined similarly. Define the *n^a-nested snake like* ordering as follows. Fix a real number a such that $0 < a < 1$ and divide the rows of the mesh into consecutive strips of width n^a. Index the processors in each strip according to the column-major snake, ordering the columns from left to right and from right to left in alternating n^a-strips. This defines two snake-like structures: there is a snake inside each strip, and the ordering of columns inside the strips makes another snake (see Figure 1). Consider the leftmost and rightmost $n^a \times n^a$ square blocks inside each strip. The combined rightmost blocks from the strips 1 and 2, 3 and 4, and so on, together with the combined leftmost blocks from strips 2 and 3, 4 and 5, and so on, form the *joining blocks*. The *joining indexing* of joining blocks is inherited from the nested snake-like ordering (see Figure 2). The algorithm below uses the vertical n^a-strips which are blocks of consecutive columns of width n^a.

Algorithm *Nested Snake*

1. Sort the vertical n^a-strips in the row-major order.
2. Sort the horizontal n^a-strips in the order inherited from the n^a-nested snake-like order.
3. Sort the joining blocks in the joining orders.
4. If the keys are not sorted then invoke any linear sorting algorithm.

We shall show that the above algorithm can be implemented so that its expected performance is $2n$. First observe that Steps 1 and 2 each can be performed in time $n + o(n)$. To this end use the algorithm of Schnorr and Shamir [9] which sorts a rectangular grid of processors with x rows and y columns in the snake-like row-major ordering in time $x + 2y + o(x + y)$. Since in our case $y = n^a$, we obtain time $n + o(n)$. Step 3 can be performed in time $O(n^a)$ using any linear-time sort. In the next section we show that the input is already sorted after steps 1 through 3 with an overwhelming probability .

3 Probabilistic Analysis of Nested Snake

First we shall show that after Step 1 most of the keys will be in their proper vertical strips with large probability.

In the estimation of distributions of keys among the vertical strips we bound the real distributions by related binomial distributions. This enables us to use the Chernoff bounds on the tails of a binomial distribution. The basic idea behind this aproximation is the following. Suppose that we select a key, say x, and consider all the keys smaller than x in the input. We need bounds on the distribution of these keys among the vertical n^a-strips, more precisely, the difference between the maximal and minimal numbers of keys in two n^a-strips. The keys may be placed in the strips one by one as follows: put the consecutive key in one of the strips, selecting a strip with probability proportional to the number of vacant places in it.

Instead of this process, which models the true distribution, we consider the following approximation: each strip may be selected with the same probability. Observe that the true distribution is more balanced: placing a key in the strip decreases the probability that the next key will be placed there. Therefore the estimates obtained in this section by way of binomial distributions are also valid for the distributions given by random permutations.

The Chernoff bounds are the following: if X is a random variable equal to the number of successes in n Bernoulli trials, each with probability p of success then

$$\mathrm{Prob}(X \geq (1+\epsilon)np) \leq e^{-\epsilon^2 np/3}, \text{ and}$$

$$\mathrm{Prob}(X \leq (1-\epsilon)np) \leq e^{-\epsilon^2 np/2},$$

for each $0 \leq \epsilon \leq 1$. For a proof and more references see the recent exposition paper by Hagerup and Rüb [3] (cf. [1]). In the sequel we use the notation $\log n$ for $\log_2 n$ and $\ln n$ for the natural logarithm of n.

Lemma 1 *Suppose that $k = nh$ keys are distributed independently and with equal probability among n^{1-a} vertical strips, where $0 < a < 1$ and $1 \leq h \leq n$. Then, for each $0 < c < a$, $1 - \frac{a-c}{2} < b < 1$ and sufficiently large n, with probability at least $1 - e^{-n^c h}$ the number S of keys in any vertical strip satisfies the inequalities*

$$n^a(h - (1-a)n^b \log n) \leq S \leq n^a(h + (1-a)n^b \ln n).$$

Proof. First divide the keys into two parts taking each key either to the eft or to the right with probability 1/2. Then again divide each part into two halves. After $\log n^{1-a}$ steps and the total of $n^{1-a} - 1$ divisions the keys will be in n^{1-a} parts.

Consider the first division. From the Chernoff bounds the number X of keys in one part is at most $\frac{1}{2}(1+\epsilon)k$ with probability $\geq 1 - e^{-\epsilon^2 k/4}$. Hence the inequality $\frac{1}{2}(1-\epsilon)k \leq X \leq \frac{1}{2}(1+\epsilon)k$ holds with probability $\geq 1 - 2e^{-\epsilon^2 k/6}$.

Each key takes part in at most $\log(n^{1-a}) = (1-a)\log n$ divisions. Repeating the above argument we obtain certain bounds on the number S of elements in each vertical strip. They hold with the respective probabilities which we estimate later on. Iterating the Chernoff formula we obtain

$$S \leq \frac{k(1+\epsilon)^{\log(n^{1-a})}}{2^{\log(n^{1-a})}} = \frac{nh(1+\epsilon)^{(1-a)\log n}}{n^{1-a}} = hn^a(1+\epsilon)^{(1-a)\log n}.$$

Set $\epsilon = n^{b-1}$ for $1 - \frac{a-c}{2} < b < 1$. Observe that

$$(1+n^{b-1})^{(1-a)\log n} = n^{(1-a)\log(1+n^{b-1})} \leq n^{(1-a)n^{b-1}} =$$

$$= \exp((1-a)n^{b-1}\ln n) \leq 1 + (1-a)n^{b-1}\ln n$$

for sufficiently large n. Hence

$$S \leq hn^a(1 + (1-a)n^{b-1}\ln n) \leq n^a(h + (1-a)n^b \ln n).$$

Similarly we have the reverse inequality $S \geq hn^a(1-\epsilon)^{(1-a)\log n}$. Since $(1-\epsilon)^{(1-a)\log n} \geq 1 - \epsilon(1-a)\log n$, setting $\epsilon = n^{b-1}$ we get

$$S \geq n^a(h - (1-a)n^{b-1}h\log n) \geq n^a(h - (1-a)n^b \log n).$$

The probability that the above inequalities hold for each strip can be bounded from below by the product $\prod_{i=1}^{n^{1-a}-1}(1 - 2e^{-\epsilon^2 y_i/6})$, where y_i is the lower bound on the number of keys in the ith division given by the Chernoff formula and depending on the "depth" of this division. Since $(1-\epsilon)^{(1-a)\log n} \geq \frac{1}{2}$ for sufficiently large n, we may use the inequality $y_i \geq \frac{1}{2}hn^a$.

Therefore the overall probability is at least

$$(1 - 2e^{-n^{2(b-1)+a}h/12})^{n^{1-a}} \geq 1 - 2n^{1-a}e^{-n^{2b-2+a}h/12} \geq 1 - e^{-n^c h}$$

for sufficiently large n. □

Using the above lemma we can estimate the number of keys which will be in their proper vertical strips after Step 1. Take nh smallest keys for $h = n/2$, and distribute them randomly among vertical strips. There will be $n^a(h \pm (1-a)n^b \log n)$ keys in each strip with the respective probability given by the above lemma. Fill the remaining places in the strips with larger keys. After sorting the vertical strips most of the keys will fall properly either in the lower or upper half, the remaining keys lying in the horizontal "dirty" strip of width $2(1-a)n^b \log n$ in the middle of the mesh, with the respective probability. Cut off all the keys which fall into this middle horizontal strip after sorting the vertical n^a-strips, and for the remaining upper and lower parts repeat the same operation. Again we obtain two dirty strips, this time however of width at most twice as large as the previous one, each width being a valid bound with the probability equal to the square of the probability given by lemma 1.

Repeating this a number of times we obtain that the maximal width of any dirty strip is at most

$$2(1-a)n^b \log n \log(n^{1-a}) = 2n^b((1-a)\log n)^2$$

with the probability

$$\geq (1 - e^{-n^c})^{n^{1-a}} \geq 1 - n^{1-a}e^{-n^c} \geq 1 - e^{-n^d}$$

for some $0 < d < c$ and sufficiently large n. The number of keys in each dirty strip is

$$2nn^b((1-a)\log n)^2 = 2n^{1+b}((1-a)\log n)^2 < n^{2a}$$

for $b < 2a - 1$ and sufficiently large n. These many keys will be moved to joining blocks by Step 2 and then put in the proper order by Step 3. This completes the proof of the following theorem.

Theorem 1 *For each rational constant a such that $\frac{1}{2} < a < 1$ there is a constant $d > 0$ such that Nested Snake sorts a random sequence of keys in the $n \times n$ mesh of processors in the n^a-nested snake-like order within $2n + o(n)$ steps with the probability at least $1 - e^{-n^d}$.*

4 Lower bounds

In this section we present some lower bounds on sorting algorithms which use randomness as a resource.

Theorem 2 *Any algorithm sorting $n \times n$ mesh must have at least $2n - o(n)$ expected behaviour on randomly ordered inputs.*

Proof. We estimate the probability that the maximum distance bound is $2n - o(n)$.

Select a number a such that $0 < a < 1$, and consider the $n^a \times n^a$ block of processors in the left-upper corner of the mesh. (Some conditions on number a will be given later.) Let X be the set of keys which will be moved to this block after sorting is completed. A random distribution of the input keys among the processors can be obtained by first placing the elements of X randomly among the processors, and then similarly distributing the remaining keys. The distribution of X can be done as follows: first put the keys one by one, each into one of the four quadrants, selecting a quadrant with the probability proportional to the number of vacant places. Then continue repeating this operation for smaller quadrants untill all the keys are assigned to processors. Observe that if the keys selected for some square portion of the mesh are divided into four parts, then each key falls into the lower-right quadrant with probability at least $1/5$.

From the Chernoff bound we see that there are at least $(1-\epsilon)n^{2a}/5$ keys in the right-lower quadrant with probability $\geq 1 - e^{-\epsilon^2 n^{2a}/10}$, for each $0 \leq \epsilon \leq 1$. Let $0 < b < 1$ and consider the first $k = (1-b)\log n$ stages of distributing the keys into quadrants. Combining the Chernoff bounds we have that there are at least $n^{2a}(\frac{1-\epsilon}{5})^k$ keys in the $n^b \times n^b$ right-lower corner of the mesh with some probability P, where

$$P \geq \prod_{i=0}^{k-1}(1 - e^{-\epsilon^2 n^{2a}(\frac{1-\epsilon}{5})^i/10}).$$

Transform $n^{2a}(\frac{1-\epsilon}{5})^k$ to obtain $n^{2a}(\frac{1-\epsilon}{5})^{(1-b)\log n} = n^{2a+(1-b)\log(\frac{1-\epsilon}{5})}$

It follows from straightforward calculations that is $b > 1 - 2/\log 5$ then there are numbers a and ϵ between 0 and 1 such that the inequality $2a + (1-b)\log(\frac{1-\epsilon}{5}) = c > 0$ holds. Therefore with probability

$$P \geq (1 - e^{-\epsilon^2 n^{2a}(\frac{1-\epsilon}{5})^k/10})^k \geq 1 - ke^{-\frac{\epsilon^2}{10}n^c} \geq 1 - e^{n^d},$$

for some $d > 0$ and sufficiently large n, there are at least n^c keys from X in the right-lower $n^b \times n^b$ corner of the mesh. Their distance to the left-upper $n^a \times n^a$ corner is at least $2n - (n^a + n^b) = 2n - o(n)$. □

Theorem 1 combined with Theorem 2 show that $2n$ is the optimal expected-time complexity for sorting random sequences. This is in sharp contrast with the deterministic case where $3n$ is the best known time behaviour of any sorting algorithm, and $2.27n$ is the largest known lower bound for all indexings.

5 Remarks

It is an interesting open problem if there is a randomized algorithm, that is an algorithm using a random sequence of bits, which does better than expected $3n$ time for all inputs. The following result provides some information concerning randomized sorting in the snake-like row-major order.

Theorem 3 *For each randomized algorithm sorting the $n \times n$ mesh of processors in the snake-like row-major order there is an input for which the expected time of the algorithm is at least $2.5n$.*

Proof. The proof will be given in the final version of the paper. □

References

[1] D. Angluin, and L.G. Valiant, Fast probabilistic algorithms for Hamiltanian circuits and matchings, J. Comput. System Sci. 18 (1979) 155-193.

[2] B.S. Chlebus, and M. Kukawka, A guide to sorting on mesh-connected processor arrays, Computers and Artificial Intelligence, to appear.

[3] T. Hagerup, and C. Rüb, A guided tour of Chernoff bounds, report, 1989.

[4] Y. Han, and Y. Igarashi, Time lower bounds for parallel sorting on a mesh connected processor array, Proceedings AWOC'88, pp. 434-439.

[5] Y. Han, Y. Igarashi, and M. Truszczynski, Indexing schemes and lower bounds for sorting on a mesh-connected computer, report, 1988.

[6] M. Kunde, Bounds for l-selection and related problems on grids of processors, Proceedings Parcella'88, Akademie-Verlag, Berlin, 1988, pp. 298-307.

[7] M. Kunde, Lower bounds for sorting on mesh-connected architectures, Acta Informatica 24 (1987) 121-130.

[8] Y. Ma, S. Sen, and I.D. Scherson, The distance bound for sorting on mesh connected processor arrays is tight, Proceedings FOCS 1986, pp. 255-263.

[9] C.P. Schnorr, and A. Shamir, An optimal sorting algorithm for mesh-connected computers, Proceedings STOC 1986, pp.255-263.

[10] C.D. Thompson, and H.T. Kung, Sorting on a mesh connected parallel computer, Comm. ACM 20 (1977) 263-271.

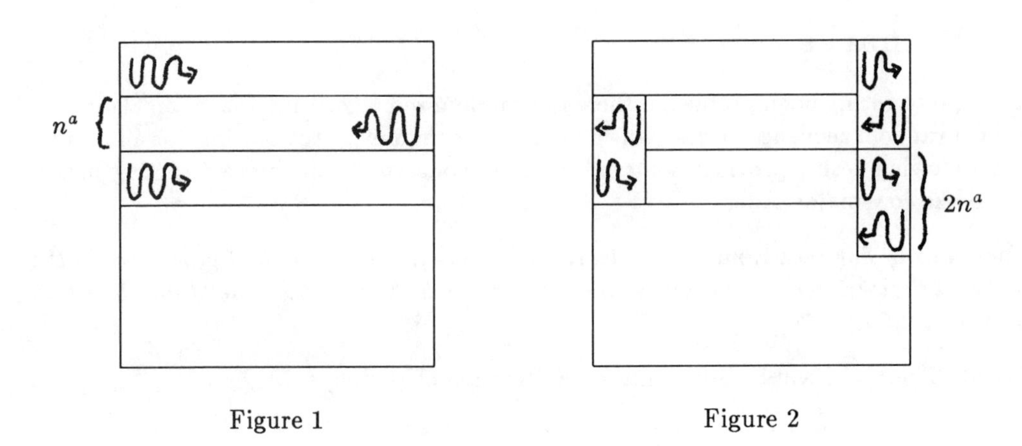

Figure 1

Figure 2

Local Insertion Sort Revisited

Jyrki Katajainen*, Christos Levcopoulos, and Ola Petersson
Department of Computer and Information Science
Linköping University
S-581 83 Linköping, Sweden

Abstract

Two measures of presortedness, *Dist*(X) and *logDist*(X), are presented. The measures are motivated by a geometric interpretation of the input sequence X. Using these measures, an exact characterization of the local insertion sort algorithm of Mannila is achieved. Variants of local insertion sort, using many fingers (i.e., pointers into a finger search tree) instead of only one, are suggested. The number can either be fixed beforehand or dynamic, i.e., vary during the sorting process. Different strategies for moving the fingers yield different algorithms. Some general conditions which all strategies should satisfy are stated, and two such strategies are given. Furthermore, a method for designing dynamic algorithms that allocate an optimal number of fingers is provided. For any specific strategy satisfying our conditions, this method yields an optimal algorithm.

Key words: sorting algorithm, measures, presortedness, local insertion sort, finger search trees, geometric interpretation.

1 Introduction

A *measure of presortedness* is an integer function on a permutation π of a totally ordered set that reflects how much π differs from the total order. Examples of presortedness measures include the number of runs or inversions. The term presortedness was coined by Mehlhorn [Meh79], who used the number of inversions in the file as a measure. Mehlhorn [Meh84a, Section III.5.3] also gave an algorithm, A-Sort, which is efficient with respect to this measure. Intuitively, an algorithm is efficient with respect to a measure of presortedness if it sorts all permutations, but performs particularly well on permutations that have a high degree of presortedness. Mannila [Man85] formalized the concept of presortedness and gave algorithms that are efficient with respect to several measures. Recently, Estivill-Castro and Wood [EW87], Skiena [Ski88], and Levcopoulos and Petersson [LP89] have considered some other measures. Cook and Kim [CK80] have studied the problem empirically.

*On leave from Department of Computer Science, University of Turku, Lemminkäisenkatu 14A, SF-20520 Turku, Finland. The research of this author was supported by the Academy of Finland.

This paper includes a description of two measures of presortedness, motivated by a geometric interpretation of the input. The measures, called $Dist(X)$ and $logDist(X)$, are used to characterize the performance of the local insertion sort algorithm of Mannila [Man85]. In Section 2 we recall local insertion sort. We also state the precise definitions of the measures, study some of their basic properties, and analyse the performance of local insertion sort as a function of the presortedness measures. A brief comparison between the new measures and other measures proposed in the literature is made. Our basic observation is that some of the previous measures are included in the measure $Dist(X)$, and that the above mentioned characterization of local insertion sort is not possible with the previous ones.

In Section 3 we present some modifications of the local insertion sort algorithm. The asymptotic running time of the algorithms can never be slower than that of local insertion sort, and the class of input sequences which can be sorted fast will increase considerably. The idea is to implement the local insertion sort algorithm with a finger search tree and maintain many fingers simultaneously, instead of only one, as done originally. Different strategies for moving the fingers are discussed, and some general conditions, which should hold for any reasonable strategy, are stated. A combination of a method similar to the *least recently used*-heuristic (see e.g., [PS86, Section 6.6]), and the idea of always moving the closest finger, turns out to be a promising heuristic. The case when the number of fingers vary during the sorting process is explored as well. We provide a method for designing such dynamic algorithms which, regardless of strategy used for moving the fingers, allocates an optimal number of fingers and sorts in optimal time.

In Section 4 we discuss the results and outline some possible extensions.

2 Local Insertion Sort

In this section, we first recall the local insertion sort algorithm proposed by Mannila [Man85]. Its implementation, by finger trees, is briefly discussed. We then state the definitions of the presortedness measures, $Dist(X)$ and $logDist(X)$, and point out their geometric motivation. In Section 2.3 the running time of local insertion sort, as a function of these measures, is explored. Finally, we make some observations regarding the relation between the new measures and other known measures of presortedness.

2.1 The Algorithm

Let $X = \langle x_1, \ldots, x_n \rangle$ be a *sequence* of n elements drawn from some totally ordered set, that is, for all $i, j \in \{1, \ldots, n\}$, $x_i \leq x_j$ or $x_j \leq x_i$. For simplicity, we assume that the elements are integers.

Insertion sort algorithms insert the elements, one by one, starting with x_1, into a sorted sequence, formed by the already inserted elements. When the already inserted elements are stored in an array, we obtain straight insertion sort (or binary insertion sort) (see any standard textbook on sorting and searching, for example [Knu73, Section 5.2.1]). An asymptotically more efficient way to implement the already inserted elements is by using search trees. It was proposed by Mannila [Man85] that when the input sequence is presorted, one should use so-called *finger search trees* (finger trees) [GMPR77] (see also [TVW88, Appendix]). When a finger search tree is used to implement the already inserted elements, the insertion sort algorithm is called *Local Insertion Sort* (LIS).

One way of implementing a *finger tree* is by a level-linked (2–4)-tree. Each node of the tree except the leaf nodes, has 2, 3 or 4 child nodes. The level linking means that each node contains pointers to the left and right neighbouring nodes on the same level. In addition, every node contains a pointer to its father. The elements are stored at the leaf level and a finger is a pointer to a leaf node. Each internal node v contains the smallest and largest elements of the subtree rooted at v. We use one special finger; the *latest insertion finger*, through which the tree is accessed. When inserting x_i, we must first search for the appropriate slot for x_i starting from the node pointed at by the latest insertion finger, and proceeding upwards until an interval inside which the new element resides is found. The search for the exact position then continues downward in this subtree as an ordinary range search. Excluding the search time, the total time needed for performing n insertions, i.e., the time needed for node splittings, is $\Theta(n)$. A search for a slot which is d positions away from the latest insertion finger takes time $\Theta(1 + \log d)$[1] (for details, see [Meh84a, Section III.5.3.3]). LIS is thus adaptive, in the sense that if consecutive elements are close to each other in the sorted output, the sorting process will be fast.

2.2 Measures of Presortedness

In this subsection we define two measures of presortedness and study their basic properties.

We begin with some preliminary definitions. For two sequences, $X = \langle x_1, \ldots, x_n \rangle$ and $Y = \langle y_1, \ldots, y_m \rangle$, their *catenation* XY is the sequence $\langle x_1, \ldots, x_n, y_1, \ldots, y_m \rangle$. Furthermore, let $|X|$ denote the *length* of X and $|S|$ the *cardinality* of a set S. If $Z = \langle x_{f(1)}, x_{f(2)}, \ldots, x_{f(m)} \rangle$ and $f : \{1, \ldots, m\} \to \{1, \ldots, n\}$, $m \leq n$, is injective and monotonically increasing then Z is called a *subsequence* of X. In particular, Z is a *consecutive subsequence* of X if there exists an i, $1 \leq i \leq n - m + 1$, such that $Z = \langle x_i, \ldots, x_{i+m-1} \rangle$.

Mannila [Man85] proposed some general conditions which any measure of presortedness should satisfy. Our approach will be a bit looser. Instead of demanding that some conditions (axioms) are satisfied, we say that a function $m : S^{<N} \to N$, where N denotes the set of natural numbers, and $S^{<N}$, the set of all finite sequences of elements taken from some totally ordered set S, is a *measure of presortedness* if it in some intuitive way quantifies disorder among the input sequence to be sorted.

Now consider the pairs of consecutive elements in X. Each such pair $\{x_i, x_{i+1}\}$, $i \in \{1, \ldots, n-1\}$, can be interpreted as a vertical line segment in the plane, with endpoints (i, x_i) and (i, x_{i+1}) (see Figure 1a). To define a measure of presortedness, we simply count the total number of *proper* intersections that the aforementioned line segments have with the horizontal line seqments $\{(j, x_i) \mid i \leq j \leq n\}$,

[1] $\log n$ is defined as $\max\{1, \log_2 n\}$

$i \in \{1, \ldots, n\}$. An equivalent way of saying this is to map each element x_i into the point (i, x_i) and count the number of points properly contained inside the rectangles $(-\infty, i) \times (\min\{x_i, x_{i+1}\}, \max\{x_i, x_{i+1}\})$, $i \in \{1, \ldots, n-1\}$ (see Figure 1b). In this way sorting can be interpreted as a special geometric problem; a restricted intersection problem or a range search problem. Both of the above problems are well studied in computational geometry. (An interested reader should consult, for example [Meh84b], [PS85], [Ede87].)

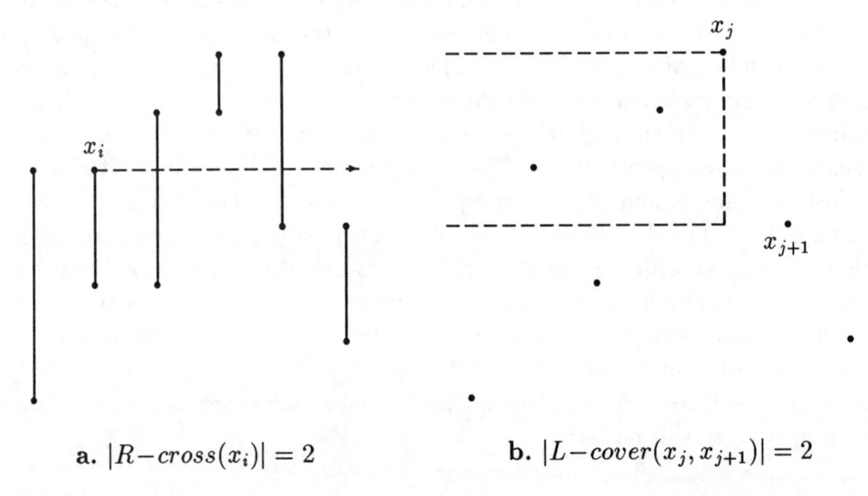

Figure 1. *Geometric interpretation, illustrating the definitions.*

We continue by defining the new measures of presortedness, formally.

Definition 2.1 Let $X = \langle x_1, \ldots, x_n \rangle$ be a sequence. For each $i \in \{1, \ldots, n\}$, let $R(ight){-}cross(x_i) = \{\{x_j, x_{j+1}\} \mid i < j < n \text{ and } \min\{x_j, x_{j+1}\} < x_i < \max\{x_j, x_{j+1}\}\}$.

As suggested in [Man84] we define the following measure of presortedness.

Definition 2.2 For a sequence X, let $Dist(X) = \sum_{i=1}^{|X|} |R{-}cross(x_i)|$.

Definition 2.3 Let $X = \langle x_1, \ldots, x_n \rangle$ be a sequence. For each $i \in \{1, \ldots, n-1\}$ let $L(eft){-}cover(x_i, x_{i+1}) = \{x_j \mid 1 \leq j < i \text{ and } \min\{x_i, x_{i+1}\} < x_j < \max\{x_i, x_{i+1}\}\}$.

For an illustration of the definitions, we refer again to Figure 1. As mentioned above, both $R{-}cross(x_i)$ and $L{-}cover(x_i, x_{i+1})$ can be used to express the measure of presortedness:

Fact 2.4 For any sequence X,

$$Dist(X) = \sum_{i=1}^{|X|} |R{-}cross(x_i)| = \sum_{j=1}^{|X|-1} |L{-}cover(x_j, x_{j+1})|.$$

To motivate the name $Dist(X)$, let us shortly turn our attention to LIS. When x_{j+1} is inserted, the finger is moved a distance corresponding to $|L\text{–}cover(x_j, x_{j+1})|$. $Dist(X)$ thus expresses the total distance covered by the finger, when LIS is applied to sort X.

To get an *exact* characterization of the performance of LIS we define the following measure.

Definition 2.5 For a sequence $X = \langle x_1, \ldots, x_n \rangle$, let

$$logDist(X) = \sum_{i=1}^{n-1} \log(1+ \mid L-cover(x_i, x_{i+1}) \mid).$$

Some basic properties of $Dist(X)$ and $logDist(X)$ are stated in the following

Theorem 2.1 For $X = \langle x_1, \ldots, x_n \rangle$ and $Y = \langle y_1, \ldots, y_m \rangle$,

1. $Dist(X) = 0$ if X is sorted in ascending or descending order;
2. For any element x, $Dist(\langle x \rangle X) \leq (|X| - 1) + Dist(X)$;
3. $Dist(X) \leq (n-1)(n-2)/2$;
4. If $m = n$ and $x_i \leq x_j \Leftrightarrow y_i \leq y_j$, for all $i, j \in \{1, \ldots, n\}$, then $Dist(X) = Dist(Y)$;
5. If Y is a subsequence of X then $Dist(Y) \leq Dist(X)$;
6. $logDist(X) = O(n \cdot \log(1 + Dist(X)/n))$.

Proof The first three conditions are obvious. Moreover, the input sequence given in Figure 2a shows that the upper bound in the third condition is the best that could be obtained. Let us next verify the other conditions.

4. Suppose, without loss of generality, $Dist(X) > Dist(Y)$. Then there must be an index i, $1 \leq i \leq n$, such that $\mid R-cross(x_i) \mid > \mid R-cross(y_i) \mid \geq 0$. Furthermore, there exists $\{x_j, x_{j+1}\} \in R-cross(x_i)$, such that $\{y_j, y_{j+1}\} \notin R-cross(y_i)$. By applying the definition of $R-cross(x_i)$ we get

$$\begin{aligned} \{x_j, x_{j+1}\} \in R-cross(x_i) &\Leftrightarrow \min\{x_j, x_{j+1}\} < x_i < \max\{x_j, x_{j+1}\} \\ &\Leftrightarrow \min\{y_j, y_{j+1}\} < y_i < \max\{y_j, y_{j+1}\} \\ &\Leftrightarrow \{y_j, y_{j+1}\} \in R-cross(y_i) \end{aligned}$$

and we have obtained a contradiction.

5. The sequence X is obtained from Y by insertions of elements. Since new elements, obviously, cannot decrease $Dist(Y)$, it follows that $Dist(Y) \leq Dist(X)$.

6. Let $r_i = 1+ \mid L-cover(x_i, x_{i+1}) \mid$, $i \in \{1, \ldots, n-1\}$. The claimed property follows directly from the fact that the geometric mean can never be larger than the arithmetic mean. That is, for any positive numbers r_i, $i \in \{1, \ldots, m\}$, $\prod_{i=1}^{m} r_i^{1/m} \leq (\sum_{i=1}^{m} r_i)/m$. Namely,

$$logDist(X) = \sum_{i=1}^{n-1} \log(r_i) = \log(\prod_{i=1}^{n-1} r_i)$$

$$
\begin{aligned}
&= (n-1)\cdot \log((\prod_{i=1}^{n-1} r_i)^{1/(n-1)}) \\
&\leq (n-1)\cdot \log(\frac{1}{n-1}\cdot \sum_{i=1}^{n-1} r_i) \\
&= O(n\cdot \log(1+\frac{Dist(X)}{n})).
\end{aligned}
$$

∎

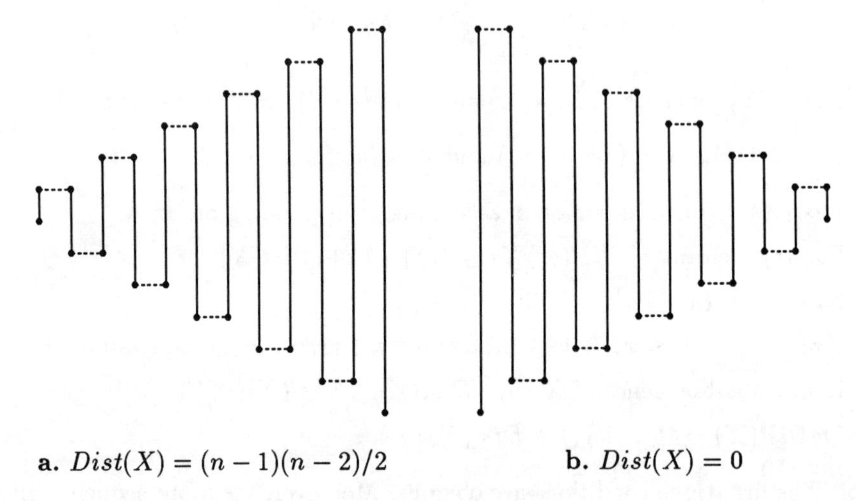

a. $Dist(X) = (n-1)(n-2)/2$ **b.** $Dist(X) = 0$

Figure 2. **(a)** *A hard instance for* LIS. **(b)** *An easy instance for* LIS.

Even though $Dist(X)$ intuitively is a measure of presortedness, it does not satisfy all Mannila's axioms (cf. [Man85]). The axiom which is not valid for $Dist(X)$ is the following: For $X = \langle x_1, \ldots, x_n\rangle$ and $Y = \langle y_1, \ldots, y_m\rangle$, if $x_i \leq y_j$, for all $i \in \{1, \ldots, n\}$, and $j \in \{1, \ldots, m\}$ then $measure(XY) \leq measure(X) + measure(Y)$. It should, however, be observed that we could have defined the measure $Dist(X)$ by considering only the crossings with the pairs $\{x_j, x_{j+1}\}$ where $x_j > x_{j+1}$. Let $RD(own){-}cross(x_i)$ denote the set of all such crossings for the element x_i. This measure would satisfy all Mannila's axioms. Furthermore, it is easy to show that, for any x_i,

$$|R{-}cross(x_i)| \leq 2\cdot |RD{-}cross(x_i)| + 1.$$

This means that the measure obtained would be almost the same as $Dist(X)$. The main reason for using the present definitions for the measures is that these give a nice characterization of LIS, as will be seen in the next subsection.

2.3 Analysis

The performance of LIS is characterized more precisely in the following

Theorem 2.2 LIS uses $\Theta(n + log Dist(X))$ time to sort a sequence $X = \langle x_1, \ldots, x_n\rangle$ of n elements. To sort any sequence with $Dist(X) = p$, LIS requires $\Theta(n + n\log(p/n))$ time.

Proof The execution time of LIS is (see [Man85])

$$T(n) = c_1 \cdot n + \sum_{i=1}^{n-1} c_2 \cdot (1 + \log(1 + |L{-}cover(x_i, x_{i+1})|)) = \Theta(n + logDist(X)).$$

From Theorem 2.1, it directly follows that $logDist(X) = O(n \cdot \log(1 + Dist(X)/n))$ for any sequence X of length n. Hence, $O(n + n\log(p/n))$ is an upper bound, for LIS. On the other hand, $\Omega(n)$ is a trivial lower bound for the execution time. The tightness follows if we can show that for any $p \geq 2(n-1)$, there exists an instance $Y = \langle y_1, y_2, \ldots, y_n \rangle$, such that $Dist(Y) = \Theta(p)$, and the algorithm requires $\Omega(n \log(p/n))$ time. This follows easily from the fact that the sum

$$\sum_{i=1}^{n-1} \log(1 + |L{-}cover(x_i, x_{i+1})|)$$

is maximized when the sets $L{-}cover(x_i, x_{i+1})$ are of almost equal size. For simplicity, we can assume that $(n-1)$ divides p; let $h = p/(n-1)$. Consider the sequence $Y = \langle y_1, \ldots, y_n \rangle$, where

$$\begin{aligned} y_{2i} &= i & &\text{for } i = 1, 2, \ldots, \lfloor n/2 \rfloor \\ y_{2i-1} &= i + h & &\text{for } i = 1, 2, \ldots, \lceil n/2 \rceil. \end{aligned}$$

It is easily verified that $p/2 \leq Dist(Y) < p$, for $n > 5$. For this sequence, it holds that $|L{-}cover(y_j, y_{j+1})| \geq \lfloor h/2 \rfloor$, for any $j = h, \ldots, n$. Hence,

$$logDist(Y) \geq (n-h)\log(1 + \lfloor \frac{h}{2} \rfloor) > \frac{n}{2} \log \frac{h}{2} = \Omega(n \log(\frac{p}{n}))$$

where the last inequality follows from the fact that $h < n/2$. (Otherwise we would have $p > (n-1)(n-2)/2$, contradicting property 3 in Theorem 2.1.) ∎

Let M be any measure of presortedness and A a sorting algorithm which uses $T_A(X)$ comparisons to sort a sequence X. Mannila [Man85] defined that A is *M-optimal* if, for some $c > 0$, we have for all $X = \langle x_1, x_2, \ldots, x_n \rangle$:

$$T_A(X) \leq c \cdot \max\{|X|, \log(|below(X, M)|)\},$$

where $below(X, M) = \{\pi \mid \pi \text{ is a permutation of } \{1, \ldots, n\} \text{ and } M(\pi(X)) \leq M(X)\}$.

Lemma 2.3 For a sequence X of length n,

$$\log(|below(X, Dist)|) = \Omega(n \log(\frac{Dist(X)}{n})).$$

Proof Similar to the lower bound proof in [LP89]. ∎

Thus, we have

Theorem 2.4 LIS is *Dist*-optimal.

Mannila [Man85] proved that LIS is M-optimal for many measures of presortedness. These include $Inv(X)$ (the number of inversions), $Runs(X)$ (the number of maximal ascending consecutive subsequences, less one), and $Rem(X)$ (the minimum number of elements that have to be removed to leave a sorted sequence). It could be proven that LIS is M-optimal for the measures $Par(X)$ [EW87] and $Osc(X)$ [LP89].

Of course, different measures consider different sequences as presorted. A natural question to ask is whether there is a measure that, in some sense, includes some of the other measures. In fact, we can prove that, if a sequence is presorted according to $Inv(X)$, $Runs(X)$, $Par(X)$, or $Osc(X)$, it is also presorted according to $Dist(X)$. In particular, any *Dist*-optimal sorting algorithm is also M-optimal, for these measures. (Mannila [Man85] proved this result for the measures $Inv(X)$ and $Runs(X)$.) However, $logDist(X)$ is superior to $Rem(X)$ as well [Man85].

As a final remark, we want to emphasize that not even $Dist(X)$ is sufficient to characterize the performance of LIS, for all instances. Namely, for some instance X, the actual running time $\Theta(n + logDist(X))$ can be much smaller than $\Theta(n + n\log(Dist(X)/n))$. Hence, the only measure that can be used to characterize the *instance complexity* of LIS is $logDist(X)$.

3 Improved Algorithms

In this section we propose two modifications of LIS. The first idea is to use many finger trees and one finger in each tree (Subsection 3.1). It is proven that this can yield asymptotic speedup. Then, we maintain many fingers, instead of only one, pointing to the bottom level of a single finger search tree. This will speed up the process of inserting a new element into its proper position. In Subsection 3.2 we consider the case when the number of fingers is fixed beforehand. Different strategies for moving the fingers are discussed, and some general properties for such strategies are proposed. Finally, in Subsection 3.3, we present a method for designing algorithms which maintain the fingers dynamically.

3.1 Many Finger Trees

Consider the input sequence given in Figure 2a. The sequence consists of two interleaved sequences; one sorted in ascending order and the other sorted in descending order. Let $X = \langle x_1, x_2, \ldots, x_n \rangle$ be a sequence, of positive integers, sorted in ascending order. The instance $Z = \langle x_1, -x_1, x_2, -x_2, \ldots, x_n, -x_n \rangle$ will be a difficult instance for LIS, since $logDist(Z) = \Theta(n \log n)$. On the other hand, the reversed sequence $\bar{Z} = \langle -x_n, x_n, \ldots, -x_2, x_2, -x_1, x_1 \rangle$ (see Figure 2b) is an easy instance for the algorithm, since $logDist(\bar{Z}) = 0$.

According to the above discussion, it is a natural idea to start LIS simultaneously from both ends of the sequence. Actually, we *run two algorithms in parallel* by performing one step of the algorithms alternatively. Each version of LIS maintains its own finger tree. At some moment, all elements are inserted into some finger tree. Then, we have two sorted subsequences that we merge to obtain the final output. Since merging requires only linear time, the sorting time of the new algorithm is

$O(n + \min\{logDist(X), logDist(\bar{X})\})$ for a sequence X of length n. Asymptotically, this is never worse than that of LIS. Moreover, observe that the bound is not tight since the sequence $\bar{Z}Z$ will be sorted in time $\Theta(n)$, while the above bound gives $\Theta(n \log n)$. Hence, the speedup can be asymptotically significant. However, in practice this new algorithm might be about two times slower, since we have to maintain two finger trees instead of only one.

3.2 Many Fingers in One Finger Tree

Let $X = \langle x_1, x_2, \ldots, x_n \rangle$ be a sequence of positive integers, already sorted in ascending order. To sort the shuffled sequence $\langle x_1, -x_1, x_2, -x_2, \ldots, x_n, -x_n \rangle$ (see Figure 2a) will take $\Theta(n \log n)$ time for LIS with one finger. By using *two fingers*, in one finger tree, the same sequence can be sorted in linear time provided that we always *move the finger nearest* (in case of a tie, we move the left one) to the latest insertion position. The other finger is held where it was before the insertion. Let MN denote this heuristic.

More generally, we can use $k \geq 1$ fingers (where k is fixed), and any heuristic for moving the fingers. Given an element x, which is to be inserted, we find the finger that points to the smallest element greater than x. This finger, together with its predecessor, defines an interval in the finger tree in which x resides. Starting at these two fingers, we search in parallel (one step alternatively) for the proper position of x. When k fingers are available, let d_x^k denote the distance from the nearest finger to the position into which x is inserted. Now, the work of finding the right position of x in the finger tree is proportional to $\log(d_x^k + 1)$.

When x has been inserted we move some finger and let it point at x; which one depends on the heuristic. There is no such thing as a best heuristic, valid for all applications. Rather, the heuristic should be choosen depending on what kind of existing order is expected. We next suggest some general properties that should be valid for any reasonable heuristic.

Property 1 Nothing can be lost if the number of fingers is increased; that is, for any sequence X,

$$\sum_{x_i} \log(d_{x_i}^{k+1} + 1) \leq \sum_{x_i} \log(d_{x_i}^k + 1).$$

We also require that, at least for some input, the speedup should be asymptotically significant:

Property 2 There exists a sequence X of length n, such that

$$\sum_{x_i} \log(d_{x_i}^k + 1) \notin O(\sum_{x_i} \log(d_{x_i}^{k+1} + 1)).$$

Let us assume that during the first k insertions no fingers are moved, but they are initialized to point at the k inserted elements. Further, let $\text{LIS}_H^k(X)$ denote LIS applied to X, when using k fingers and heuristic H, and let $T(\text{LIS}_H^k(X))$ be the time complexity of $\text{LIS}_H^k(X)$ (including the overhead for maintaining the fingers; when k is fixed this will be linear).

It is easy to prove that Property 1 holds for the heuristic MN, since $d_x^{k+1} \leq d_x^k$. This follows by observing that if $k+1$ fingers are in use, k of them will point at the same elements as will all k fingers, if just k fingers are in use. Further, the sequence X given in Figure 3 shows that Property 2 holds, as well. Note that the existence of a sequence X such that $T(\mathrm{LIS}_H^k(X)) \notin O(T(\mathrm{LIS}_H^{k+1}(X)))$ implies Property 2. If k fingers are available, one finger oscillates over $\Theta(n)$ elements, while $k-1$ fingers stay motionless. Thus $T(\mathrm{LIS}_{MN}^k(X)) = \Theta(n \log n)$, while $T(\mathrm{LIS}_{MN}^{k+1}(X)) = \Theta(n)$. Observe that $T(\mathrm{LIS}_{MN}^k(X)) = T(\mathrm{LIS}_{MN}^1(X))$, indicating that the fingers are not always effectively used in MN.

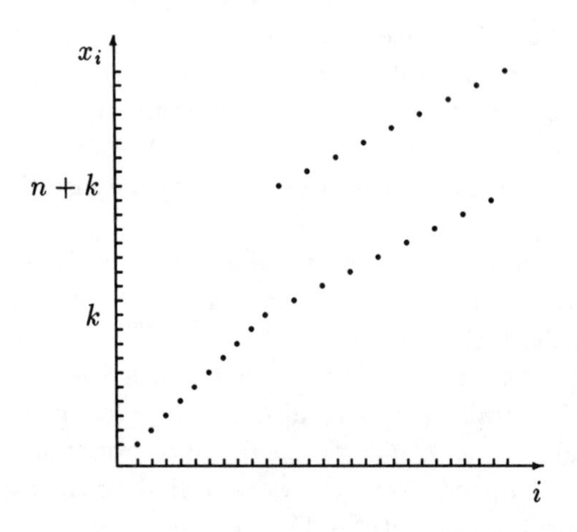

Figure 3. *An instance showing that Property 2 holds for the heuristic* MN; $X = \langle 1, \ldots, k-1, k, n+k, k+1, n+k+1, \ldots, n+k-1, 2n+k-1 \rangle$.

We remind the reader of the similarity between choosing which finger to move and page replacement heuristics in a virtual memory implementation (see e.g. [PS86, Section 6.6]). This leads us to consider another heuristic: *Least Recently Moved* (LRM). As the name indicates, in LRM the finger that has not been moved for the longest period of time is set to point at the latest element inserted.

By applying the same argument as for MN, it can be proven that $d_x^{k+1} \leq d_x^k$ for LRM as well, and hence Property 1 is fulfilled. The sequence Y in Figure 4 shows that Property 2 holds. In spite of this, the performance of LRM is not always satisfactory; in fact, it can be too greedy. When sorting Y with k fingers, all fingers are tricked together, covering only small portion of the finger tree, even if they are not needed there. Then, an element which resides far away from this area causes an expensive insertion. Observe that $T(\mathrm{LIS}_{LRM}^k(Y)) = \Theta(n \log n)$, while $T(\mathrm{LIS}_{MN}^2(Y)) = \Theta(n)$!

From the above example, we can learn the following: As long as the new positions are found *reasonably fast* it would be wise having the remaining fingers spread out, rather than gathered together. In fact, this can be achieved, at least to some extent, if a hybrid of LRM and MN is used: whenever the distance from the nearest finger is less than some *threshold* $Near(n,k)$, then MN is applied, otherwise LRM is applied. For example, we can have $Near(n,k) = c \cdot (n/k)$, for some constant $c > 0$. With a proper choice of $Near(n,k)$, this hybrid guarantees that LRM is invoked sometimes,

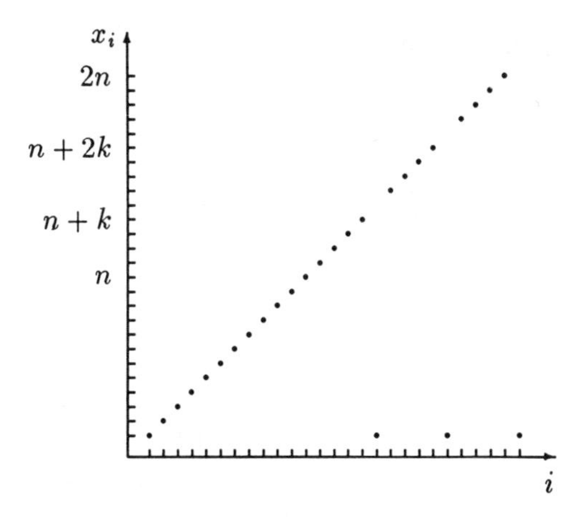

Figure 4. *The sequence* $Y = \langle y_1, \ldots, y_{2n+(n/k)} \rangle$, *where* $y_i = 1$ *if* $i \geq n$ *and* $i - n \bmod k \neq 0$ *and* $y_i = i$ *otherwise, shows that Property 2 holds for* LRM.

preventing situations like the one given in Figure 3, where most fingers stay motionless when they are actually needed somewhere else in the tree.

If $Near(n, k)$ is non-decreasing with respect to k, we can prove that Property 1 and Property 2 hold for this hybrid of LRM and MN. It should however be noted that Property 1 need not hold if $Near(n, k)$ is an increasing function on k.

A crucial question when implementing LIS with k fingers is how these are stored. The efficiency depends highly on the strategy used for moving the fingers. Let us first consider the MN heuristic, and let x be the element to be inserted. It turns out that the fingers can be stored in an array, sorted with respect to the elements pointed at by the fingers. By a binary search, we find the finger that points to the smallest element greater than x, in $O(\log k)$ time. With this finger, together with its predecessor, we find the proper position for x within the finger tree. When this position has been found, and x has been inserted, the nearest finger is moved to point at x. The array of fingers remains sorted, since two fingers can never cross. When k fingers are used, the insertion of x takes $O(\log k + \log(d_x^k + 1))$ time. Hence,

$$T(\mathrm{LIS}_{MN}^k(X)) = \sum_{x_i} \Theta(\log k + \log(d_{x_i}^k + 1)) = \Theta(n \log k + \sum_{x_i} \log(d_{x_i}^k + 1)).$$

If another strategy for moving the fingers is used, it might not be possible to store them in a sorted array; at least if reasonable, i.e., logarithmic, access and update time is required. In the case of LRM, or the hybrid of LRM and MN, the fingers can be stored in a balanced search tree that supports nearest neighbour queries, insertions and deletions. In combination, a move-to-front list can be used to support the LRM-part of the heuristic. Hence, the overhead is logarithmic per element for these heuristics, too.

3.3 Dynamic Algorithms

A natural idea is to let the number of fingers vary, instead of keeping it fixed. Let H be a heuristic fulfilling Property 1. For a sequence X, we say that k_{opt} fingers is optimal for $\text{LIS}_H(X)$, if $T(\text{LIS}_H(X)) = \Omega(T(\text{LIS}_H^{k_{opt}}(X)))$, for whatever number of fingers is available. Denote by $g_H(k)$ the worst case time complexity of an operation on the data structure storing the fingers, when heuristic H is used. Then, clearly, $T(\text{LIS}_H^{k_{opt}}(X)) = O(n \cdot g_H(k_{opt}))$. The idea is to balance the cost of the operations on the data structure storing the fingers and the cost of searching for the proper positions in the finger tree, and thereby match this bound.

The method proposed here is based on a well known dynamization technique by Overmars [Ove83], called *global rebuilding.* We start by allocating $k_1 = 2$ fingers and then repeatedly increase the number, if it turns out that more fingers are needed. The amount of increase depends on $g_H(k)$. Let $T(k_i)$, $i = 1, 2, \ldots, max$, denote the time spent when k_i fingers are available. The algorithm is obviously optimal if

$$\sum_{i=1}^{max} T(k_i) = O(n \cdot g_H(k_{opt})).$$

This is achieved if

1. $T(k_i) \leq T(k_{i+1})/c$, for some constant $c > 1$, and
2. $T(k_{max}) = O(n \cdot g_H(k_{opt}))$

since then

$$\sum_{i=1}^{max} T(k_i) = \sum_{i=0}^{max-1} \frac{T(k_{max})}{c^i} \leq O(n \cdot g_H(k_{opt})) \cdot \sum_{i=0}^{max-1} \frac{1}{c^i} = O(n \cdot g_H(k_{opt})).$$

Let $f_H(k)$ be a function that satisfies the conditions 1 and 2 above. (For example, if $g_H(k) = \log k$ then $f_H(k) = k^2$ is sufficient, or if $g_H(k) = \log\log k$, then $f_H(k) = 2^{\log^2 k}$ is sufficient.) The algorithm is as follows:

$k \leftarrow 2$;
while not sorted **do**
begin
  Run $n \cdot g_H(k)$ steps of $\text{LIS}_H^k(X)$;
  $k \leftarrow f_H(k)$
end.

By the above discussion, the number of steps used by our dynamization algorithm is $\Theta(T(\text{LIS}_H^{k_{opt}}(X)))$, and hence we conclude

Theorem 3.1 For any heuristic H fulfilling Property 1, the above method yields a dynamic variant of LIS_H, which sorts in optimal time.

4 Concluding Remarks

The main contribution of this paper is twofold. First, we introduced a geometric characterization of the performance of LIS. Second, this characterization enhanced our understanding of the algorithm and, in this way we were able to present some modifications of the basic algorithm, improving its performance.

Usually LIS is implemented by using a finger tree with one (insertion) finger. Previously, it was well known that almost sorted files can be sorted efficiently by the algorithm. We proposed two measures of presortedness that can be used to characterize the performance of the algorithm. The measure $Dist(X)$ represents the running time when we want to sort *any* sequence X with $Dist(X) = p$. The measure $logDist(X)$ will tell the performance for *any particular instance* X. Even if we have two sequences X and Y, with $Dist(X) = Dist(Y)$, the sorting times for these sequences can differ from each other considerably. In the extreme case, the sequence X can be sorted in linear time (i.e., $logDist(X) = \Theta(n)$) whereas the sorting of Y could require $\Omega(n \cdot \log(Dist(Y)/n))$ time (i.e., $logDist(Y) = \Omega(n \cdot \log(Dist(Y)/n))$).

We also proposed two variants of LIS. First, it was pointed out that asymptotical speedup could be achieved by using more than one finger tree. Then we explored the idea of maintaining many fingers in the same finger tree. The number of fingers can either be fixed beforehand or vary during the sorting process. There are many possibilities for moving the fingers, each one giving a different algorithm. Two properties that should hold for any reasonable heuristic for moving the fingers were stated, and two such heuristics were presented. In the first heuristic (MN) the finger closest to the latest inserted element is moved. In LRM the least recently moved finger follows the inserted element. A combination of these two heuristics seems promising. Finally, we proposed a method for dynamizing the algorithms, in the sense that they allocate new fingers when needed during the sorting process. For any specific heuristic these algorithms allocate an optimal number of fingers and sort in optimal time.

Numerous interesting open questions still remain unanswered.

- Can one find some geometric applications of practical importance where some special instances could be solved efficiently in a manner similar to LIS? For example, it is easily shown that some special line-segment intersection problem and range search problem can be solved faster than the corresponding unrestricted problems.

- When the number of fingers is fixed, is it possible to preprocess X in linear time to get some information about high activity areas in the finger tree? This information could then serve as a guide for the finger moving strategy.

- Recently Skiena [Ski88] proposed a measure of presortedness similar to a dynamic variant of our multi-tree approach. However, it is still possible to find sequences that can be sorted efficiently by LIS, but not by his approach and vice versa. An interesting question is whether there exists a *natural* common characterization, which includes all sequences that are preordered in some intuitive sense.

- Can one find a more natural dynamic algorithm, that allocates fingers when needed and deallocates when they are not needed any more? To be able to do this efficiently probably needs some preprocessing.
- The design of parallel sorting algorithms that utilize presortedness is an area deserving more attention (see e.g., [AI88] and [LP88]).

Acknowledgements

We would like to thank Heikki Mannila for his helpful comments on the subject of this paper.

References

[AI88] T. Altman and Y. Igarashi. *Roughly sorting, sequential and parallel approach.* Technical Report CS–88–7, Gunma University, Department of Computer Science, Kiryu, Japan, 1988.

[CK80] C.R Cook and D.J. Kim. Best sorting algorithms for nearly sorted lists. *Communications of the ACM*, 23(11):620–624, 1980.

[Ede87] H. Edelsbrunner. *Algorithms in Combinatorial Geometry.* Springer-Verlag, Berlin/Heidelberg, F.R.Germany, 1987.

[EW87] V. Estivill-Castro and D. Wood. *A new measure of presortedness.* Research Report CS–87–58, University of Waterloo, Department of Computer Science, Waterloo, Canada, 1987.

[GMPR77] L.J. Guibas, E.M. McCreight, M.F. Plass, and J.R. Roberts. A new representation of linear lists. In *Proc. 9th Annual ACM Symposium on Theory of Computing*, pages 49–60, 1977.

[Knu73] D.E. Knuth. *The Art of Computer Programming, Vol. 3: Sorting and Searching.* Addison-Wesley, Reading, Mass., 1973.

[LP88] C. Levcopoulos and O. Petersson. An optimal parallel algorithm for sorting presorted files. In *Proc. 8th Conference on Foundations of Software Technology and Theoretical Computer Science*, pages 154–160, Springer-Verlag, 1988.

[LP89] C. Levcopoulos and O. Petersson. Heapsort—adapted for presorted files. In *Proc. 1989 Workshop on Algorithms and Data Structures*, Springer-Verlag, 1989. To appear.

[Man84] H Mannila. *Implementation of a sorting algorithm suitable for presorted files.* Technical Report, Department of Computer Science, University of Helsinki, Finland, 1984.

[Man85] H. Mannila. Measures of presortedness and optimal sorting algorithms. *IEEE Transactions on Computers*, C-34(4):318–325, 1985.

[Meh79] K. Mehlhorn. Sorting presorted files. In *Proc. 4th GI Conference on Theoretical Computer Science*, pages 199–212, Springer-Verlag, 1979.

[Meh84a] K. Mehlhorn. *Data Structures and Algorithms, Vol 1: Sorting and Searching.* Springer-Verlag, Berlin/Heidelberg, F.R.Germany, 1984.

[Meh84b] K. Mehlhorn. *Data Structures and Algorithms, Vol. 3: Multidimensional Searching and Computational Geometry.* Springer-Verlag, Berlin/Heidelberg, F.R.Germany, 1984.

[Ove83] M.H. Overmars. *The Design of Dynamic Data Structures. Lecture Notes in Computer Science 156*, Springer-Verlag, Berlin/Heidelberg, F.R.Germany, 1983.

[PS85] F.P. Preparata and M.I. Shamos. *Computational Geometry: An Introduction.* Springer-Verlag, New York, N.Y., 1985.

[PS86] J.L. Peterson and A. Silberschatz. *Operating System Concepts.* Addison Wesley, Reading, Mass., second edition, 1986.

[Ski88] S.S. Skiena. Encroaching lists as a measure of presortedness. *BIT*, 28:775–784, 1988.

[TVW88] R.E. Tarjan and C.J. Van Wyk. An $O(n \log\log n)$-time algorithm for triangulating a simple polygon. *SIAM Journal on Computing*, 17(1):143–178, 1988.

Packet Routing on Grids of Processors

Manfred Kunde*
Institut für Informatik, Technische Universität München
Arcisstr. 21, D-8000 München 2, West Germany

Abstract The problem of packet routing on $n_1 \times \ldots \times n_r$ mesh-connected arrays or grids of processors is studied. For two-dimensional grids a deterministic routing algorithm is given for $n \times n$ meshes where each processor has a buffer of size $f(n) < n$. It needs $2n + O(n/f(n))$ steps on grids without wrap-arounds. Hence it is asymptotically optimal and as good as randomized algorithms routing data only with high probability. Furthermore it is demonstrated that on r-dimensional cubes of processors packet routing can be performed by asymptotically $(2r-2)n$ steps which is faster than the running times of so far known randomized algorithms and of deterministic algorithms.

1. Introduction

It has turned out that in the performance of parallel computation data movement plays an important role. We present routing algorithms on grids where the number of parallel data transfers is asymptotically minimal or nearly minimal.

An $n_1 \times \ldots \times n_r$ mesh-connected array or grid is a set $mesh(n_1, \ldots, n_r)$ of $N = n_1 n_2 \ldots n_r$ identical processors where each processor $P = (p_1, \ldots, p_r)$, $0 \leq p_i \leq n_i - 1$, is directly interconnected to all its nearest neighbours only. A processor $Q = (q_1, \ldots, q_r)$ is called nearest neighbour of P if and only if the distance between them is exactly 1. For a grid without wrap-around connections the distance is given by $d(P,Q) = |p_1 - q_1| + \cdots + |p_r - q_r|$. For grids with wrap-around connections we define $d_{wrap}(P,Q) = |(p_1 - q_1) \bmod n_1| + \cdots + |(p_r - q_r) \bmod n_r|$, where $-b/2 \leq a \bmod b \leq b/2$ is assumed for arbitrary integers a and b. The control structure of the grid of processors is thought to be of the MIMD type. A main restriction is that each processor can send data only to its nearest neighbours during one clock period, and that each processor has only a limited number of registers for data.

For the permutation routing problem N elements or packets are loaded in the N processors. Each packet has a destination address specifying the processor to which it has to be sent. Different packets have different addresses. The routing problem is to transport each packet to its address. Thus the problem is described by the pair $(mesh, address)$ where address is a one-to-one mapping from mesh

* This work was supported by the Siemens AG, München

onto itself. Very often the routing problem is solved by sorting algorithms. For the sorting problem we assume that N elements from a linearily ordered set are initially loaded in the N processors, each receiving exactly one element. The initial loading of a processor P is denoted by $cont_0(P)$. The processors are thought to be indexed by a certain one-to-one mapping g from $mesh(n_1, \ldots, n_r)$ onto $\{0, \ldots, N-1\}$. (Compare Figure 1.) With respect to this function the sorting problem is to move the i-th smallest element to the processor indexed by $i-1$ for all $i = 1, \ldots, N$. Hence a sorting problem is described by the triple $(mesh, cont_0, g)$. Note that the routing problem $(mesh, address)$ can be solved by a sorting algorithm for the problem $(mesh, goaddress, g)$ for which the initially contents is given by $cont'_0(P) = g(address(P)) = g(Q)$ for all $P \in mesh$, meaning that the contents of P must be transported to $Q = address(P)$.

Figure 1 Mesh–connected arrays and index functions

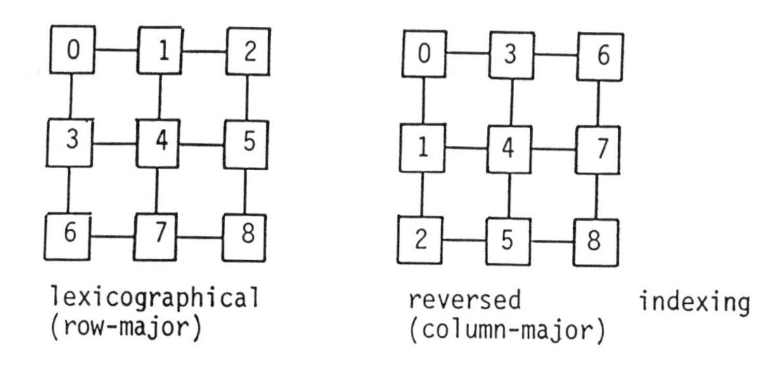

In order to get lower and upper bounds for sorting and routing on meshes the possible steps for computations must be described very carefully. We will use two models. The first one allows interchanges of data between two directly neighboured processors and data shifts on cycles of processors. The interchange of data may be caused by a comparison or not. After each parallel step exactly one element (packet) is in each processor. In the second model all operations of the first model are allowed. However, packets may be stored in a processor until a limited buffer is filled.

For 2-dimensional $n \times n$ meshes without wrap-arounds several sorting algorithms have been proposed for the first model. The fastest ones need about $3n + O(low\ order\ term)$ steps for snake-like indexing [TK,SS] or blockwise snake-like indexing [MSS] which turns out to be asymptotically optimal for such indexings [SS,Ku1]. For wrap-around meshes a lower bound of $3n/2$ can be shown for arbitrary indexings [Ku3]. The fastest algorithms for snake-like indexings need about $5n/2$ steps [Ku2] and $2n$ steps for blockwise snake-like indexing [MSS].

The best algorithm for an $n \times \cdots \times n$ cube, $r \geq 3$, needs only $(2r-1)n +$ *low order term* steps for sorting with respect to snake-like indexing [Ku2, Ku4] which is asymptotically optimal for this type of indexing [Ku1]. For meshes with wrap-around-connections the lower bound is $(r-1/2)n$ and the so far best algorithm asymptotically needs rn steps [Ku4].

The question is, whether there are better routing algorithms than those mentioned above. For model 1 the best sorting algorithms are more or less the best routing algorithms. In the case that a bigger buffer size is given, better algorithms can be obtained. In the second section we show that on an $n \times n$ grid routing in model 2 can be done within $2n + O(n/f(n))$ steps in the case where buffer size $f(n)$ is given. Hence this routing algorithm asymptotically matches the distance bound of $2n-2$ provided that $f(n)$ is a strictly increasing function. If wrap-around-connections are given then the performance can be improved to $n+O(n/f(n))$ steps which again is asymptotically optimal. For example, for buffer size $f(n) = log\, n$ the proposed deterministic algorithm is faster than a randomized algorithm presented in [VB] and as good as that one in [KRT], both algorithms reaching their execution times with high probability only. It is a little bit worse than a recently proposed algorithm of Leigthon, Makedon and Tollis [LMT] which exactly matches the distant bound of $2n - 2$ steps with constant buffersize. However, it is open whether that algorithm fits to grids with wrap-around-connections.

In the third section we then present a generalization of the routing algorithm for an $n \times ... \times n$ $r-$dimensional cube of processors. It needs only $(2r - 2)n + 0(n/f(n)^{1/(r-1)})$ steps and $(r - 1)n + 0(n/f(n)^{1/(r-1)})$ steps if wrap-arounds are given. It is the first deterministic algorithm which is faster than the routing algorithms based on sorting [Ku2,Ku4] and the randomized algorithms of Valiant and Brebner [VB].

2. Preliminaries and basic notations

In this section we first present some basic definitions and then give some results on linear (or 1-dimensional) and on 2-dimensional grids. We will concentrate on the cases where all sidelengths are equal, that is $n_i = n$ for all $i = 1, ..., r$. The second model of computation is used where each processor is able to store a limited amount of data during the computation. For sidelength n we assume that the size of the additional buffer is $f(n)$, $2 \leq f(n) < n^{1/2}$. Typical buffer sizes are for example $log\, n$, $n^{1/4}$ or a constant c.

Indexing and intervalls of processors

Mainly two index functions are used: the *lexicographical indexing lex* which is defined by $lex(p_1, ..., p_r) = p_1 n^{r-1} + ... + p_{r-1} n + p_r$ and the *reversed indexing* given by $rev(p_1, ..., p_r) = lex(p_r, ..., p_1)$. For the rest of the paper we assume that the packets are linearily ordered with respect to the lexicographical order of their adresses, that is $address(P) \leq address(Q)$ if and only if $lex(address(P)) \leq lex(address(Q))$. For submeshes the indexings are used in the corresponding manner. Let g be any index function. Then an intervall of processors (or addresses) with respect to indexing g is the set of processors $[P, Q]_g = \{X | g(P) \leq g(x) \leq g(Q), X$ a processor$\}$. Some further abreviations are used. For all $i = 1, ..., r$ let $[a_i, b_i]$ denote an intervall of integers with $0 \leq a_i \leq b_i < n$. Then $([a_1, b_1], ..., [a_i, b_i], ..., [a_r, b_r]) = \{(p_1, ..., p_i, ..., p_r) | a_i \leq p_i \leq b_i, i = 1, ..., r\}$. If $a_i = b_i = p_i$ then only the p_i is written instead of the intervall and the intervall $[0, n - 1]$ is abbreviated by $*$. That is, an intervall of processors along the i-th is described by $(p_1, ..., p_{i-1}, *, p_{i+1}, ..., p_r) = (p_1, ..., p_{i-1}, [0, n-1], p_{i+1}, ..., p_r)$,which is called an i-tower . For fixed $r - i$ coordinates $p_{i+1}, ..., p_r$ i-dimensional subgrids are denoted by $(*, ..., *, p_{i+1}, ..., p_r)$.

For the routing algorithms presented in the following we very often need solutions for routing problems which are not permutation routing problems. That is the address function is no longer a 1-to-1 mapping. It may happen that there are $f(n)$ packets in the beginning and the same number of packets at the end in a processor. We say that an r-dimensional routing problem (with at most n^r packets) is *balanced* (with respect to the r-th axis) if and only if there are at most $b_r - a_r + 1 + f(n)$ packets with addresses in the intervall of processors $(p_1, ..., p_{r-1}, [a_r, b_r])$ for all $p_j, 0 \leq p_j < n, j = 1, ..., r-1$, and all $a_r, b_r, 0 \leq a_r \leq b_r < n$.

Routing on linear arrays

Two lemmas on strategies for routing on one-dimensional meshes are presented which are used as basic operations for routing in r-dimensional arrays.

Lemma 2.1
On a linear array with n processors let each processor contain at most one packet initially and an arbitrary number finally. Then the routing can be done in at most $n - 1$ transport steps.

Proof Use the routing strategy where a packet once started is never delayed. W.l.o.g observe only packets going from left to right. Note that packets going in the opposite direction do not interfere because of the bidirectional connections between processors. Since at the beginning at each processor there is at most one packet, it can be started and is never delayed until it reaches its destination.

Lemma 2.2
A balanced routing problem on a linear array of n processors with buffer size $f(n)$ can be solved by at most $n - 1 + f(n)$ transport steps.

Proof The routing strategy is "farthest destination first" [KRT]. Route a packet from source i to destination j (w.l.o.g. $i \leq j$). Since the problem is balanced there are at most $n - j + 1 + f(n)$ packets going to or past j. Therefore a packet at i can be delayed by at most $n - j + f(n)$ packets and reaches its destination after at most $n - j + f(n) + (j - i) = n - i + f(n)$ steps. The maximum is reached for $i = 1$.

An optimal routing algorithm for a two-dimensional grid

We will present an asymptotically optimal algorithm for routing problems which are balanced with respect to the second coordinate. The algorihm was originally constructed for permutation routing problems [Ku2]. A fundamental method for obtaining asymptotically optimal or nearly optimal routing or sorting algorithms is as follows: divide the rectangle of n^2 processors into small rectangles called blocks, do some computations on these blocks and then use the obtained information to perform the whole task. In our case we assume a block to be a $n/k \times n/k$ submesh where $4 \leq 2k \leq f(n)$.

For an easier understanding let us denote the first coordinates by an r (for row) and the second one by a c (for column). We call r the row address and c the column address of the packet. As mentioned above the addresses are ordered by the lexicographical order. That is $(r_1, c_1) \leq (r_2, c_2)$ if and only if $r_1 n + c_1 \leq r_2 n + c_2$.

Routing algorithm : Sort-and-Route

1. Sort all blocks with row-major indexing (that is lexicographical indexing)
2. In each column c, $0 \leq c \leq n-1$: transmit packets with row address r to processor in row r.
3. In each row r, $0 \leq r \leq n-1$: transmit packets with column address c to processor in column c.

In order to proof the correctness of our algorithm let us observe all packets with row address r for an arbitrary r, $0 \leq r \leq n-1$. Let the number of such packets be z. Since the problem is balanced with respect to the second coordinate, the column coordinate, z is bounded by $n + f(n)$. Take an arbitrary column of blocks. Assume that these blocks are numbered from 1 to k. Let $b(i)$ be the number of packets with row address r in block i, $1 \leq i \leq k$. After the first step of the routing algorithm the packets in block i are ordered with respect to their row addresses. Hence all packets with row address r have been sent to processors in neighbouring rows of processors in block i. Since the side length of a block is n/k that means that in each column of block i there are at most $\lceil b(i)/(n/k)) \rceil$ packets with row address r (compare Figure 2).

Figure 2 Columns of blocks after the first step of the routing algorithm

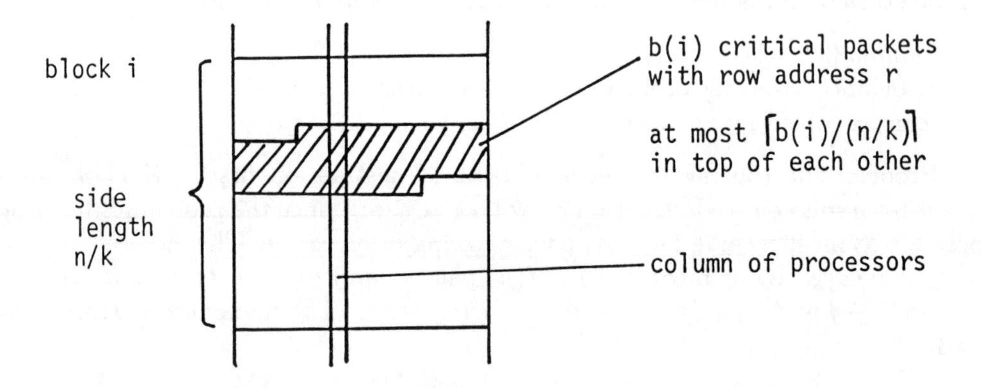

Therefore in each column of the total block column there are at most

$$x = \sum_{i=1}^{k} \lceil k \cdot b(i)/n \rceil < (k/n) \cdot \sum_{i=1}^{k} b(i) + k$$

$$\leq (k/n) \cdot z + k \leq (k/n) \cdot (n + f(n)) + k \leq 2k + 1$$

packets with row address r. Since x is an integer we get $x \leq 2k \leq f(n)$.

Hence in each column c, $0 \leq c \leq n-1$, of the total mesh there are at most $f(n)$ packets with row address r. These at most $f(n)$ packets are then transported to processor (r, c) which can store all these packets because its buffer is large enough. By lemma 2.1 this can be done within n transport steps. Since the problem was balanced we now have in every row a balancėd linear routing proplem. In the last step the packets are then distributed to their destination processor, which costs at most $n + f(n)$ transport steps by lemma 2.2.

Since $f(n) \leq n^{1/2}$ we have proven the following theorem

Theorem 1   An balanced routing problem on an $n \times n$ grid can be solved by $2n + O(n/f(n))$ steps.

Note that in the case of wrap-arounds the last two steps of the algorithm can be performed together in n steps. Hence in this case only $n + O(n/f(n))$ steps are needed.

It is easily seen that the method can be applied to $a \times b$ meshes. If the buffersize is given by a function $g(a, b)$, the complexity is then $a + b + O((a + b)/g(a, b))$ for meshes without wrap-arounds and $(a+b)/2 + O((a+b)/g(a, b))$ in the wrap-around case. Note that for both types of meshes the distance bound asymptotically is matched if the buffersize is given as a strictly increasing unbounded function. That means that in all these cases the method is asymptotically optimal. Hence this deterministic algorithm is asymptotically faster than the randomized algorithm in [VB] and as good as the randomized algorithm in [RT]. However, the randomized algorithms are oblivious, while the algorithms of this paper are non-oblivious. Recently Leigthon, Makedon and Tollis [LMT] gave a non-oblivious algorithm needing only $2n - 2$ steps for $n \times n$ meshes with constant buffersize. However, it is not clear whether the distance bound of $2n - 2$ can be reached for balanced problems.

Furthermore, it should be pointed out that the routing algorithm only needs $2d_{max} + O(n/f(n))$ steps, if one knows in advance the maximum distance $d_{max} = \{d(P, address(P)) \mid P \in mesh\}$ for a problem $(mesh, address)$. Hence for local routing problems with small d_{max} the routing algorithm can make use of the locality.

3. A fast routing algorithm for r-dimensional grids, $r \geq 3$

In this section we show how the ideas of the last section can be used to obtain fast routing algorithms on grids with more than two dimensions. At the beginning we will briefly concentrate on a three-dimensional $n \times n \times n$ cube of processors in order to illustrate some basic ideas. The cube of n^3 processor can be viewed as collection of n planes (where $plane(h) = \{(r, c, h) \mid 0 \leq r, c \leq n-1\}$ for $h = 0, ..., n-1$). In each plane we may consider the *projected routing* problem $(plane(h), address_{1,2})$ with $address_{1,2}((r_1, c_1, h_1)) = (r_2, c_2)$ where $(r_2, c_2, h_2) = address(r_1, c_1, h_1)$. Usually the projected routing problems are neither permutation nor balanced routing problems. The main idea for the following algorithm is to rearrange the packets in the cube in such a way that in each plane we have a balanced routing problem.

Sort-and-Route for 3-dimensional cubes of processors

1. Rearrange packets such that in each plane the corresponding projected routing problem becomes a balanced routing problem.(Rearrange phase)
2. In each plane solve the balanced routing problem by the 2-dimensional Sort-and-Route.
3. For all $r, c, 0 \leq r, c \leq n-1$, transport packets in $tower(r,c) = \{(r,c,h) \mid h = 0, ..., n-1\}$ to their destination processors. (Correction phase)

Let us assume that the first stage of the algorithm has been performed correctly. Then in each plane the projected routing problem has become a balanced one and can therefore be solved by the second stage. After this stage an arbitrary processor $P = (r,c,h)$ only contains packets with adresses of type $(r,c,j), 0 \leq j < n$. That means packets have arrived positions correct up to the third coordinate and in each tower of processors there are exactly n packets. Hence all the linear routing problems in a tower can be performed in parallel by n transport steps (as shown in lemma 2.2).

Figure 3 Partitioning of the cube into blocks

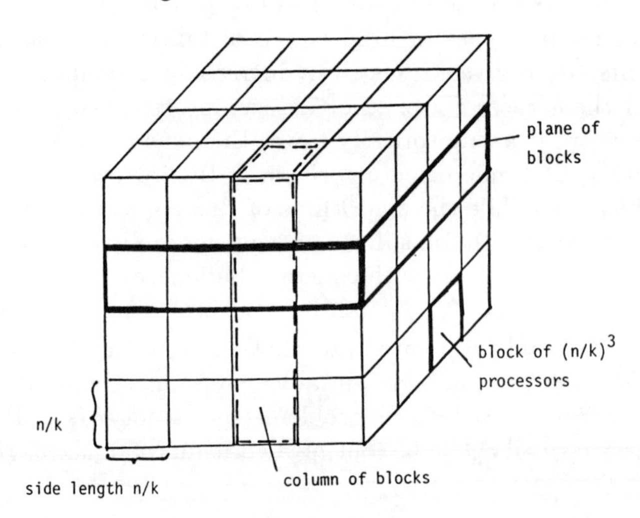

The r-dimensional problem, $r > 3$, will be solved in a similiar manner. The r-dimensional Sort-and-Route has to solve balanced routing problems on all $(r-1)$-dimensional subgrids by the $(r-1)$-dimensional Sort-and-Route. The transport of packets has to be done in parallel to the r-th axis as well as the shift operation of the corresponding r-dimensional rearrange algorithm presented in the following. Before we can describe the rearrange algorithm in its general form some definitions must be given. For arbitrary $s, 2 \leq s \leq r$, s-dimensional subcubes with sidelength n are considered. For example, if the last $r-s$ coordinates are fixed then we have a s-dimensional subcube of type $(*, ..., *, p_{s+1}, ..., p_r)$. For the rest of this section a block parameter k is assumed which is restricted to $2 \leq k \leq (f(n))^{1/(r-1)} \leq n^{1/r}$ and where n/k^2 is an integer. For this integer k the s-dimensional cube of processors is divided into k^s small cubes called blocks. The blocks are given by

$B(k_1,...,k_s) = [k_1 n/k, (k_1+1)n/k - 1] \times .... \times [k_s n/k, (k_s+1)n/k - 1]$ for all $k_i, 0 \le k_i \le k-1$, $i = 1,...,s$. That is each block contains $(n/k)^s$ processors. For a fixed k_s, $0 \le k_s \le k-1$, the union of blocks $\bigcup_{i=1}^{s-1} \bigcup_{k_i=0}^{k-1} B(k_1,...,k_i,...,k_s)$ is called the (k_s+1)-th plane of blocks and for fixed $k_i, i = 1,...,s-1$ the set $\bigcup_{k_s=0}^{k-1} B(k_1,...,k_{s-1},k_s)$ is called a column of blocks. (Compare Figure 3). Note that there are exactly k^{s-1} columns of blocks. Assume that these columns of blocks are numbered from 1 to k^{s-1} by any fixed numbering. Then let B_{ij} denote that block which is in the j-th column of blocks and in the i-th plane of blocks.

As in the last section the addresses are lexicographically ordered. That is $(p_1,...,p_s) \le (q_1,...,q_s)$ iff $lex(p_1,...,p_s) \le lex(q_1,...,q_s)$.

Rearrange algorithm : *Rearrange*$(s, mesh)$

1. In each block sort the packets with respect to indexing *rev*.
2. Shift packets along the s-axis: for all $p_i, i = 1,...,s, 0 \le p_i < n$, move contents of processor $(p_1,...,p_s)$ to processor $(p_1,...,p_{s-1},(p_s + (p_1 \ mod \ k)n/k) \ mod \ n)$.
3. In each block sort packets with respect to lexicographical indexing *lex*.

Lemma 3.1

Let *mesh* be an arbitrary s-dimensional submesh. For an arbitrary intervall of addresses $[P,Q]_{lex}$ let z be the number of packets in *mesh* with addresses in $[P,Q]_{lex}$. Then after applying the rearrange algorithm along the s-axis in every $(s-1)$-dimensional submesh $(*,...,*,p_s)$, $0 \le p_s < n$, there are at most $z/n + k + k^{s-1}$ packets with adresses in $[P,Q]_{lex}$.

Proof Let us call packets with adresses in $[P,Q]_{lex}$ critical packets. For a fixed $j, 1 \le j \le k^{s-1}$ let z_{ij} denote the number of critical packets in a block B_{ij}. Note that after the first step of the rearrange algorithm there are at most two 1-towers $(*,p_2,...,p_s) \cap B_{ij}$ and $(*,q_2,...,q_s) \cap B_{ij}$ which may contain both critical and non-critical packets. All the other 1-towers in the block contain either critical or non-critical packets. For these clean towers the shift distributes the critical packets uniformly to the k different blocks in the j-th column of blocks.

Now have a closer look at the at most two dirty towers. In one tower the critical packets are gathered at one side , and in the other they are on the other side. Hence there is at most one dirty intervall of length k in each of the dirty towers. (Compare Figure 4.) Again by the sorting in one intervall of length k the critical packets are gathered at the left side, and in the other they are on the right side. Hence after the shift operation the critical packets from dirty k-intervalls differ at most by one in the different blocks of a column of blocks. That is each block B_{hj} now contains either $\lfloor z_{ij}/k \rfloor$ or $\lceil z_{ij}/k \rceil$ critical packets from block B_{ij}. Let after shifting a_{hj} denote the number of critical packets in block B_{hj}. Then for all $h = 1,...,s$

$a_{hj} \le \sum_{i=1}^{k} \lceil z_{ij}/k \rceil$.

After the sorting with respect to lexicographical indexing there are at most two dirty s-towers within each block. Hence in an arbitrary $(s-1)$-dimensional submesh $(*,...,*,p_s)$, where $(*,...,*,p_s) \cap B_{hj} \ne \emptyset$, there are either $\lfloor a_{hj}/(n/k) \rfloor$ or $\lceil a_{hj}/(n/k) \rceil$ critical packets. Therefore in each $(s-1)$-dimensional submesh

$(*,...,*,p_s)$ there are at most

$$\sum_{j=1}^{k^{s-1}} \lceil (1/(n/k)) \sum_{i=1}^{k} \lceil z_{ij}/k \rceil \rceil \leq \sum_{j=1}^{k^{s-1}} \lceil (k/n) \sum_{i=1}^{k} (z_{ij}/k + 1) \rceil$$

$$\leq \sum_{j=1}^{k^{s-1}} \sum_{i=1}^{k} (z_{ij}/n + k/n) + k^{s-1} \leq z/n + k^{s+1}/n + k^{s-1}$$

critical packets. Since $k^s \leq n$ and $3 \leq s$ the lemma is proven.

Figure 4 Clean and dirty s-towers

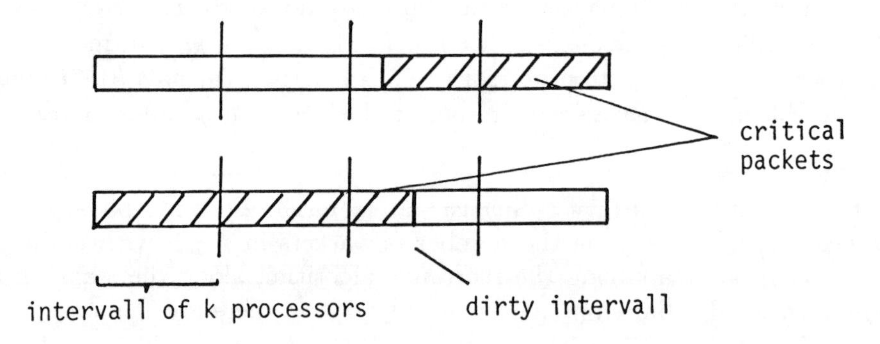

We are now ready to formulate the algorithm *Sort – and – Route* for r-dimensional cubes of processors.

Algorithm *Sort – and – Route* for r-dimensional cubes

for $s = r$ **downto** 3 **do**
begin { rearrange phase s }
for all $j = s+1, ..., r$, for all $p_j = 0, ..., n-1$ do in parallel:
$Rearrange(s, (*, ..., *, p_{s+1}, ..., p_r))$
end
for all $j = 3, ..., r$, and for all $p_j = 0, ..., n-1$ do in parallel
$Sort - and - Route(2, (*, *, p_3, ..., p_r))$
for $s = 3$ **to** r **do**
begin { correction phase s }
for all $j = 1, ..., r, j \neq s$, for all $p_j = 0, ..., n-1$ do in parallel:
in all s-towers $(p_1, ..., p_{s-1}, *, p_{s+1}, ..., p_r)$)transport packets to their correct s-address.
end

Lemma 3.2
For all intervalls of packets with addresses in $(p_1, ..., p_{i-1}, [a_i, b_i], *, ..., *)$, $2 \leq i \leq r$, after rearrange phases $r, ..., r-j$, $j = 0, ..., r-i$, there are at most $(b_i - a_i + 1)n^{r-i-j-1} + k + k^{r-1}$ packets in each hyperplane $(*, ..., *, p_{i+1}, ..., p_r)$.

Proof Before the rearrange phases start exactly $z = (b_i - a_i + 1)n^{r-i}$ critical packets have addresses in intervall $(p_1, ..., p_{i-1}, [a_i, b_i], *, ..., *)$ since the original problem is a permutation routing problem. For $j = 0$, that is after rearrange phase r, by lemma 3.1 we know that at most $z/n+k+k^{r-1} = (b_i-a_i+1)n^{r-i-1}+k+k^{r-1}$ lie in each $r-1$-dimensional submesh $(*, ..., *, p_r)$.

Assume that the lemma is true for all $h, h = 0, ..., j < r-i$. Then by induction hypothesis after rearrange phase $r-j$ there are at most $(b_i-a_i+1)n^{r-i-j-1}+k+k^{r-1}$ critical packets in each hyperplane $(*, ..., *, p_{i+1}, ..., p_r)$. Hence after rearrange phase $r-(j+1) = r-j-1$ by lemma 3.1 there are at most

$$((b_i - a_i + 1)n^{r-i-j-1} + k + k^{r-1})/n + k + k^{r-(j+1)-1}$$

$$= (b_i - a_i + 1)n^{r-i-(j+1)-1} + (k + k^{r-1})/n + k + k^{r-(j+1)-1}$$

$$\leq (b_i - a_i + 1)n^{r-i-(j+1)-1} + k + k^{r-1}$$

packets in each hyperplane $(*, ..., *, p_{i+1}, ..., p_r)$. Q.e.d.

Lemma 3.3

In all s-dimensional submeshes $(*, ..., *, p_{s+1}, ..., p_r)$:

- before correction phase s all packets have arrived at processors which are correct up to the first $s-1$ coordinates.
- after correction phase s all packets have arrived at processors which are correct up to the first s coordinates.

Proof Note that

(1) after rearrange phase $j, 3 \leq j \leq r$, a packet never leaves that $(j-1)$-dimensional submesh it was transported to before correction phase j starts.

That means that after rearrange phase 3 and before correction phase 3 for all $i, i = 3, ..., r$ and for all $p_i, 0 \leq p_i \leq n-1$ in each 2-dimensional submesh $(*, *, p_3, ..., p_r)$ by lemma 3.2 there are at most $b_2 - a_2 + 1 + k + k^{r-1}$ packets with addresses in $(p_1, [a_2, b_2], *, ..., *)$ for arbitrary p_1 and $a_2, b_2, 0 \leq a_2 \leq b_2 < n$. Hence the routing problems in each 2-dimensional submesh $(*, *, p_3, ..., p_r)$ are balanced (with respect to the second coordinate) and can be solved by the 2-dimensional Sort-and-Route. Hence before correction phase 3 all packets have arrived at processors which are correct up to the first two coordinates. Furthermore, in each 3-dimensional submesh $(*, *, *, p_4, ..., p_r)$ again by lemma 3.2 there are at most $b_3 - a_3 + 1 + k + k^{r-1}$ packets with addresses in $(p_1, p_2, [a_3, b_3], *, ..., *)$ for arbitrary $p_1, p_2, a_3, b_3, 0 \leq a_3 \leq b_3 < n$. Hence after the two-dimensional Sort-and-Route and before correction phase 3 all the 3-dimensional submeshes are balanced (with respect to the third coordinate). Therefore in every 3-tower $(p_1, p_2, *, p_4, ..., p_r)$ there are only packets with addresses correct up to the first two coordinates, that are addresses of type $(p_1, p_2, *, ..., *)$, and they are uniformly distributed. Hence in each 3-tower a balanced 1-dimensional problem is to be solved. By lemma 2.2 this can be done by $n + O(k^{r-1})$ transport steps. After routing along the 3-towers packets have arrived at processors which are correct up to the first three coordinates.

The argumentation for arbitrary $s, s \geq 4$, is analoguous as before and follows by induction. Assume that after correction phase $s-1$ and before correction phase s all packets have arrived at processors which are correct up to the first $s-1$ coordinates.

By lemma 3.2 and (1) we know that for all $i, i = s+1, ..., r$, and for all $p_i, 0 \leq p_i \leq n-1$, in each s-dimensional submesh $(*, ..., *, p_{s+1}, ..., p_r)$ there are at most $b_s - a_s + 1 + k + k^{r-1}$ packets with addresses in $(p_1, ..., p_{s-1}, [a_s, b_s], *, ..., *)$ for arbitrary $p_1, ..., p_{s-1}, a_s, b_s, 0 \leq a_s \leq b_s < n$. Hence all the s-dimensional submeshes are balanced (with respect to the s-th coordinate). Therefore in every s-tower $(p_1, ..., p_{s-1}, *, p_{s+1}, ..., p_r)$ there are only packets with addresses correct up to the first $s-1$ coordinates, that are addresses of type $(p_1, ..., p_{s-1}, *, ..., *)$. Hence in each s-tower a balanced 1-dimensional problem is to be solved. By lemma 2.2 this can be done by $n + O(k^{r-1})$ transport steps. After routing along the s-towers packets have arrived at processors which are correct up to the first s coordinates. Q.e.d

For the complexity of the algorithm note that sorting of the blocks with sidelength n/k can be performed by $O(n/k))$ steps. The shift operation costs n steps in a cube without wrap-arounds. Therefore each of the $r-2$ rearrange phases costs at most $n + O(n/k)$, and all these phases sum up to $(r-2)n + O(n/k)$. The two-dimensional Sort-and-Route needs in total $2n + O(n/k)$ steps. Each of the $r-2$ correction phases needs at most $n + O(k^{r-1})$ transport steps, which is bounded by $n + O(n/k)$ in the case of $k^r \leq n$. Then in total the r-dimensional Sort-and-Route needs at most $2(r-2)n + 2n + O(n/k) = 2(r-1)n + O(n/k)$ transport steps. Since $k^{r-1} \leq f(n)$ we can formulate the following theorem

Theorem 2

The permutation routing problem on an r-dimensional cube with sidelength n and buffersize $f(n), f(n) \leq n^{(r-1)/r}$, can be solved within
$2(r-1)n + O(n/f(n)^{1/(r-1)})$ transport steps.

Hence Sort-and-Route beats randomized routing algorithms [VB] as well as deterministic algorithms based on sorting [Ku2,Ku4] both approaches needing at least $(2r-1)n$ steps.

Furthermore note that only half of the number of steps is necessary if wrap-around connections are given.

It is not hard to see that the method can be extended to grids with arbitrary side lengths. For an $n_1 \times ... \times n_r$ grid the number of parallel data transport steps is then asymptotically $n_1 + n_2 + 2(n_3 + ... + n_r)$. Note that for $r \geq 3$ there is a certain asymmetry with respect to different dimensions. For example, for a $2^{1/3}n \times 2^{1/3}n \times 2^{-2/3}n$ mesh with n^3 processors asymptotically $2(2^{1/3}n + 2^{-2/3}n) = 54^{1/3}n < 3.84n$ steps are only needed instead of $4n$ for an 3-dimensional cube.

At the end of this section let us remark that the algorithm $Sort-and-Route$ is a so-called uniaxial algorithm. That is, at each time step all processors communicate only along one coordinate axis. Hence the algorithm is suitable for routing algorithms solving multi-packet problems by the help of overlapping techniques [KT].

4. Conclusion

In this paper we presented routing algorithms for r-dimensional $n \times \cdots \times n$ meshes with additional buffer for each processor. If the buffersize is $f(n)$ and wrap-around connections are given, then the algorithm needs $(r-1)n$ parallel data movements (neglecting low order terms) and for grids without wrap-arounds the method works with $(2r-2)n$ steps. For 2-dimensional meshes the routing algorithms match the distance bound, and therefore are asymptotically optimal. For $r \geq 3$ the routing algorithms are faster than the so far known deterministic routing methods based on sorting algorithms which need processors with buffersize of only one packet. Although the routing problem seems to be easier than the sorting problem, it is still an open question whether sorting can be solved as fast as routing on meshes with additional buffer. One result of this paper is that in the 2-dimensional case routing on meshes with additional buffer can be done faster than sorting on meshes without additional buffer. Another open question is whether the distance bound can be matched by any routing algorithm for r-dimensional grids with $r \geq 3$.

References:

[KRT] Krizanc, D.,Rajasekaran, S., Tsantilas, Th.: Optimal routing algorithms for mesh-connected processor arrays. Proceedings AWOC'88.In: Reif, J.H. (ed.),Lect. Notes Comp. Sci.,vol 319,pp. 411-422,Berlin:Springer 1988

[KT] Kunde, M., Tensi, T.: Multi-packet-routing on mesh connected arrays. To appear in: Proceedings of ACM Symposium on Parallel Algorithms and Architectures SPAA89, June 1989.

[Ku1] Kunde, M.: Lower bounds for sorting on mesh-connected architectures. Acta Informatica 24, 121-130 (1987).

[Ku2] Kunde, M.: Optimal sorting on multi-dimensionally mesh-connected computers. Proceedings of STACS 87. In: Brandenburg, F.J., et al. (eds.), Lect. Notes Comp. Sci., vol. 247, pp. 408-419, Berlin: Springer 1987

[Ku3] Kunde, M.: Bounds for l-section and related problems on grids of processors. Proceedings of PARCELLA 88. In: Wolf, G., Legendi, T., Schendel, U. (eds.), Mathematical Research, vol. 48, pp. 298-307, Akademie Verlag, Berlin, 1988

[Ku4] Kunde, M.: Routing and Sorting on Mesh-Connected Arrays. Proceedings AWOC'88. In: Reif, J.H. (ed.),Lect. Notes Comp. Sci.,vol 319,pp. 423-433,Berlin:Springer 1988

[LMT] Leighton, T., Makedon, F., Tollis, I.G.: A $2n-2$ algorithm for routing in an $n \times n$ array with constant size queues. To appear in: Proceedings of ACM Symposium on Parallel Algorithms and Architectures SPAA89, June 1989.

[MSS] Ma, Y., Sen, S., Scherson, I.D.: The distance bound for sorting on mesh-connected processor arrays is tight. Proceedings FOCS 86, pp. 255-263

[RT] Rajasekaran, S., Tsantilas, Th.: An optimal randomized routing algorithm for the mesh and a class of efficient mesh-like routing networks. 7th Conf. on Found. of Software Technology and Theoret. Comp. Science, Pune, 1987, pp. 226-241

[SS] Schnorr, C.P., Shamir, A.: An optimal sorting algorithm for mesh-connected computers, pp. 255-263. Proceedings STOC 1986. Berkley 1986

[TK] Thompson, C.D., Kung, H.T.: Sorting on a mesh-connected parallel computer. CACM **20**, 263-271 (1977)

[VB] Valiant, L.G., Brebner, G.J.: Universal schemes for parallel communication. Proceedings STOC 81, pp. 263-277.

OPTIMAL PARALLEL COMPUTATIONS FOR HALIN GRAPHS

Krzysztof Diks & Wojciech Rytter

Instytut Informatyki
Uniwersytet Warszawski, PKiN, 8p., pok. 850
00-901 Warszawa, Poland

Abstract.

Optimal parallel algorithms are given for two hard problems (the Hamiltonian cycle and the travelling salesman problem) restricted to graphs having a simple structure - Halin graphs. These problems were previously investigated for Halin graphs from the sequential point of view [1,5,6]. The travelling salesman problem (the computation of the shortest Hamiltonian cycle) for the Halin graph is interesting because such a graph can contain an exponential number of Hamiltonian cycles. Two tree-oriented algorithmic techniques are used: computation of products for paths of the tree (which gives $\log_2 n$ time algorithm for the Hamiltonian cycle) and a special parallel pebble game (giving $\log_2 n$ time for the travelling salesman problem).

1. Introduction.

By an optimal parallel algorithm we mean here an algorithm working in logarithmic time using $O(n/\log n)$ processors. Our model of parallel computations is the parallel random access machine without write conflicts, so called CREW PRAM (see [2] for an exact definition). Optimal parallel computations in the graph theory are mostly for problems on trees. We demonstrate techniques for the design of such optimal algorithms on an example class of tree-structured graphs: Halin graphs. These graphs are a simple variation of trees and in the same time some problems for them are nontrivial. We consider the Hamiltonian cycle problem and the travelling salesman problem. The nontriviality of these problems follows from the fact that the number of distinct Hamiltonian cyles in such graphs can be exponential. We show an application of the Euler tour technique, parallel prefix computation and the tree contraction method. All these methods can be implemented optimally on CREW PRAM, see [2,3].

A Halin graph (sometimes called a skirted tree) is a planar graph which consists of a tree T with no vertices of degree two and a circuit C (called the skirt) composed precisely of all the leaf vertices of T. In any planar embedding of the Halin graph, C forms the boundary of some face. Without loss of generality we can always take this to be the external face. We presume a natural representation for a Halin graph as follows. For each vertex v we have an adjacency list in which the neighbours of v are ordered in their, say clockwise, rotational occurence about v in the planar embedding of the graph. Given an adjacency list representation and the circuit C, we can easily obtain the representation used here. One of the most useful and general techniques for parallel computations on trees is the Euler tour technique due to Tarjan and Vishkin [7]. Using this technique a tree T of a given Halin graph can be rooted and for each vertex we can know its father and its sons. Let H=(T,C) be an arbitrary Halin graph where T is its interior tree and C is its skirt. We root the tree T at an arbitrary vertex of the circuit C. We will denote the root by *root* and its only son by *root1*. For each internal node v of T we will denote the subgraph of G induced by the subtree (of T) rooted at this node by G_v (see Fig. 1).

2. Hamiltonian cycles in Halin graphs.

We start from the Hamiltonian cycle problem. Let H=(T,C) be a given Halin graph, *root* a root of T (*root* is on C) and *root1* its only son. Let us consider some vertex v of the tree different from the root and consider the subgraph G_v. It is easy to observe that each Hamiltonian cycle have only three possibilities to enter and to leave the subgraph G_v. We can traverse this subgraph from the left to the top or from the right to the top or from the left to the right (see Fig. 2). These three different directions will be denoted respectively by d1={left,top}, d2={right,top}, d3={left,right}}. We want to find a Hamiltonian cycle containing both circuit edges incident to the root. It means that such Hamiltonian cycle must traverse the subgraph rooted at the only son of the root from the left to the right. Our goal will be to compute for each remaining vertex v the proper direction of traversing G_v (with respect to the whole graph). These directions will be values of the function val:T-->{d1,d2,d3}. This function plays a crucial role in the construction of the Hamiltonian cycle. The example of Fig. 1 will be used to illustrate the algorithm of calculating values of the function val whilst at the same time we indicate how the algorithm works for the general case (see also Figures 3, 4, 5).

At the first step we define the function dep_{e_v} :{d1,d2,d3}-->{d1,d2,d3}, such that dep_{e_v}(val(father(v))=val(v), where e_v=(v,father(v)) for each v≠*root* ,*root1* . This function defines the initial values of function val at each vertex v≠*root* ,*root1* with respect to the values of this function at father(v). At each vertex v we store the tabular representation of dep_{e_v}. The table records the value of val(v) for each possible value of this function at father(v). For example at the vertex v1 of the graph in Fig. 1 $dep_{e_{v1}}$ is :

d1	d2	d3
d3	d2	d1

. Here the first row stores arguments (the possible values of function val at father(v1) while the second row stores the corresponding values of val(v1). The evaluation of this function is of constant time because the number of values and arguments is constant.

In order to obtain at each vertex the final value of the function val we must compose functions dep_{e_v} for each vertex on the unique path in T from v to *root1*.

Let v=v1,v2, . . . vk=*root1*be consecutive vertices on the path from the vertex v to *root1*. Then val(v)=$dep_{e_{v1}}$ *$dep_{e_{v2}}$ * ... *$dep_{e_{vk}}$ (val(root1)), where val(root1)=d3. Composing functions dep_{e_v} is a simple matter because this operation is associative and it can be done optimally in logtime using the parallel prefix computation. It should be clear that computed directions uniquely determine edges of the Hamiltonian cycle.

Theorem 1

A Hamiltonian cycle of a given n-vertex Halin graph can be found in log*n* time using *n*/log*n*processors of CREW PRAM. □

3. The travelling salesman problem.

To solve the travelling salesman problem we use essentially the same technique as in [3]. A weighted Halin graph H=(T,C) (the meaning of T and C is the same as in the previous section) is a Halin graph in which each edge has associated a non-negative real number called its weight. The goal is to find in H the shortest Hamiltonian cycle (i. e. with the minimum sum of the edge weights). Assume that a Halin graph is represented in the same way as before.

We introduce a parallel pebble game on binary trees [2] because our algorithm is a variant of this game on weighted Halin graphs. This game is a kind of the tree contraction method. Within the game each vertex v of the tree T points to some vertex denoted by cond(v). At the outset of the game cond(v)=v for all v. During the game the pairs (v,cond(v)) can be thought of as additional edges. Another notion we shall require is that of "pebbling" a vertex. Within our application of the pebble game, "pebbling" a vertex denotes the fact that in the current state of the game it is enough information to evaluate the length of the shortest Hamiltonian cycle of the subgraph rooted there. At the outset of the game only the leaves of the tree are therefore pebbled. We say that a vertex is "active" if and only if cond(v)≠v.

The three operations activate, square, and pebble are components of a "move" within the game and are defined as follows:

```
activate:
  for each nonleaf vertex v in parallel do
  begin
    if v is not active and precisely one of its sons is pebbled
    then cond(v) becomes the other son;
    if v is not active and both sons are pebbled then cond(v) becomes
    one of the sons arbitrarily;
  end;

square:
  for each vertex v in parallel do cond(v):=cond(cond(v));

pebble:
  for each vertex in parallel do
    if cond(v) is pebbled then pebble v;
```

Now we define one (composite) move of the pebbling game to be a sequense of individual operations (activate; square; square; pebble) in that order. Then the following lemma provides a key result.

Lemma 1

Let T be a binary tree with n leaves. If initially only the leaves are pebbled then only $n/\log n$ processors are sufficient to pebble the root of T in $O(\log n)$ moves of the pebbling game.

□

For the proof see [2]. The main difference between operations described above and the operations activate, square and pebble we will introduce for our algorithm is that now each vertex v≠*root* has, not at most, but at least two sons.

Let H=(T,C) be a weighted Halin graph and G_{root1} a subgraph rooted at *root1*. As before the set directions={d1,d2,d3} where d1={left,top}, d2={right,top}, d3={left,right} consists of all possible directions of traversing the subgraph of G_{root1} by suitable Hamiltonian paths. We use the parallel pebble game to evaluate the length of the shortest Hamiltonian path in the subgraph G_{root1} for each of these three directions. Simultaneously we will determine edges of these paths in essentially the same fashion as in the previous algorithm. Given the three shortest Hamiltonian paths in G_{root1} it is a simple matter to find the shortest Hamiltonian cycle of the Halin graph H with respect to the weights of the edges outgoing from the root.

Let us define a function $cost_v$:directions-->R which for each vertex v gives the lengths of the shortest Hamiltonian paths in the subgraph rooted at v for all three directions d1,d2,d3. At each vertex v we store a tabular representation of $cost_v$. Within our algorithm the evaluation of values of this function for a given vertex corresponds to pebbling this vertex. At the outset of the game only the leaves of T1 are pebbled and for each leaf l $cost_l$(di)=0, i=1,2,3.

In this application of the pebbling game we add to the operation activate a recalculation of the function cost. If at some state of the game at least at two sons of vertex v, values of the functon cost haven't been calculated yet, then we say that this vertex is not active and then cond(v)=v. However it may be the case that only one son of v ,say w, hasn't evaluated values of the function cost. During the operation activate cond(v):=w and then a function $ccost_v$:$directions^2$-->R is constructed, which relates the value of the function $cost_v$ to the value of $cost_w$. More precisely values of the function $ccost_v$ store the length of the shortest Hamiltonian path of the subgraph rooted at v for each of directions d1,d2,d3 provided that the subgraph rooted at w is traversed by such a path also in one of the three directions and under the assumption that $cost_w$(di)=0 for i=1,2,3. Let us note that values of the function ccost we can store in a table of the constant size. If at some state of the game the values of the function cost at each son of the vertex v are known then $cost_v$ may be evaluated at hand. In our algorithm the following recalculation of the function ccost is done during the operation square. Suppose that cond(vi)=vj and cond(vj)=vk for some vertices vi,vj,vk. Then for instance $ccost_{vi}$(d1,d2)=min($ccost_{vi}$(d1,d)+$ccost_{vj}$(d,d2)), where d$\in$directions.

Pebbling a vertex within our algorithm corresponds to computing values of the function cost at that vertex. If cond(vi)=vj and vj is pebbled (i. e. $cost_{vj}$ is known) then we can compute the $cost_{vi}$ in the following way: $cost_{vi}$(dk)=min($ccost_{vi}$(dk,d)+$cost_{vj}$(d)), k=1,2,3 and d$\in$directions.

Note that in this application of the pebble game to each of original operations we add a recomputation of the functions defined for all vertices of the subgraph G_{root1}. It is not difficult to see that such a recomputation in the case of "square" and "pebble" requires a constant time. The only real difficulty might arise from the cost of evaluating the function cost in the operation activate since in the worst case this cost could be O(logn) for graphs with large degree (in a straightforward implementation of 'activate'). The cost of activate is constant for graphs with degree bounded by a constant. For general graphs one can make a kind of binarization. The details are very technical. The tree contraction (and the parallel pebble game) can be implemented by an optimal parallel algorithm, see [3]. Together this proves the following:

Theorem 2

The travelling salesman problem in the Halin graph can be solved in O(log*n*) time using *n*/log*n* processors. □

4. Closing remarks.

The presented methods can be used generally for each tree structured graph (where by tree structured we mean decomposable, partial k-tree or of bounded tree-width). However Halin graphs are most suitable for demonstrating the techniques of efficient parallelization for problems on tree-structured graphs.

Bibliography

1. G. Cornnuejols, D.Nadoff, W.Pullybank. Halin graphs and the travelling salesman problem. Math.Progr. 26 (1983) 287-294
2. A.Gibbons, W.Rytter. Efficient parallel algorithms. Cambridge University Press (1988)
3. A. Gibbons, W. Rytter. Optimal parallel algorithms for dynamic expression evaluation and context-free recognition. Information and Computation 81 (1989), 32-45.
4. W.Rytter. Fast parallel omputations for some dynamic programming problems. Theor.Computer Science (1988)
5. M.Syslo. NP-complete problems on some tree structured graphs. WG'83 (ed.M.Nagl).
6. M.Syslo, A.Proskurowski. On Halin graphs, in M.Borowiecki, J.W.Kennedy, M.Syslo (Editors), Graph Theory-Lagow 1981, LN in Maths, Springer-Verlag, Berlin-Heidelberg, 1983.
7. R.E.Tarjan,U.Vishkin, An Efficient Parallel Biconnectivity Algorithm, SIAM J.Comput. 14:4(1985).

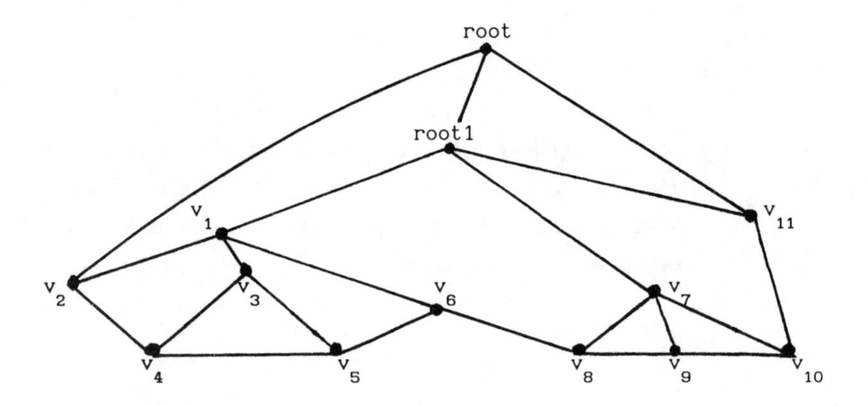

Fig. 1 An example of a Halin graph.

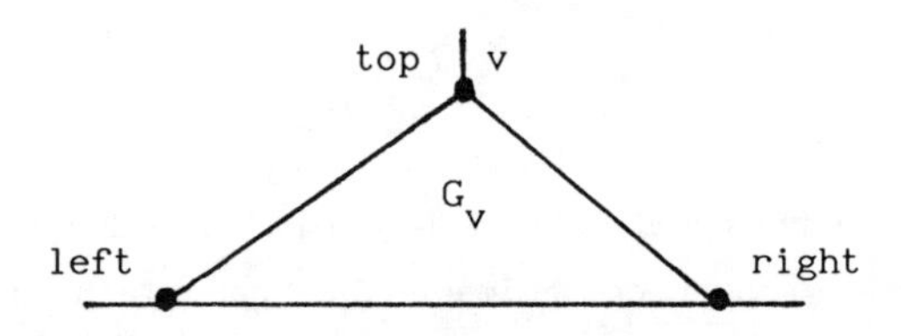

Fig. 2 A subgraph rooted at v.

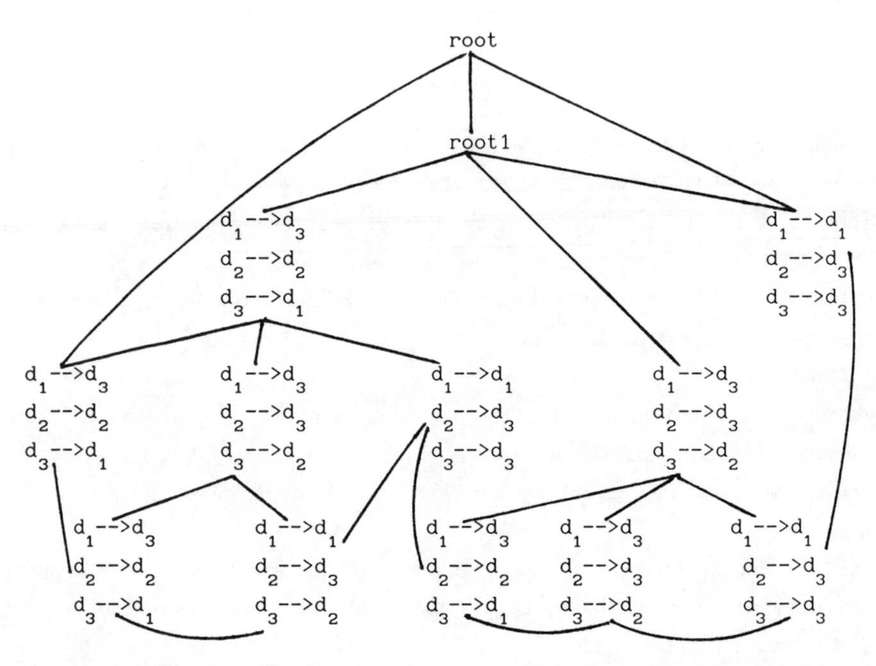

Fig. 3 The Halin graph of fig. 1 with computed values of the function dep_{e_v} at each vertex $v \neq$ root, root .

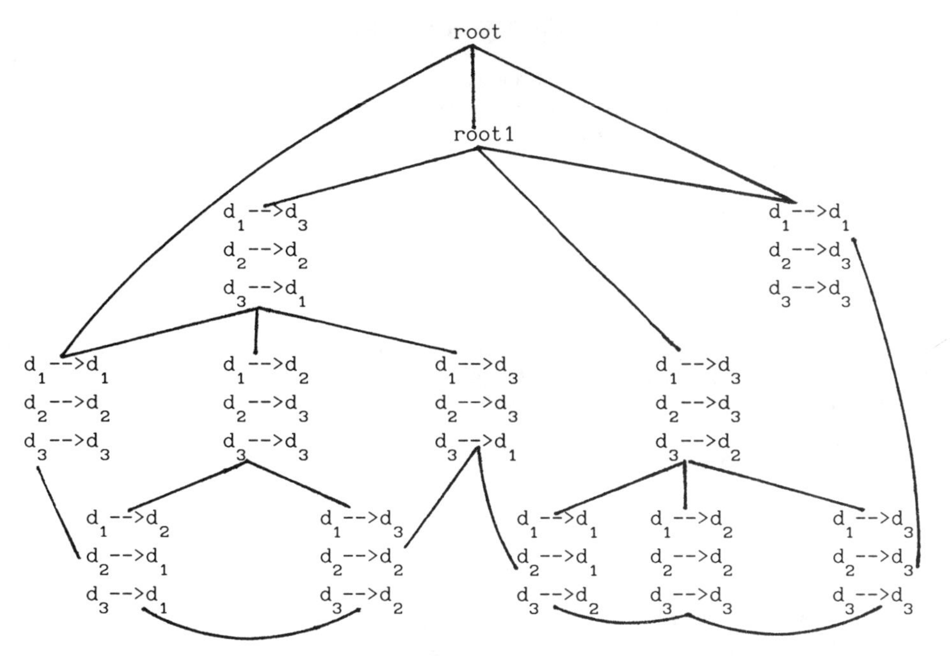

Fig. 4 The Halin graph of Fig. 1 with computed compositions of functions dep on paths.

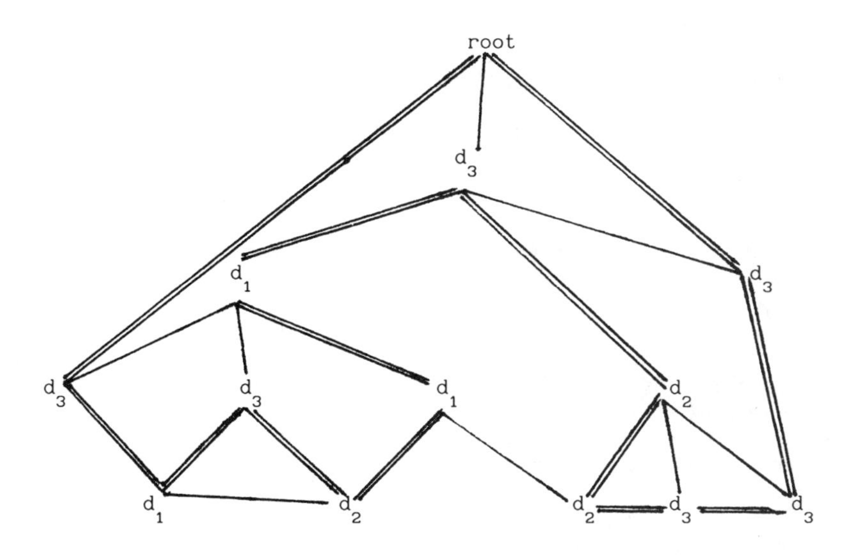

Fig.5. The Halin graph of Fig. 1 with computed values of the function val with respect to val(root1)=d_3 and the suitable Hamiltonian cycle consisting of marked edges.

OPTIMAL PARALLEL ALGORITHMS FOR b-MATCHINGS IN TREES

Constantine N.K. Osiakwan and Selim G. Akl
Department of Computing and Information Science
Queen's University, Kingston, Ontario, Canada K7L 3N6.

Abstract

We present adaptive parallel algorithms for $b-$matchings in trees. The algorithms are designed using the exclusive-read exclusive-write parallel random-access machine (EREW PRAM) model of parallel computation. For a tree of n vertices, the algorithms run in $O(n/p + \log n)$ time using p processors ($p \leq n$). When $p \leq n/(\log n)$, the algorithms are cost optimal.

Ordinary matching problems are special cases of $b-$matching problems. The best previously known parallel algorithm for ordinary matching in trees runs in $O(\log n)$ time using $O(n)$ processors on the EREW PRAM model. Our algorithms achieve the same time complexity with fewer processors.

Keywords: $b-$matching, matching, tree, parallel algorithm, EREW PRAM, postorder numbering, minimum set cover, maximum independent set.

1. Introduction

We are given a tree $T=(V_T, E_T)$ of n vertices. Let $deg(v)$, $v \in V_T$, be the degree of vertex v and $b(v)$ be a given bound such that $0 \leq b(v) \leq deg(v)$. Further, let $M \subseteq E_T$. If

$$|\{ \{u,v\} : \{u,v\} \in M\}| \leq b(v) \text{ for all } v \in V_T,$$

in other words, the number of edges in M incident on v is at most $b(v)$ for all vertices v in V_T, then M is referred to as a *$b-$matching* of T.

The following problems arise from the definition of a $b-$matching:

(i) Determine a $b-$matching M such that the cardinality of M is maximum. This problem is referred to as the *maximum cardinality $b-$matching problem*.

(ii) Suppose that a real-valued positive weight $\omega(\{u,v\})$ is associated with each edge $\{u,v\} \in E_T$ of the tree. The problem of finding a $b-$matching M, such that

$$\sum_{\{u,v\} \in M} \omega(\{u,v\})$$

is maximum, is known as the *maximum weight $b-$matching problem*.

(iii) A problem related to problem (ii) is to determine a maximum cardinality $b-$matching M such that

$$\sum_{\{u,v\} \in M} \omega(\{u,v\})$$

is maximum. Let us refer to this problem as the *maximum weight maximum cardinality $b-$matching problem*.

To clarify the differences between these $b-$matchings, consider the weighted tree in Figure 1. The bound on each of the eleven vertices is the integer (in brackets) shown next to the vertex. A maximum cardinality $b-$matching of the tree in Figure 1 is shown in Figure 2, where only edges in the $b-$matching are displayed. A maximum weight $b-$matching and the maximum weight maximum cardinality $b-$matching are shown in Figures 3 and 4 respectively. The total weight of the maximum weight $b-$matching in Figure 3 is 119 and the maximum weight maximum cardinality $b-$matching in Figure 4 has weight 74.

Goodman, Hedetniemi and Tarjan [1976] observed that the maximum weight maximum cardinality $b-$matching problem is reducible to the maximum weight $b-$matching problem. Let ψ be the sum of the weights of all the edges in the tree T. We can scale the weights of the edges as follows:

$$\omega'(\{u,v\}) = 1 + \frac{\omega(\{u,v\})}{2\psi} \quad \text{for } \{u,v\} \in E_T.$$

This work was supported by the Natural Sciences and Engineering Research Council of Canada under grant A3336.

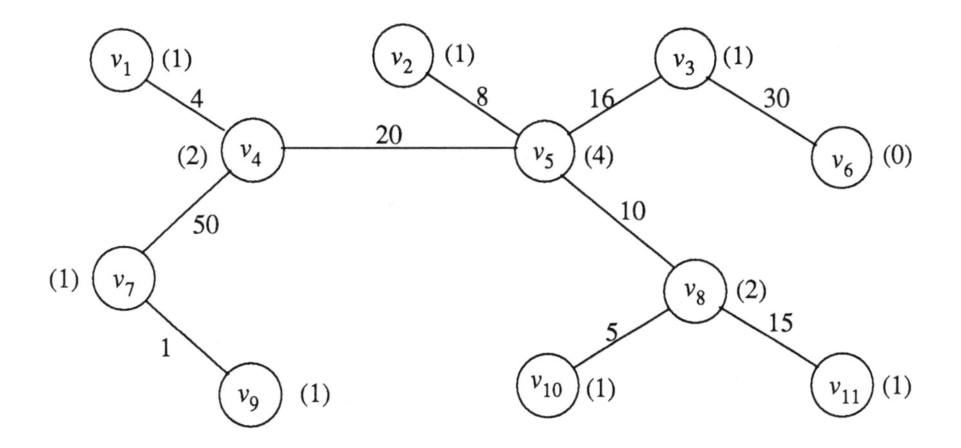

Figure 1. A Weighted Tree with Bounds on the Vertices.

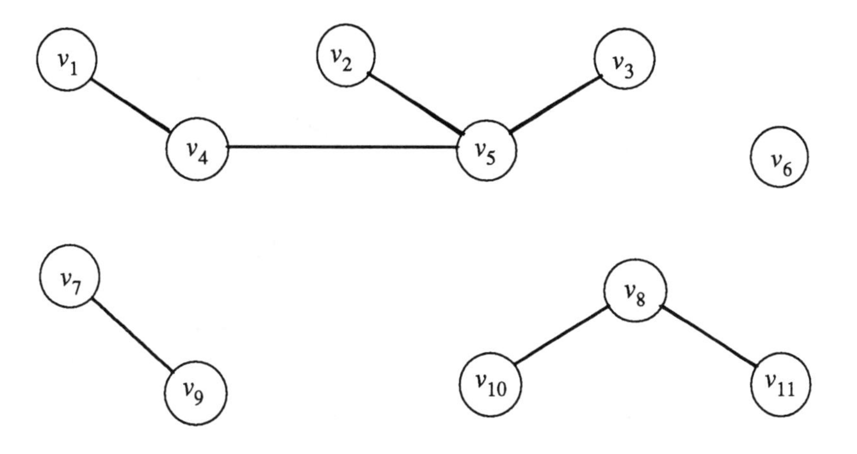

Figure 2. A Maximum Cardinality b-Matching for the Tree in Figure 1.

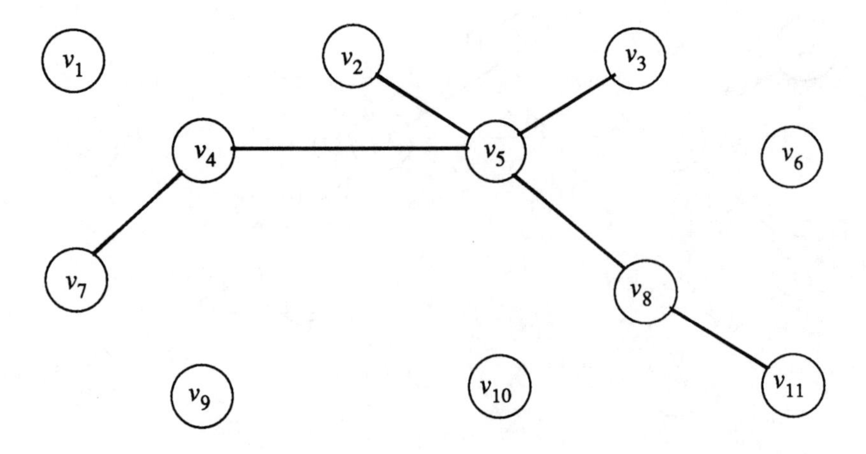

Figure 3. A Maximum Weight b-Matching for the Tree in Figure 1.

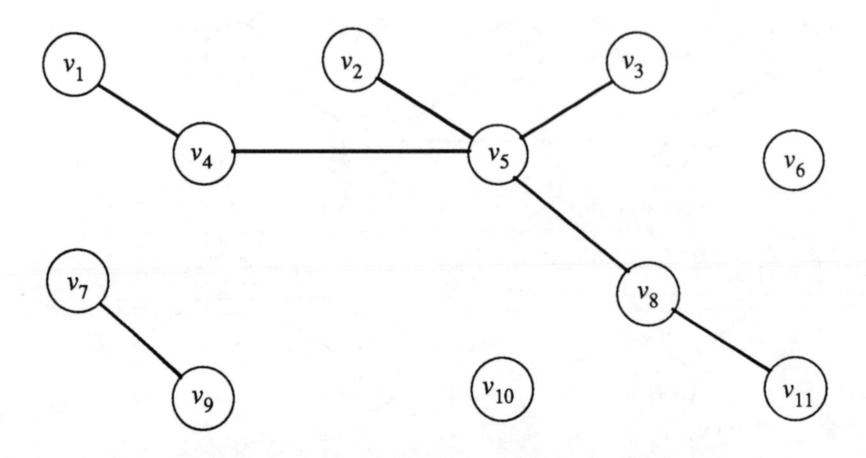

Figure 4. A Maximum Weight Maximum Cardinality b-Matching for the Tree in Figure 1.

This implies

$$\sum_{\{u,v\}\in T} (\omega'(\{u,v\}) - 1) = \frac{1}{2}.$$

By this transformation, if M is any $b-$matching of T, then M is a $b-$matching such that $\sum_{\{u,v\}\in M} \omega'(\{u,v\})$ is maximum if and only if M is a maximum cardinality $b-$matching such that $\sum_{\{u,v\}\in T} \omega(\{u,v\})$ is maximum. A method for solving the maximum weight $b-$matching problem can be used to solve the maximum weight maximum cardinality $b-$matching problem. We will therefore not consider the maximum weight maximum cardinality $b-$matching problem further.

The ordinary matching problems (sometimes referred to as 1-matching problems) are special cases of $b-$matching problems (Berge [1973]), where all the bounds for the vertices are unity. It is also known that $b-$matching problems in arbitrary graphs are reducible to the ordinary matching problems (Lawler [1976], Papadimitriou and Steiglitz [1982]). Efficient algorithms (Gabow [1976], Gabow [1974], Lawler [1976]), based on the algorithms by Edmonds (Edmonds [1965a], Edmonds [1965b]), exist to solve the ordinary matching problems in arbitrary graphs. Linear time sequential algorithms for $b-$matchings in trees were presented by Goodman, Hedetniemi and Tarjan [1976].

A parallel algorithm for ordinary maximum weight matching in trees was first presented by Pawagi (Pawagi [1987]). The algorithm runs in $O(\log^2 n)$ parallel time with $O(n)$ processors using the concurrent-read exclusive-write (CREW) PRAM model of parallel computation. A direct simulation of this algorithm on the exclusive-read exclusive-write (EREW) PRAM model runs in $O(\log^3 n)$ time. Recently, an $O(\log n)$ parallel algorithm for ordinary maximum cardinality matching in trees was obtained by He and Yesha [1988]. He and Yesha designed their algorithm to use $O(n)$ processors on the EREW PRAM model. (For definitions of these models of parallel computation, see Akl [1989].)

In this paper we present $b-$matching algorithms that achieve a running time of $O(n/p + \log n)$ on the EREW PRAM model using p processors. When all the bounds are set to 1, the $b-$matching algorithms solve the ordinary maximum cardinality and maximum weight matching problems in trees. By comparison with the algorithms mentioned in the previous paragraph, our algorithms solve a more general problem, are adaptive (since they use $p \leq n$ processors), and cost optimal when $p \leq n/\log n$. Notice that when $p = n/\log n$, the algorithms run in $O(\log n)$ time which is faster than the algorithm of Pawagi, while using fewer processors than that of He and Yesha.

In Section 2, we discuss the internal representation of the tree for the algorithms. The significance of postorder numbering of the vertices of the tree and a description of a parallel algorithm for postorder numbering in trees are also given in Section 2. Section 3 presents sequential and parallel algorithms for maximum cardinality $b-$matching in trees. Sequential and parallel algorithms for maximum weight $b-$matchings in trees are described in Section 4. Some applications of $b-$matching algorithms are mentioned in Section 5. A summary of the paper is given in Section 6 together with a list of open problems.

2. Postorder Numbering

We first describe the representation of the tree. The tree is assumed to be represented by a set of vertices $V_T = \{v_1, v_2, \ldots, v_n\}$ in the array VERTICES of length n. The adjacency lists (lists of incident edges) of the vertices are sequentially stored in the array EDGES of length $2n-2$. Each undirected edge $\{v_i, v_j\}$ is represented by two directed edges (v_i, v_j) and (v_j, v_i). An edge (v_i, v_j) is said to be incident on vertex v_i. Let ADJACENCY(v_i) denote the set of edges incident on vertex v_i. The elements of ADJACENCY(v_i) are stored in consecutive locations of EDGES. For each vertex v_i there are pointers to the respective locations in the array EDGES where the adjacency list for vertex v_i begins and ends. Also for each directed edge (v_i, v_j), there is a pointer to its reversal (directed) edge (v_j, v_i). An example of the array EDGES for the tree in Figure 1 is shown in Figure 5(a).

In solving the maximum cardinality $b-$matching problem, Goodman, Hedetniemi and Tarjan [1976], observed

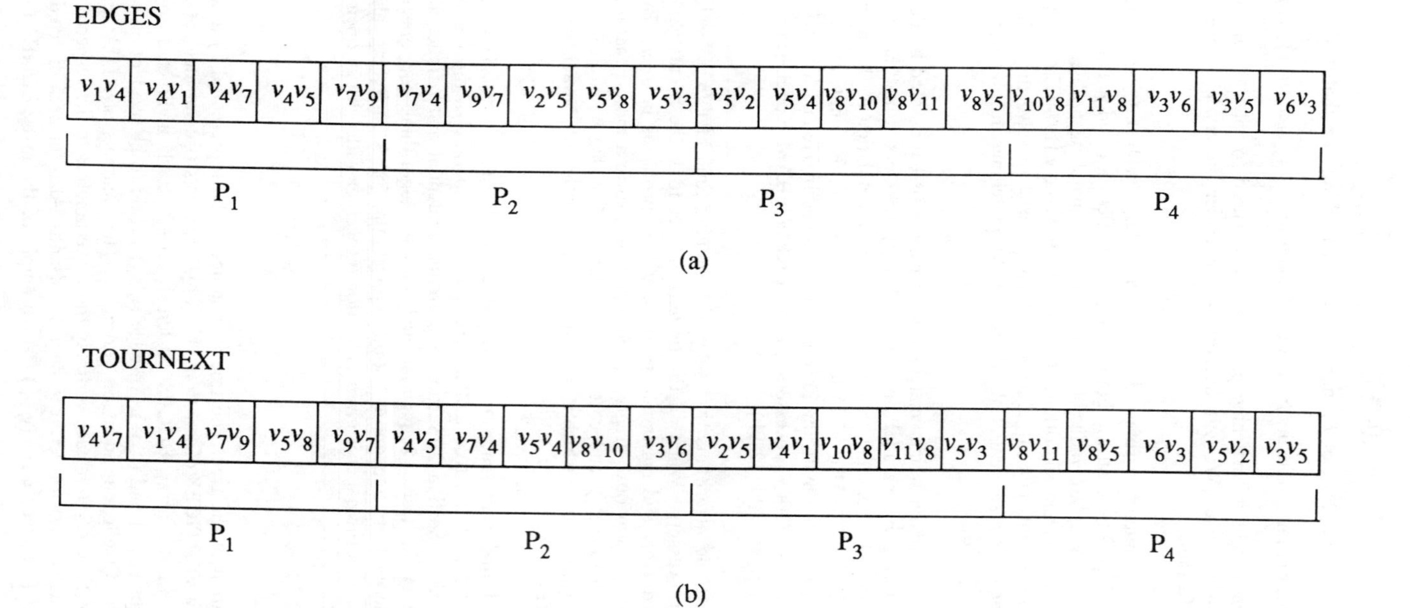

Figure 5. (a) A Representation of the Edges of the Tree in Figure 1 and (b) the Corresponding Representation of the Euler Tour.

that: if an edge $\{v_i, v_j\}$ is the only edge incident on v_i in the tree, with $b(v_i) > 0$ and $b(v_j) > 0$, then there is a maximum cardinality $b-$matching which contains the edge $\{v_i, v_j\}$.

Their algorithm for computing the maximum cardinality $b-$matching is based on this observation. Starting from an empty $b-$matching, the algorithm basically identifies any vertex v_i with degree one that has $b(v_i)$ greater than zero. The only edge $\{v_i, v_j\}$ incident on v_i is added to the $b-$matching (provided $b(v_j)$ is also greater than zero) and the degrees of the vertices v_i and v_j are each decreased by one. The bounds for v_i and v_j are each decreased by one as well. However, if $b(v_j)$ is zero the edge $\{v_i, v_j\}$ is deleted from the tree and $deg(v_j)$ is decreased by one. Also if v_i has degree one and $\{v_i, v_j\}$ is an edge such that $b(v_i)$ is zero but $b(v_j)$ is greater than zero, then $\{v_i, v_j\}$ is removed from the tree and $deg(v_j)$ is decreased by one. This process is repeated until all the vertices are considered. In order to efficiently process the vertices in the way described, a bijection

$$\beta : \{1, 2, \ldots, n\} \longleftrightarrow \{v \in V_T\}$$

is defined such that

$$\pi(v) = \{w \mid (v,w) \in E_T \text{ and } \beta^{-1}(w) > \beta^{-1}(v)\}$$

has no more than one element. For each vertex $v \in V_T$, $P(v)$ is defined to be the unique vertex of $\pi(v)$ if $|\pi(v)| = 1$ and 0 otherwise. Once β has been defined, the vertices are assumed to be identified by number (i.e. $\beta(v) = v$ and therefore $\beta^{-1}(v) = v$).

One such bijection is postorder numbering of vertices of the tree with respect to an arbitrary root. Here $P(v)$ is the parent of vertex v when the tree is traversed in postorder manner (Knuth [1976], Tarjan [1972]). Postorder numbering satisfies the conditions of the bijection as stated above, and therefore guarantees that the maximum cardinality matching is obtained in linear time. In what follows, we recap the sequential linear time postorder numbering of vertices in a graph, discuss a parallel algorithm for postorder numbering due to Tarjan and Vishkin [1984], and show how an optimal algorithm can be derived from their algorithm.

2.1. Sequential Postorder Numbering Algorithm

Let a tree T and a distinguished vertex r, referred to as the *root* of the tree, be given. Postorder numbering is defined recursively as follows. Let $T_1, T_2, \cdots, T_k$ be the subtrees whose roots $r_1, r_2, \cdots, r_k$ are the children of r. Visit the subtrees $T_1, T_2, \cdots, T_k$, in that order, in postorder manner assigning consecutive numbers from 1 to $n-1$ to the vertices in the order in which they are visited. Visit the root r of the tree T and assign it the number n.

Tarjan [1972] presents a sequential technique for such numbering of the vertices of T. Starting from the root, all vertices adjacent to it are visited. When visiting a vertex v, the vertices adjacent to v are postorder numbered. This algorithm, called PO-NUMBER, is given below together with procedure POSTORDER which it calls. The algorithm traverses each edge in the tree exactly twice. Since there are $2n-2$ edges in the tree, postorder numbering is achieved in $O(n)$ time. A postorder numbering of our sample tree in Figure 1 using this algorithm is shown in Figure 6. The root is vertex v_4, and the postorder numbers are shown next to the vertices.

```
Algorithm PO-NUMBER
    for all vertices u do
        NUMBER[u] ← 0
    end for
    i ← 0
    POSTORDER(root)
end PO-NUMBER
procedure POSTORDER(u)
    for v such that {u,v} is an edge and v is not visited do
        POSTORDER(v)
        i ← i + 1
        NUMBER[u] ← i
    end for
end POSTORDER
```

2.2. Parallel Algorithm for Postorder Numbering

Tarjan and Vishkin [1984] gave a parallel algorithm for computing postorder numbers of vertices in a tree. The algorithm assumes the EREW PRAM model of parallel computation and executes in $O(\log n)$ time using $O(n)$ processors. We discuss how their algorithm can be modified to execute on the same model in $O(n/p + \log n)$ time using p processors.

Our representation of the tree satisfies the representation of graphs assumed by Tarjan and Vishkin [1984]. They refer to the first edge on the incidence list of a vertex v_i in a tree T, as TREEADJ(v_i). For each directed edge (v_i, v_j), the edge that follows it on the adjacency list of vertex v_i (if any) is referred to as TREENEXT[(v_i, v_j)].

Tarjan and Vishkin assigned a processor to each vertex v_i and a processor to each directed edge (v_i, v_j). A circularly linked list corresponding to an Euler tour of the directed version of the tree is constructed. For each edge (v_i, v_j) the next edge TOURNEXT[(v_i, v_j)] in the tour is TREENEXT[(v_j, v_i)] if the latter is not *null*, TREEADJ(v_j) otherwise. This construction takes O(1) parallel time in their case or $O(n/p)$ time using p processors, by assigning $(2n-2)/p$ edges to a processor.

In Figure 5(a), the edges of our sample tree are distributed among four processors. The array TOURNEXT of the tree, computed by these four processors, is given in Figure 5(b). Consider the computation of TOURNEXT[(v_1, v_4)]. TREENEXT[(v_4, v_1)] is the edge (v_4, v_7), and therefore TOURNEXT[(v_1, v_4)] is (v_4, v_7). For the edge (v_4, v_5), TREENEXT[(v_5, v_4)] is *null*. So we let TOURNEXT[(v_4, v_5)] be equal to TREEADJ(v_5), which is (v_5, v_8). TOURNEXT for other edges are computed similarly.

The Euler tour corresponds to the order of advancing and retreating along edges during a depth-first search traversal of the tree, starting at an arbitrary vertex (see Figures 7 and 8). To root the tree, the Euler tour is broken at an arbitrary edge causing some edge, say (v_i, v_j), to be the first edge on the list. The vertex v_i becomes the root of the tree. The broken list is referred to as the *traversal list*. In the worst case, this can be achieved in $O(\log n)$ time using $O(n)$ processors. Using p processors, this can be done in $O(\log p)$ time, since all processors need to know where the circular list is broken. To do this,

(a) the first processor breaks the list at an arbitrary edge, and
(b) makes known to all other processors, the point of breakage by a broadcast.

Let the edge used to visit a vertex after the first time in an Euler tour be called a *retreat edge*. Suppose we number the retreat edges from 1 to $n-1$ as we encounter them in the Euler tour. We observe that the number, k, of a retreat (directed) edge (v_i, v_j) is the postorder number of the vertex v_i. This implies that if we are able to identify edges which are retreat edges, then we can easily postorder number the vertices.

To identify the retreat edges, Tarjan and Vishkin number the edges of the traversal list from 1 to $2n-2$ in traversal order in $O(\log n)$ time with $O(n)$ processors using a *doubling* technique to compute, in effect, for each edge (v_i, v_j) the number of edges before (v_i, v_j) from the beginning of the traversal list. The *doubling* technique is the same as the one used to compute prefix sums (Akl [1985]). An array RANK[(v_i, v_j)] is initialized to 1 for each edge (v_i, v_j). The TOURNEXT pointers of the traversal list are then reversed, and a list ranking (prefix sums) computation performed on RANK using the reversed TOURNEXT in the computation. Tarjan and Vishkin observed that of the two edges (v_i, v_j) and (v_j, v_i), the lower-numbered one corresponds to an advance from v_i to v_j along the tree edge $\{v_i, v_j\}$ and the higher-numbered one to a retreat from v_j to v_i along $\{v_i, v_j\}$. Thus using the edge numbers, each directed edge can be marked as either an advance edge or a retreat edge. For each vertex v_j other than the root, there is exactly one retreat edge (v_i, v_j), the parent $P(v_i)$ of v_i in the tree is v_j.

Once the retreat edges are identified, we can then number the vertices in postorder much as was done for the edge numbers. The only difference is that we re-initialize RANK[(v_i, v_j)] to 1 if (v_i, v_j) is a retreat edge and 0 otherwise. The root gets the postorder number n.

An optimal parallel algorithm for computing list ranking is given in Cole and Vishkin [1988], [1986]. The algorithm runs on the EREW PRAM in $O(\log n)$ time using $n/\log n$ processors. Using Brent's theorem

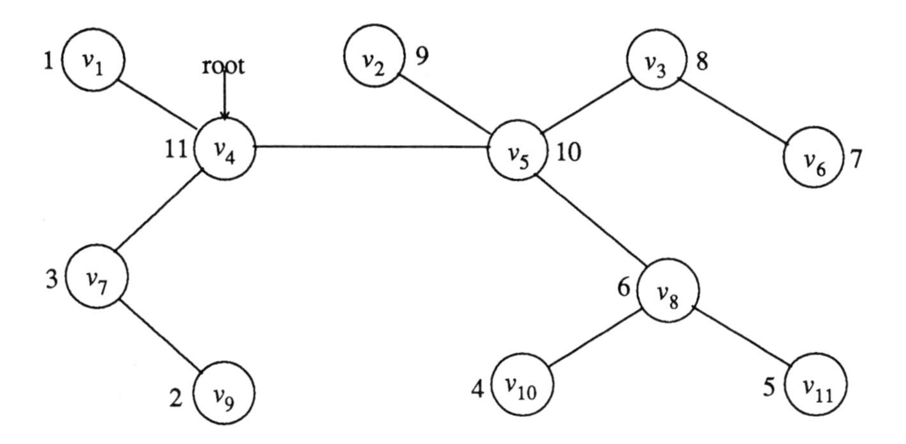

Figure 6. A Postorder Numbering of the Tree in Figure 1.

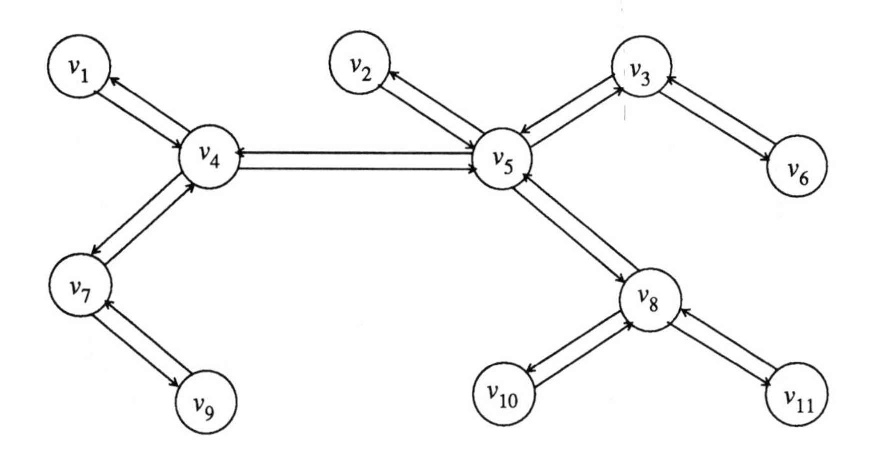

Figure 7. A Directed Graph of the Tree in Figure 1.

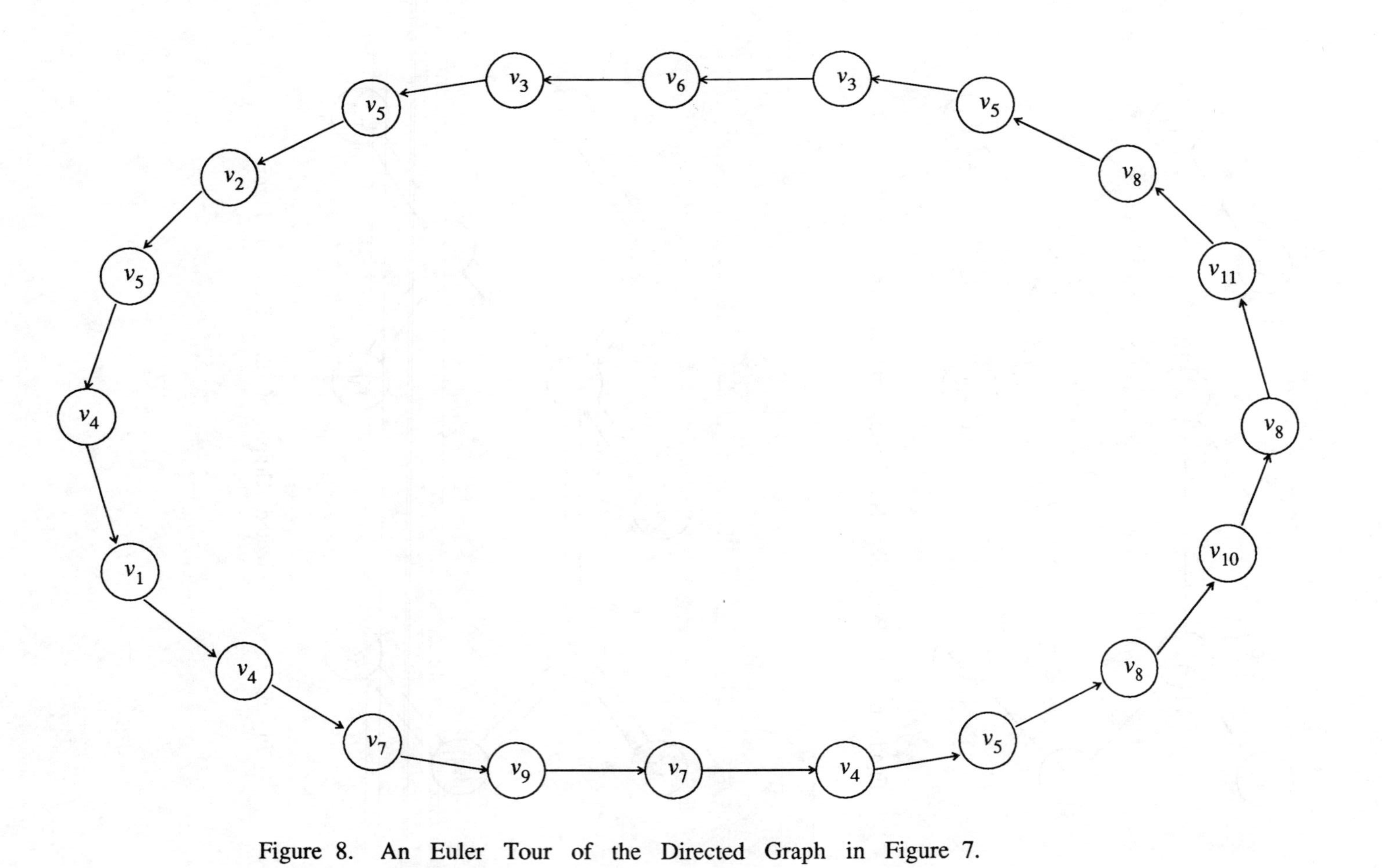

Figure 8. An Euler Tour of the Directed Graph in Figure 7.

(Brent [1974]) the algorithm of Cole and Vishkin can be implemeted to run in $O(n/p+\log n)$ time using p processors. The algorithm of Cole and Vishkin can be used for the list ranking computation.

Once the postorder numbers are computed, we can replace each occurrence of a vertex by its postorder number, retaining an inverse map to restore the original vertex names when computation is complete (for each number i, we remember VERTEX(i), the vertex v_k with postorder number i). We summarize the above discussion in the high level algorithm POSTORDERNUMBER.

Theorem 2.1.
Given an undirected tree, the vertices of the tree can be numbered in postorder manner in $O(n/p + \log n)$ time in parallel using p $(\leq n)$ processors.
Proof.
Steps 0, 1 and 8 of the algorithm POSTORDERNUMBER are executable in constant time. It takes $O(\log p)$ time to execute Step 2. Steps 3, 5, 7, and 9 will each take $O(n/p)$ time. The algorithm of Cole and Vishkin can be used to process Steps 4 and 6. For $p \leq n$, these will take $O(n/p + \log n)$ time. Thus the running time of the modified parallel algorithm for postorder numbering runs in $O(n/p + \log n)$ time using p processors. □

Algorithm POSTORDERNUMBER
  Step 0:
    Assign $(2n-2)/p$ edges to a processor

  Step 1:
    Processor P_1 arbitrarily chooses an edge (v_i, v_j) as the break point of the circular list

  Step 2:
    for all processors **do in parallel**
      broadcast edge (v_i, v_j) to all processors
    end for

  Step 3: {Determine the retreat edges}
    for all processors **do in parallel**
      (a) reverse the links, TOURNEXT, of your sub-list
      (b) set RANK[(v_i, v_j)] to 1, for each edge (v_i, v_j) on your sub-list
    end for

  Step 4:
    for all processors **do in parallel**
      perform a list ranking on RANK using the reversed TOURNEXT
    end for

  Step 5:
    for all processors **do in parallel**
      for each edge (v_i, v_j) **do**
        if (RANK[(v_j, v_i)] > RANK[(v_i, v_j)])
          then
            (a) set RETREAT[(v_j, v_i)] to *true*
            (b) set RANK[(v_j, v_i)] to 1
          else
            (a) set RETREAT[(v_j, v_i)] to *false*
            (b) set RANK[(v_j, v_i)] to 0
        end if
      end for
    end for

```
Step 6: {Compute postorder numbers}
    for all processors do in parallel
        perform a list ranking on RANK using the already reversed links of TOURNEXT
    end for

Step 7:
    for all processors do in parallel
        for each edge (v_i ,v_j ) do
            if (RETREAT[(v_i ,v_j )] = true ) then
                set NUMBER[v_i ] to RANK[(v_i ,v_j )]
            end if
        end for
    end for

Step 8:
    Assign n /p vertices to a processor

Step 9:
    for all processors do in parallel
        for each vertex v_i do
            (a) j ← NUMBER[v_i ]
            (b) set VERTEX(j ) to v_i
        end for
    end for
```

To illustrate the postorder numbering of the tree in Figure 1, an Euler tour with vertex v_4 as the root is shown in Figure 9. The edges are numbered in the order in which they are encountered during the traversal. The circularly linked list of the tour is shown in Figure 10. The numbers by the links are the numbers associated with the edges in Figure 9. The traversal list of the tour is given in Figure 11(a). The reversed pointers of TOURNEXT, and the initial values of RANK are shown in Figure 11(b). The values of RANK after the list ranking computation on RANK are given in Figure 11(c) with each retreat edge indicated by an "x".

To compute the postorder numbers, we re-initialize the array RANK as shown in Figure 12(a). The retreat edges are still marked with an "x". Performing the list ranking computation on RANK using the already reversed pointers of TOURNEXT, we obtain the values of RANK as shown in Figure 12(b). This gives the order in which we encounter the retreat edges in the tour of the tree with vertex v_4 as the root, as shown in the traversal list in Figure 12(c). The circularly linked list of the traversal list in Figure 12(c) is given in Figure 13 with the order of encounter of the retreat edges during traversal of the tree indicated by the numbers. The tour which corresponds to the circularly linked list in Figure 13 is shown in Figure 14. For each retreat edge (v_i, v_j), the postorder number of the tail vertex v_i is the computed value of RANK associated with the edge. The postorder number of the *root* is 11. The numbering obtained at the end of this example is shown in Figure 6. In the remainder of this paper, "number" will refer to the postorder number of a vertex.

3. Maximum Cardinality $b-$Matching

The sequential algorithm of Goodman, Hedetniemi and Tarjan [1976], first computes the numbers of the vertices, obtaining a parent $P(u)$ for each vertex u. The vertices are then referred to by their numbers. An array S of length n is used in the computation of the $b-$matching. At completion of the computation of the $b-$matching, an edge $\{u, P(u)\}$ is in the $b-$matching if the value of $S(u)$ is *true*. Another array COUNT, also of length n, is used to indicate the number of edges incident on a vertex that are currently in the $b-$matching.

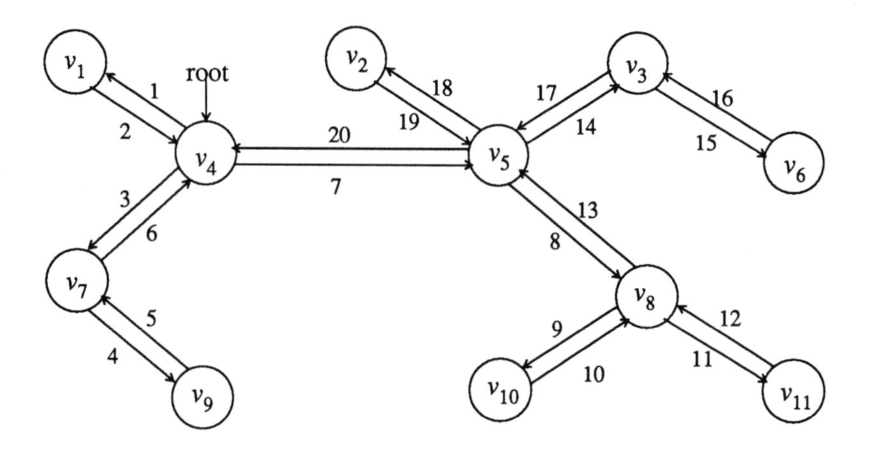

Figure 9. Order of the Euler Tour in Figure 8.

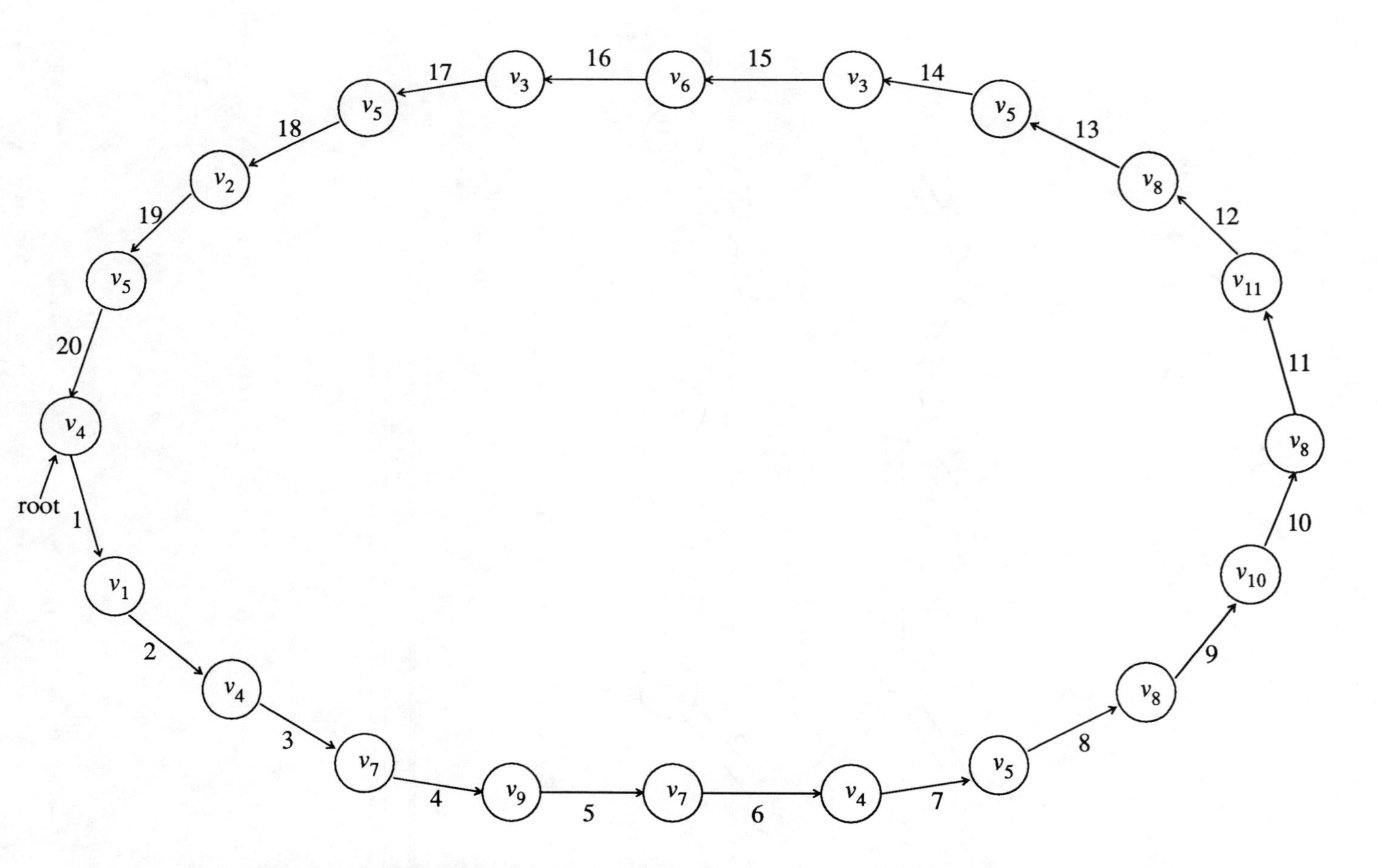

Figure 10. A Circularly Linked List Representation of the Euler Tour in Figure 9.

(a)

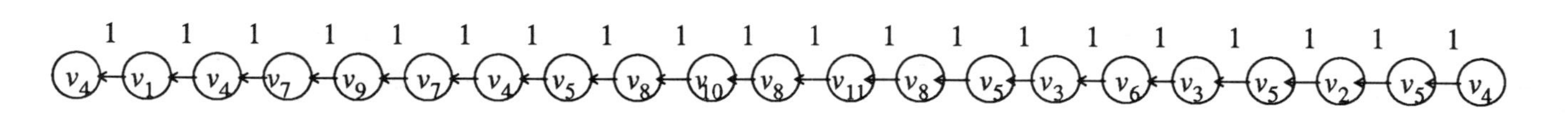

(b)

(c)

Figure 11. Computation of Retreat Edges of the Euler Tour in Figure 9 in Parallel.
(a) Traversal List of the Circularly Linked List in Figure 10,
(b) Reversal of the Links TOURNEXT and Initialization of RANK,
(c) Values of RANK after the List Ranking Computation.

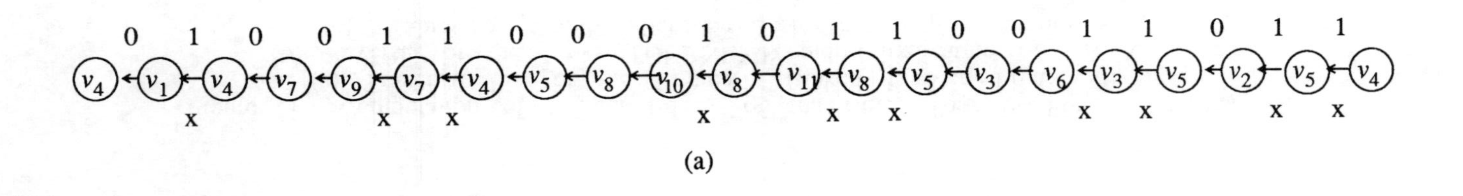

(a)

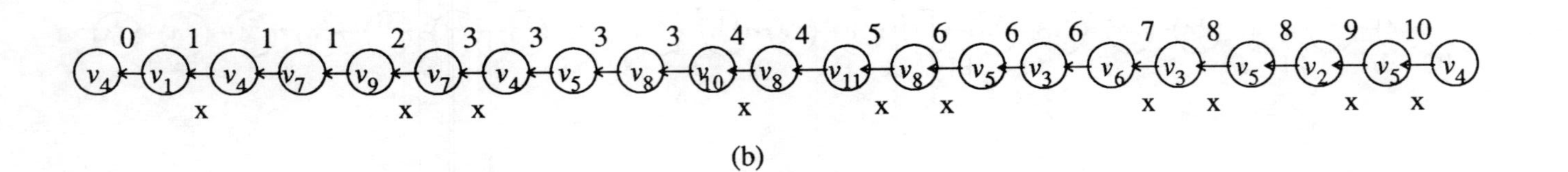

(b)

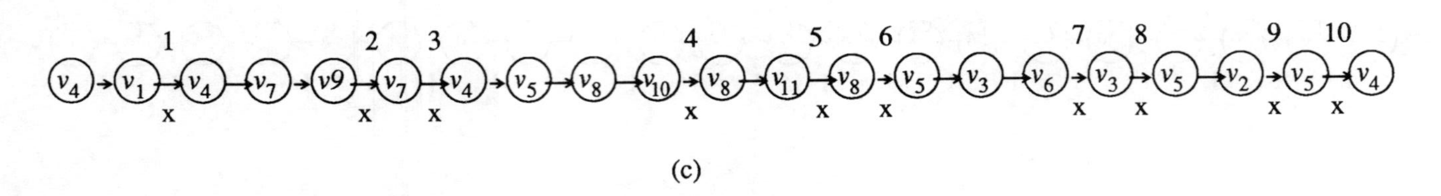

(c)

Figure 12. Computation of Order of Encounter of Retreat Edges in the Traversal in Figure 11 in Parallel.
(a) Re-initialization of RANK to compute Postorder Numbers,
(b) Values of RANK after List Ranking Computation,
(c) Values of RANK of Retreat Edges.

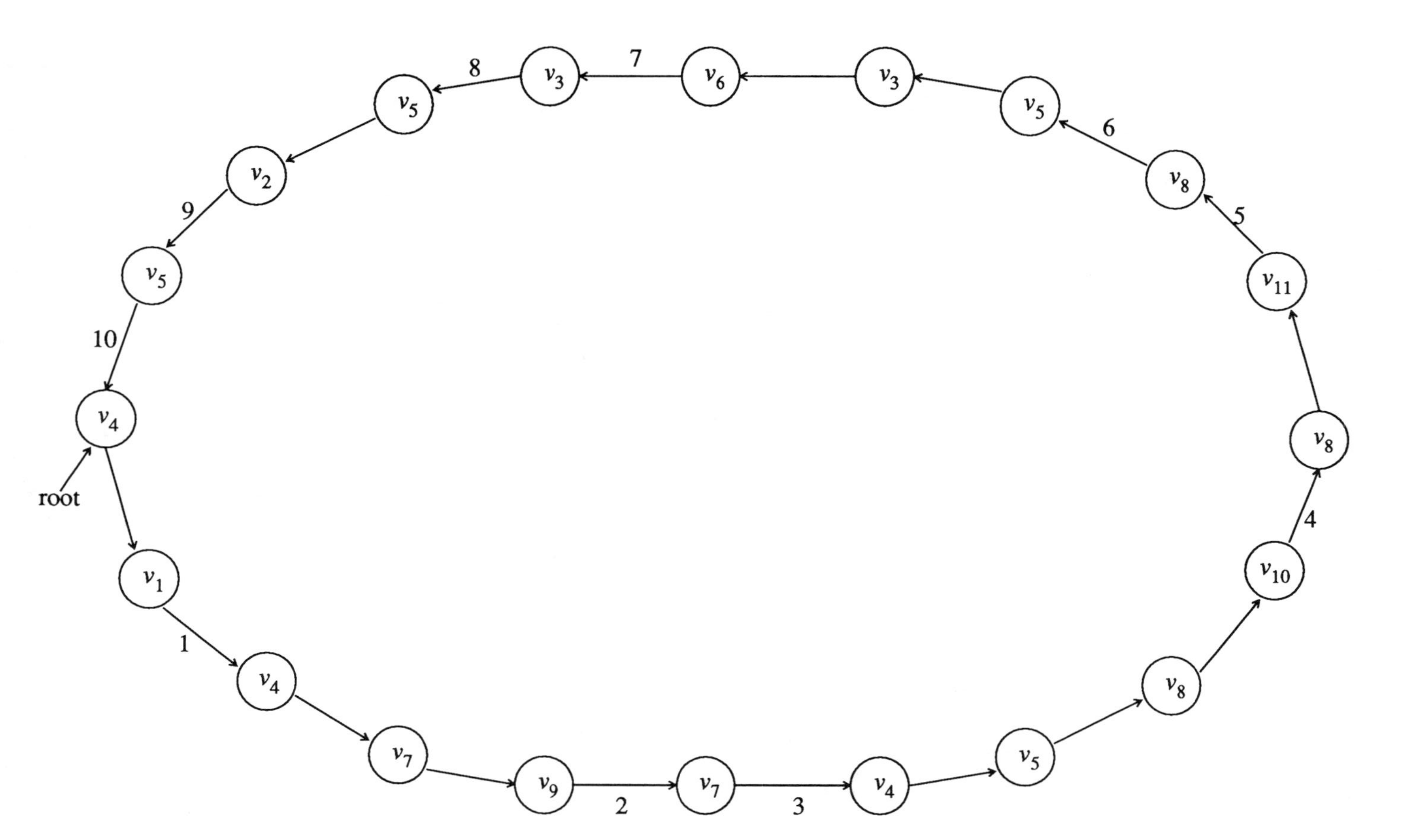

Figure 13. A Circularly Linked List Representation of the Traversal in Figure 12(c).

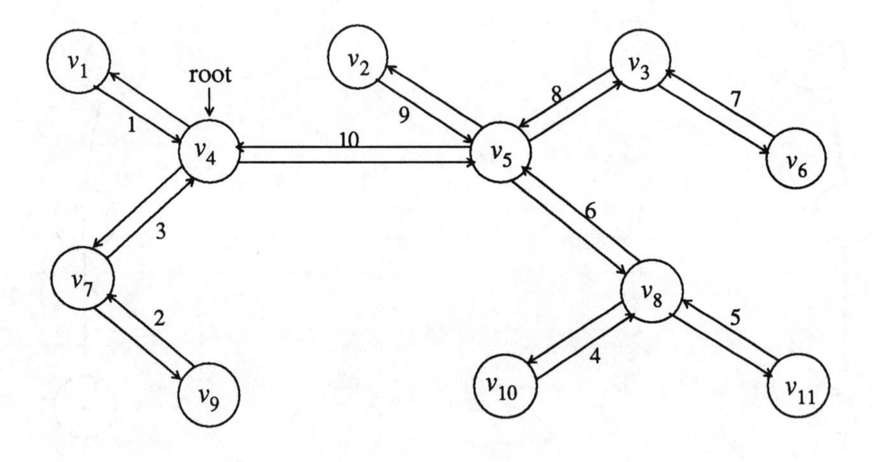

Figure 14. The Order of Encounter of Retreat Edges in the Traversal in Figure 12(c).

The algorithm initializes COUNT for all vertices to zero. A dummy vertex 0 is introduced to be the parent of vertex n. By convention, $P(n)=0$ and therefore COUNT(0) and $b(0)$ are initialized to 0. For each vertex u starting from the vertex with number one to the vertex with number n, it is determined whether the vertex u and its parent $P(u)$ have enough of their incident edges in the $b-$matching. If u has $b(u)$ matched edges incident on it or $P(u)$ has $b(P(u))$ matched edges incident on it, then the edge between the vertex u and its parent $P(u)$ is not to be in the $b-$matching and $S(u)$ is set to *false*. Otherwise the edge $\{u, P(u)\}$ is to be in the $b-$matching and $S(u)$ is set to *true*. Then COUNT(u) and COUNT(P(u)) are each incremented by one. The algorithm is given as algorithm TREEMATCH below.

Algorithm TREEMATCH

```
Step 0:
    construct the postorder numbers of the vertices of the tree using algorithm PO-NUMBER.

Step 1:
    for u ← 1 to n do
        COUNT(u) ← 0
    end for
    COUNT(0)  ← 0
    b(0)      ← 0

Step 2:
    for u ← 1 to n do
        if ((COUNT(u)=b(u)) or (COUNT(P(u))=b(P(u)))
            then S(u) ← false
            else
                S(u)    ← true
                COUNT(u) ← COUNT(u) + 1
                COUNT(P(u)) ← COUNT(P(u)) + 1
        end if
    end for
```

The correctness of the algorithm is presented in Goodman Hedetniemi and Tarjan [1976]. The running time of the algorithm is $O(n)$, since each edge is looked at only once in Step 2. The maximum cardinality $b-$matching of the tree in Figure 1, shown in Figure 2, was obtained using this algorithm. The root of the tree is chosen to be vertex v_4.

3.2. Maximum Cardinality Matching by Selection and Identification

Algorithm TREEMATCH processes the vertices from the least numbered vertex to the highest numbered vertex. Moreover, an edge $\{u, P(u)\}$ between a vertex u and its parent $P(u)$ is chosen to be in the $b-$matching only if there are fewer than $b(u)$ matched edges incident on u which are also incident on the children of u. The edges incident on vertex u which are also incident on the first $b(u)$ least numbered children, where such a child v has fewer than $b(v)$ matched edges incident on it, are chosen to be in the matching.

In considering any vertex u, therefore, we can say that an edge $\{u, v\}$ is chosen to be in the $b-$matching if

(a) vertex v is among the first $b(u)$ least numbered vertices adjacent to vertex u and vertex u is among the first $b(v)$ least numbered vertices adjacent to vertex v; or

(b) vertex v is among the first $b(u)$ least numbered vertices adjacent to vertex u, vertex u is not among the first $b(v)$ least numbered vertices adjacent to vertex v but there are fewer than $b(v)$ other edges (v, w), where v is the parent of w and w precedes u in postorder number, for which condition (a) is true.

Using this observation, we process for each vertex u the incident edges (u, v) in increasing order of the number of vertex v. For each vertex u, to identify the edges (u,v) such that the vertices v are among the first $b(u)$ least numbered adjacent vertices, we choose the edge (u,v_L) for which the number of vertex v_L is the $b(u)^{th}$ smallest of the vertices adjacent to u. We then consider all the incident edges (u,v), with the number of v less than or equal to that of v_L to be *likely* to be in the $b-$matching, indicated by setting the entry (u,v) of an array LIKELY to 1. LIKELY[(u,v)] is set to 0 if the number of v is greater than the number of v_L.

For each vertex u starting from the least numbered vertex, we traverse the adjacency list of u. For each incident edge (u,v), we check if the edge (v,u) has been marked as a *likely*. If it has, then we include the edge (u,v) in the $b-$matching. We traverse ADJACENCY(u) until we obtain $b(u)$ incident edges in the $b-$matching or we run out of incident edges to consider. We again use the array COUNT to keep a count of the number of edges incident on a vertex that are currently in the $b-$matching. The algorithm is specified below as algorithm SELECTMATCH.

Algorithm SELECTMATCH

```
Step 0:
    (a)  for u ← 1 to n do
             for all (u,v) ∈ ADJACENCY(u) do
                 LIKELY[(u,v)] ← 0
             end for
         end for
    (b)  Compute the postorder number of the vertices of the tree using the algorithm PO-NUMBER.

Step 1:
    for each vertex u ← 1 to n do
        (a)  select the incident edge (u,v_L) such that the vertex v_L is the b(u)^th least numbered
             adjacent vertex.
        (b)  traverse the adjacency list and mark the incident edge (u,v), such that v ≤ v_L as likely,
             i.e. set LIKELY[(u,v)] to 1. {These are the first b(u) edges with least numbered
             vertices adjacent to u.}
    end for

Step 2:
    for each vertex u ← 1 to n do
        COUNT(u) ← 0
    end for

Step 3:
    for each vertex u ← 1 to n do
        i ← 1
        while i ≤ deg(u) do
            let the edge (u,v) be the next edge on ADJACENCY(u)
            if ((COUNT(u) < b(u)) and (COUNT(v) < b(v)) and (LIKELY[(v, u)] = 1))
                then
                    if (edge {u,v} = {v,u} is not marked as in the b-matching) then
                        COUNT(u) ← COUNT(u) + 1
                        COUNT(v) ← COUNT(v) + 1
                        choose edge {u,v} as in the b-matching
                    end if
                else
                    if (COUNT(u) = b(u)) then i ← deg(u) end if
            end if
            i ← i + 1
        end while
    end for
```

Theorem 3.1.
Given an undirected tree and a bound $b(v)$, $0 \leq b(v) \leq deg(v)$, associated with each vertex v of the tree, a maximum cardinality $b-$matching of the tree is obtainable using the algorithm SELECTMATCH in O(n) sequential time.

Proof.
Steps 0 and 2 take O(n) time. Step 1(a) is executable in O($deg(u)$) time using the linear time algorithm of Blum et al [1973] (also see Aho, Hopcroft and Ullman [1974] and Hyafil [1976]). Step 1(b) is executable in O($deg(u)$) time. The overall execution time of Step 1 is O(n). Each iteration of the **while** loop of Step 3 is executable in O($deg(u)$) time, giving O(n) execution time for Step 3. Thus the running time of SELECTMATCH is O(n). □

3.4. Assignment of Processors to Vertices

In this subsection, we discuss how we can assign p processors to the adjacency lists of vertices so that we obtain the required execution time for the parallel $b-$matching algorithms. The distribution will be such that:

- (i) no vertex will have part of its adjacency list assigned to a processor which is assigned to the adjacency list of more than one vertex.
- (ii) if more than one processor is assigned to the adjacency list of a vertex, then none of those processors are assigned to the adjacency list of any other vertex.
- (iii) each processor has at most $(6n-6)/p$ edges assigned to it.

By conditions (i) and (ii), we mean that either a processor handles the entire adjacency list of one or more vertices or one or more processors are assigned to the adjacency list of only one vertex, but not both. Condition (iii) ensures that each processor performs no more than O(n/p) operations on its set of edges.

To get the processors assigned according to the above conditions, we initially assign $(2n-2)/p$ edges to a processor. The values n and p can be broadcast to all the processors in O($\log p$) time. We assume that the processors are identified by their numbers, 1, 2, . . . , p. Then the initial interval on EDGES for processor P_i is $\lceil (i-1)*(2n-2)/p + 1 \rceil$ to $\lfloor i*(2n-2)/p \rfloor$. Let the initial interval for processor P_i be $LEFT_i$ to $RIGHT_i$. Each processor P_i then shifts its left and right boundaries, $LEFT_i$ and $RIGHT_i$, so that the above conditions are satisfied. To do this, each processor executes the following steps in algorithm ASSIGN.

Algorithm ASSIGN

Step 0:
  $shifts \leftarrow 1$

Step 1:
  $(v_i, v_j) \leftarrow$ EDGES[$LEFT_i - shifts$]
  $(v_k, v_l) \leftarrow$ EDGES[$LEFT_i - shifts + 1$]
  if ($v_i \neq v_k$) **then** go to Step 3 **end if**

Step 2:
  if ($shifts < (2n-2)/p$) **then** $shifts \leftarrow shifts + 1$, go to Step 1 **end if**

Step 3:
  if (($v_i \neq v_k$) **and** ($shifts \leq (2n-2)/p$)) **then** $LEFT_i \leftarrow LEFT_i - shifts + 1$ **end if**

Step 4:
  $shifts \leftarrow 1$

Step 5:
  $(v_i, v_j) \leftarrow$ EDGES[$LEFT_i + shifts - 1$]
  $(v_k, v_l) \leftarrow$ EDGES[$LEFT_i + shifts$]
  if ($v_i \neq v_k$) **then** go to Step 8 **end if**

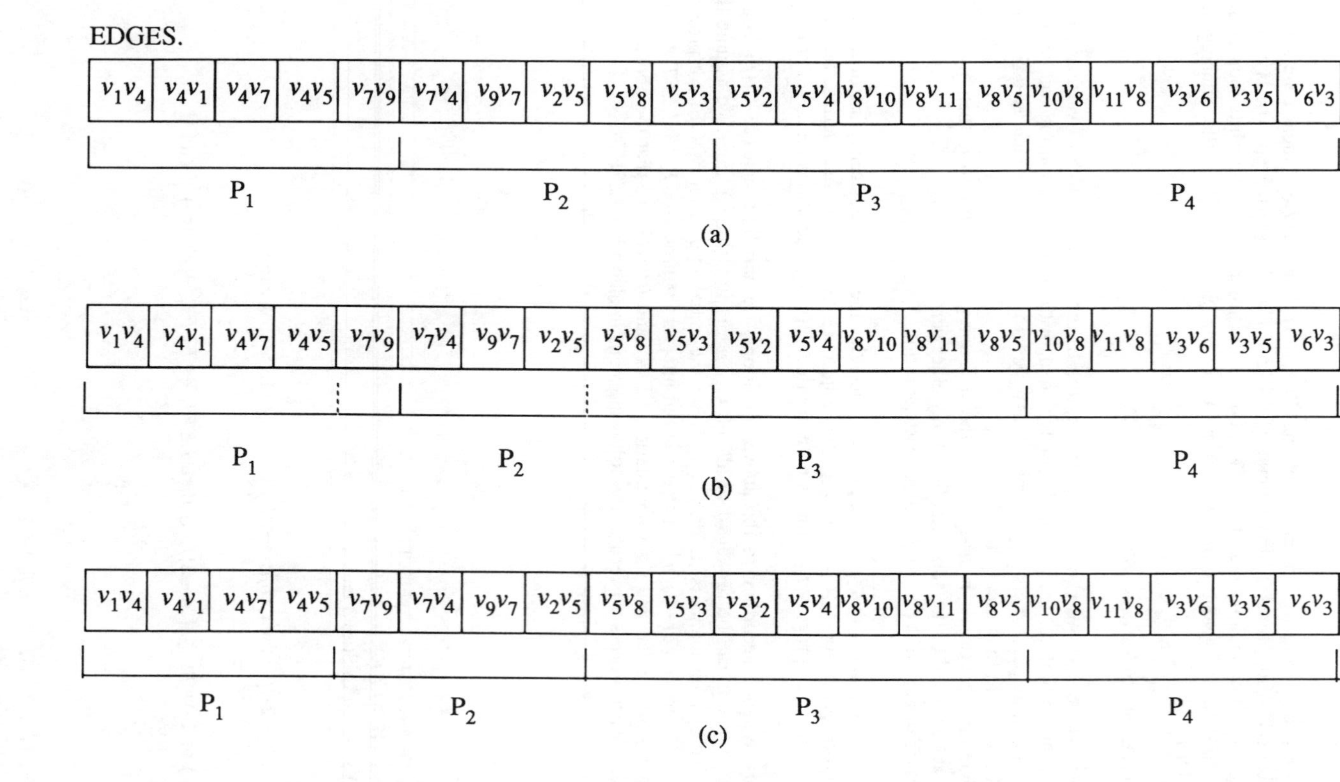

Figure 15. Assignment of Processors to Adjacency List of Vertices of the Tree in Figure 1. (a) Initial Allocation of Edges. (b) Shifting the Left Boundary to Begining of an Adjacency List. (c) Final Assignment of Adjacency Lists to Processors.

Step 6:
  if (*shifts* $< (2n-2)/p$) **then** *shifts* $\leftarrow$ *shifts* + 1, go to Step 5 **end if**

Step 7:
  if (($v_i \neq v_k$) **and** (*shifts* $\leq (2n-2)/p$)) **then** $LEFT_i \leftarrow LEFT_i$ + *shifts* $-$ 1 **end if**

Step 8:
  repeat Steps 0 to 7 replacing $LEFT_i$ by $RIGHT_i$

Theorem 3.2.
Given an adjacency list representation of a tree as described in Section 2, the adjacency lists of the vertices can be shared by p processors in $O(n/p + \log p)$ time such that

(i) no vertex will have part of its adjacency list assigned to a processor which is assigned to the adjacency list of more than one vertex.
(ii) if more than one processor is assigned to the adjacency list of a vertex, then none of those processors are assigned to the adjacency list of any other vertex.
(iii) each processor has at most $(6n-6)/p$ edges assigned to it.

Proof.
The loops of Steps 1 to 3 as well as Steps 5 to 7 will each go through at most $(4n-4)/p$ iterations when shifting a boundary. Each iteration takes constant time. Steps 0 through 7 are gone through twice. Thus we obtain an $O(n/p + \log p)$ distribution algorithm, taking into consideration the time to broadcast n and p to the processors. □

Execution of the algorithm ASSIGN on the array EDGES is shown in Figure 15. The initial allocation of the edges to the processors is shown in Figure 15(a). After executing Steps 0 through 3 to the end of the loop, we obtain the distribution shown in Figure 15(b). The left boundaries of the lists assigned to processors P_2 and P_3 are shifted as shown by the broken bars. Executing the remaining steps of the algorithm shifts the right boundaries of processors P_1 and P_2 to give the distribution as shown in Figure 15(c).

3.5. Parallelizing the Algorithm SELECTMATCH

In parallelizing algorithm SELECTMATCH, we need to determine the edge (u,v) incident to each vertex u, such that v is among the first $b(u)$ least numbered vertices adjacent to u. This requires that the selection of edge (u,v_L) for each vertex u be done in parallel. Determining which incident edge (u,v) with number of v less than or equal to that of v_L also needs to be done in parallel, after broadcasting v_L to the processors assigned to vertex u. The edges (u,v) such that NUMBER[v] is less than or equal to NUMBER[v_L] are marked as *likely* to be in the b-matching, by setting LIKELY[(u,v)] to 1. Initially, LIKELY[(u,v)] is set to 0 for all edges (u,v).

We then determine in parallel, for each vertex u, which of its adjacent vertices v has LIKELY[(v,u)] set to 1. The array SUCCESSFUL of length $2n-2$ is used for this purpose. The edges incident to the first $b(u)$ least numbered successful vertices adjacent to vertex u are chosen to be in the b-matching. The algorithm is formally given below.

Algorithm Parallel SELECTMATCH
Step 0:
  Compute the postorder numbers of the vertices of the tree using the algorithm POSTORDERNUMBER.

Step 1:
  Assign the set P_u of processors to each vertex u using the algorithm ASSIGN.

Step 2:
  Assign $deg(u)/P_u$ edges incident to vertex u to a processor of the set P_u.

Step 3:

```
for each set of processors P_u do in parallel
    for all processors P_{u,k} ∈ P_u do in parallel
        for each assigned incident edge (u,v) do
            (a) set LIKELY[(u,v)] to 0
            (b) set SUCCESSFUL[(u,v)] to 0
        end for
    end for
end for
```

Step 4:

```
for each set of processors P_u do in parallel
    (a) select the edge (u,v_L) where vertex v_L is the b(u)^th least numbered vertex adjacent to
        vertex u.
    (b) broadcast the number of v_L to all processors in P_u
    (c) for all processors in the set P_u do in parallel
            for each assigned edge (u,v) do
                if (the number of v is less than or equal to that of v_L) then
                    mark (u_i, v_j) as likely (i.e. LIKELY[(u,v)] ← 1) {Edges (u,v) with
                    LIKELY[(u,v)] set to 1 are the incident edges (u,v) for which vertex v is
                    among the b(u) least numbered vertices adjacent to vertex u}.
                end if
            end for
        end for
end for
```

Step 5:

```
for each set of processors P_u do in parallel
    for processors P_{u,k} ∈ P_u do in parallel
        for each assigned incident edge do
            if (edge (v, u) is marked as likely for vertex v, i.e. LIKELY[(v, u)]=1) then
                mark edge (u,v) as successful (i.e. SUCCESSFUL[(u,v)] ← 1)
            end if
        end for
    end for
end for
```

Step 6:

```
for each set of processors P_u do in parallel
    (a) select the edge (u,v_SL) where vertex v_SL is the b(u)^th least numbered adjacent vertex
        marked successful
    (b) broadcast the number of v_SL to all processors in P_u
    (c) for all processors in P_u do in parallel
            for each incident edge (u,v) assigned to each processor P_{u,k}∈P_u do
                if ((SUCCESSFUL[(u,v)]=1) and (v ≤ v_SL)) then
                    if (the number of u is less than that of v)
                        then mark edge (u,v) as in the b−matching end if
                end if
            end for
        end for
end for
```

Theorem 3.3.
Given an undirected tree with bounds $b(v)$, $0 \leq b(v) \leq deg(v)$, associated with the vertices of the tree, a

maximum cardinality $b-$matching of the tree can be computed in $O(n/p + \log n)$ time in parallel using p processors.

Proof.

Step 0 is the execution of the algorithm POSTORDERNUMBER which runs in $O(n/p + \log n)$ time in parallel. Step 1 is the execution of the $O(n/p + \log p)$ parallel time algorithm ASSIGN. Steps 2, 3 and 5 are each executable in $O(n/p)$ time. The select operation of Steps 4 and 6 can each be done by the optimal algorithm for parallel selection of Akl [1984] which runs in $O(n/p)$ time. The broadcast operations of Steps 4 and 6 can each be done in $O(n/p + \log p)$ time.

To see this, suppose we keep an array HEADNUM of length $2n-2$, where HEADNUM[(u,v)] is the number of the vertex v. We can construct HEADNUM by letting the first processor in P_u read the number of u and broadcast to the rest of the processors in P_u. Each processor in P_u then sets HEADNUM[(v,u)] to the number of u, for each edge (u,v) assigned to it. The time for this preprocessing step is $O(n/p + \log p)$, and can be done once the postorder numbers are computed.

We can then read the number of the head vertex v of any directed edge (u,v) and broadcast it (see Akl [1989], Akl [1985]) to the processors in P_u in $O(\log p)$ time. Steps 4(c) and 6(c) are executable in $O(n/p)$ time having done the preprocessing step as in the previous paragraph. We therefore obtain a running time of $O(n/p + \log n)$ with p processors. □

The computation of the maximum cardinality $b-$matching for the tree in Figure 1, using the postorder numbering in Figure 6, is presented in Figure 16. Figure 16(a) is the distribution of the adjacency lists among the four processors as obtained at the end of the computation of the algorithm ASSIGN by the four processors (see Figure 15). Executing the algorithm Parallel SELECTMATCH up to Step 4, we obtain the array LIKELY of Figure 16(b). The values of the array SUCCESSFUL obtained after executing Step 5 are given in Figure 16(c). The $b-$matching obtained after executing Step 6 is given in Figure 16(d). The undirected edge $\{v_i, v_j\}$ is implied to be in the $b-$matching, by the computation, when the directed edge (v_i, v_j) or (v_j, v_i) is in the $b-$matching,

4. Maximum Weight $b-$Matching

The algorithm of Goodman, Hedetniemi and Tarjan for solving the maximum weight $b-$matching problem is an extension of the algorithm TREEMATCH, for the maximum cardinality $b-$matching. This extension is based on the following observation:

> Let u be a vertex of T adjacent to at most one vertex w of degree higher than one. Let $v_1, v_2, \cdots, v_k$ be the vertices of degree one adjacent to u, in non-increasing order of $\omega\{v_i, u\}$. Then there exists some maximum weight $b-$matching containing the edges $\{v_1, u\}, \cdots, \{v_{b(u)-1}, u\}$ and in addition either $\{v_{b(u)}, u\}$ or $\{w, u\}$.

By the above observation, let a tree T' be constructed from T by deleting the edges $\{v_1, u\}, \cdots, \{v_k, u\}$. A weight function ω' can be defined on the edges of T' as

$$\omega'(\{w, u\}) = \omega(\{w, u\}) - \omega(\{v_{b(u)}, u\}) \text{ and } \omega'(\{y, z\}) = \omega(\{y, z\}) \text{ if } \{y, z\} \neq \{w, u\},$$

then any maximum weight $b-$matching of T' can be converted into a maximum weight $b-$matching of T by adding edges $\{v_1, u\}, \ldots, \{v_{b(u)-1}, u\}$ and $\{v_{b(u)}, u\}$ if $\{w, u\}$ is not already in the $b-$matching.

Let a bijection be defined on the vertices of the tree T as discussed in Section 2. The set $\chi(u)$ for each vertex u is defined as

$$\chi(u) = \{v \mid (u,v) \in \mathrm{T} \text{ and } \beta^{-1}(u) > \beta^{-1}(v) \text{ and } b(v) > 0\}.$$

When the bijection is a postorder numbering of the vertices with respect to a root, the elements of $\chi(u)$ are the children of vertex u; let us denote this set by CHILDREN(u) in such a case.

The algorithm changes the weights of the edges using the above scheme. The variable WT(u) is used to denote the current value of $\omega(\{u, \mathrm{P}(u)\})$. WT($n$) is set to 0 since the root has no parent. At termination of the algorithm, the maximum weight $b-$matching will consist of the edges $\{u, \mathrm{P}(u)\}$ such that S(u)=*true*.

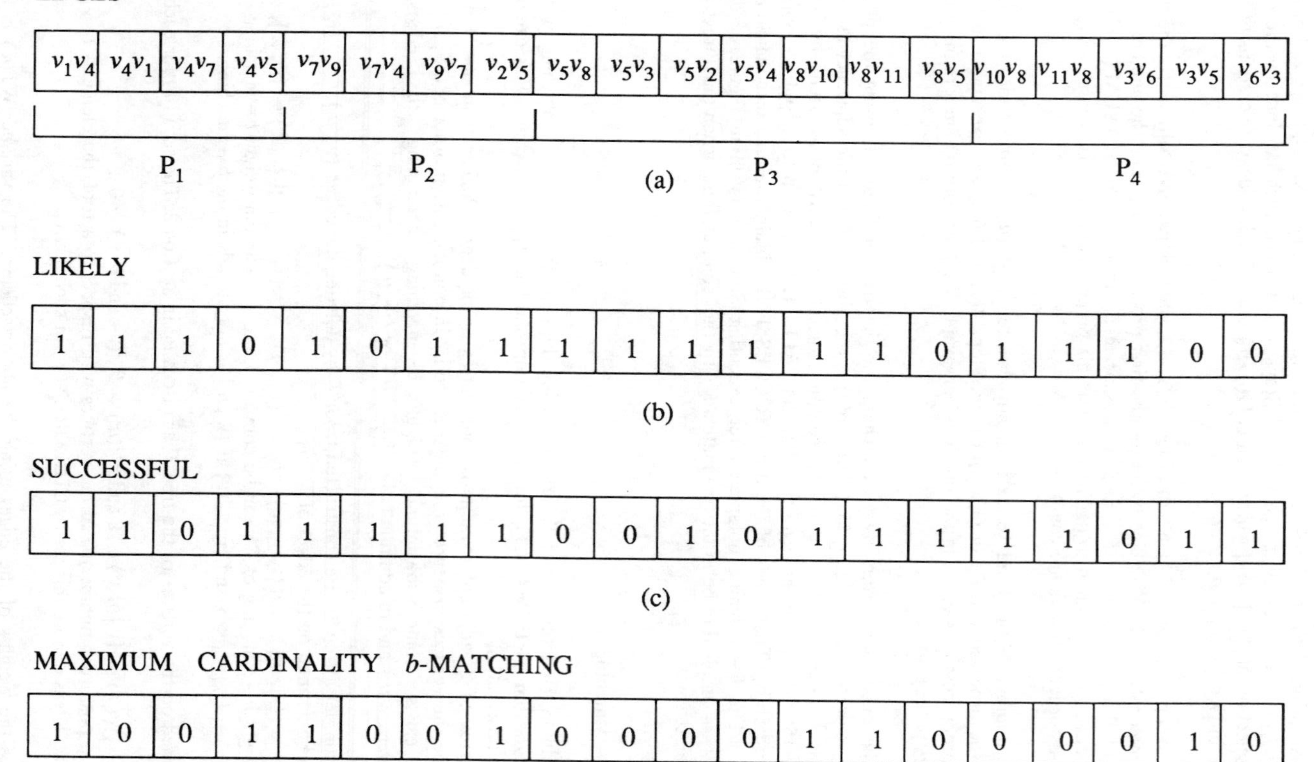

Figure 16. (a) Adjacency Lists of Vertices with assigned processors, (b) Edges likely to be in the *b*-matching, (c) Edges most likely to be in the *b*-matching, and (d) Edges in the *b*-matching.

To decide which of the edges $\{v_{b(u)},u\}$ and $\{u,w\}$ to include in the maximum weight $b-$matching, a variable PRED(u) for each vertex u is kept. PRED(u) is defined to be the vertex $v \in$ CHILDREN(u) with the $b(u)^{th}$ largest value of WT(v), if WT(u) < WT(v); otherwise PRED(u)=0. The algorithm processes vertices in order of increasing vertex number. For each vertex u, the first $b(u)-1$ heaviest incident edges are determined. For such an edge $\{u,v\}$, if u and v respectively have fewer then $b(u)$ and $b(v)$ incident edges in the matching, then the edge is chosen to be in the maximum weight $b-$matching. The edge $\{u,v_{b(u)}\}$ is chosen to be in the $b-$matching, if $\omega(\{u,\mathrm{P}(u)\}) < \omega(\{u,v_{b(u)}\})$, otherwise, $\{u,\mathrm{P}(u)\}$ is chosen. In case of a tie, the edge $\{u,v_{b(u)}\}$ is chosen. The algorithm is given as follows:

Algorithm WTREEMATCH

```
Step 0:
    Compute the postorder numbers for the vertices of the tree using the algorithm PO-NUMBER.

Step 1: {Initialization}
    for u ← 1 to n do
        WT(u) ← ω({u,P(u)})
        PRED(u) ← 0
    end for
    WT(n) ← 0
    PRED(n) ← 0

Step 2: {Main Loop}
    for u ← 1 to n do

    Step 2.1. {Select the child with b(u)^th heaviest weight.}
        find the element v_b(u) ∈ CHILDREN(u) with the b(u)^th largest value of WT(v_b(u))

    Step 2.2. {Determine which children have larger or equal weight as ω({u,v_b(u)})}
        order the elements of CHILDREN(u) so that v_b(u) occurs in the b(u)^th position and any v ∈
        CHILDREN(u) occurring before v_b(u) has WT(v) ≥ WT(v_b(u))

    Step 2.3. {Update WT(u) if necessary.}
        if (WT(u) > WT(v_b(u)))
        then
            WT(u) ← WT(u) − WT(v_b(u))
            if (WT(v_b(u)) > 0) then PRED(u) ← v_b(u) end if
            {A decision need be made whether to include the edge {u,v_b(u)} in the maximum weight
            b−matching or not.}
        else
            WT(u)      ← 0
            WT(v_b(u)) ← 0
        end if

        {Decide whether to add {u,v} to the matching.}
        flag ← true
        for each v ∈ CHILDREN(u) do
            if ((v ≠ v_b(u)) or (WT(u) = 0))
            then
                if WT(v) > 0
                    then S(v) ← flag
                    else S(v) ← false
                end if
                {Decide which of {u,P(u)} and {u,v_b(u)} to include in the matching.}
                v_t ← v
                while (PRED(v_t) ≠ 0) do
```

```
                S(PRED(v_t)) ← ¬ S(v_t)
                v_t ← PRED(v_t)
            end while
        end if
        if (v_b(u) = v) then flag ← false end if
    end for
end for
```

The correctness of this algorithm is given in Goodman, Hedetniemi and Tarjan [1976]. Step 0 and 1 execute in $O(n)$ time. Steps 2.1 and 2.2 are each executable in $deg(v)$ time using the selection algorithm of Blum et al [1973] (see also Aho, Hopcroft and Ullman [1974]). Each edge of the tree is examined once in Steps 2.3 and 2.4. Thus the running time of the algorithm is $O(n)$. The maximum weight $b-$matching in Figure 3 was obtained using algorithm WTREEMATCH on the tree in Figure 1 with the postorder numbering in Figure 6.

4.1. Maximum Weight $b-$Matching by Selection and Identification

Again, we process the vertices in increasing order of postorder number as it is done in WTREEMATCH. For each vertex, u, the vertices in ADJACENCY(u) are processed in the non-increasing order of weight. The adjacent vertices which have equal weight are processed in order of increasing number. To determine the order of processing the adjacent vertices, the edge $(u, v_{b(u)})$ which is the $b(u)^{th}$ heaviest edge incident on u is selected.

For each $v \in$ ADJACENCY(u) such that $\omega((u,v))$ is greater than $\omega((u,v_{b(u)}))$, (u, v) is chosen to be *likely* in the maximum weight $b-$matching by setting LIKELY[(u,v)] to $+1$. Adjacent vertices with $\omega((u,v))$ equal to $\omega((u,v_{b(u)}))$ have LIKELY[(u,v)] set to 0, and the others have LIKELY[(u,v)] set to -1. At this point, we have partitioned the set ADJACENCY(u) into three sets, G(u), E(u) and L(u). These are the edges with weight greater than, equal to and less than $\omega((u,v_{b(u)}))$, respectively. We will keep the number of edges in G(u) in TALLY(u), where the array TALLY is of length n.

Among the elements of E(u), the $(b(u) - |G(u)|)^{th}$ least numbered vertex v_L is selected. The elements of E(u) with number less than or equal to that of v_L have LIKELY[(u,v)] set to $+1$ and the others have LIKELY[(u,v)] set to -1. Now we have identified the incident edges (u,v) by which the edges $\{u,v\}$ are *likely* to be in the matching.

We then process the vertices in increasing order of number. For each incident edge (u,v), if u and v, respectively, have fewer than $b(u)$ and $b(v)$ incident edges in the matching, then the edge $\{u,v\}$ is chosen to be in the $b-$matching provided the edge (v, u) was marked as *likely*. The bounds for u and v are decremented by 1 and the next incident edge is considered if necessary. Otherwise the adjacency list of the next vertex is processed. We formally present the algorithm as:

Algorithm WTSELECTMATCH

Step 0:
Compute the postorder number of the vertices of the tree using the algorithm PO-NUMBER

Step 1:
for each vertex $u \leftarrow$ **1 to** n such that $b(u) > 0$ **do**
(a) select the $b(u)^{th}$ heaviest incident edge $(u, v_{b(u)})$
(b) traverse the adjacency list and set for each incident edge (u,v), such that:
(i) $\omega((u,v)) > \omega((u,v_{b(u)}))$, LIKELY[$(u,v)$] $\leftarrow +1$
(ii) $\omega((u,v)) = \omega((u,v_{b(u)}))$, LIKELY[$(u,v)$] $\leftarrow 0$
(iii) $\omega((u,v)) < \omega((u,v_{b(u)}))$, LIKELY[$(u,v)$] $\leftarrow -1$
(c) set TALLY(u) to the number of vertices marked $+1$
end for

```
Step 2:
    for each vertex u ← 1 to n  such that b(u) > 0 do
        for all the incident edges (u,v) such that LIKELY[(u,v)] = 0 do
            (a) select the incident edge (u,v_L) where vertex v_L is the (b(u)−TALLY(u))^th least
                numbered vertex
            (b) for all incident edges (u,v) such that LIKELY[(u,v)] is 0 do
                    if (the number of v is less than or equal to that of v_L)
                        then LIKELY[(u,v)] ← +1
                        else LIKELY[(u,v)] ← −1
                    end if
                end for
        end for
    end for

Step 3:
    for each vertex u ← 1 to n  do
        COUNT(u) ← 0
    end for

Step 4:
    for each vertex u ← 1 to n  do
        i ← 1
        while i ≤ deg(u) do
            let (u,v) be the next edge on the adjacency list to be processed
            if (COUNT(u) < b(u)) and (COUNT(v) < b(v)) and (LIKELY[(v,u)] =1)
            then
                if (edge {u,v}={v,u} is not marked as in the maximum weight b−matching) then
                    COUNT(u) ← COUNT(u) + 1
                    COUNT(v) ← COUNT(v) + 1
                    mark edge {u,v} as in the maximum weight b−matching
                end if
            else
                if (COUNT(u) = b(u)) then i ← deg(u) end if
            end if
            i ← i + 1
        end while
    end for
```

Theorem 4.1.
Given a weighted undirected tree, the maximum weight $b-$matching of T can be computed in time $O(n)$ using algorithm WTSELECTMATCH.
Proof.
Trivially, each step of the algorithm is executable in $O(n)$ time. Thus the running time of WTSELECTMATCH is $O(n)$. □

4.2. Parallel WTSELECTMATCH

In parallelizing the algorithm WTSELECTMATCH, we use the algorithm ASSIGN to assign vertices to processors such that each processor has no more than $(6n-6)/p$ edges assigned to it. Each vertex u will have a set of processors, P_u assigned to it.

The array LIKELY is used to identify which incident edges (u,v) are likely to be in the $b-$matching. The array SUCCESSFUL is used to indicate which reverse edges (v,u) are marked as likely to be in the $b-$matching. Out of the edges incident to u which are marked *successful*, only $b(u)$ of them will be chosen according to non-increasing order of weight. When more than one incident edge (u,v) has the same weight as the $b(u)^{th}$ heaviest edge, the edges (u,v) are processed in order of increasing number.

To determine which incident edges are likely to be in the matching, the incident edge $(u,v_{b(u)})$ with the $b(u)^{th}$ heaviest weight is selected. The weight $\omega((u,v_{b(u)}))$ is broadcast to all processors in the set P_u. Each processor in P_u then traverses its sublist comparing the weight of each of its edges (u,v) to $\omega((u,v_{b(u)}))$. LIKELY[(u,v)] is set to $+1$, 0 or -1 if $\omega((u,v))$ is greater than, equal to or less than $\omega((u,v_{b(u)}))$ respectively. We then obtain the sets G(u), E(u) and L(u) as before. The processors P_u then determine, for the elements in E(u), the edge (u,v_L) such that v_L is the $(b(u)-|G(u)|)^{th}$ least numbered vertex. The number of vertex v_L is communicated to the processors of P_u. Each processor in P_u determines which of its edges (u,v), with LIKELY[(u,v)] equal to 0, has number of v less than or equal to that of v_L and sets LIKELY[(u,v)] to $+1$, otherwise LIKELY[(u,v)] is set to -1.

Among the incident edges (u,v) each processor sets SUCCESSFUL[(u,v)] to 1 if LIKELY[(v,u)] is $+1$, else SUCCESSFUL[(u,v)] remains zero for the edge (u,v). Among the successful incident edges, $b(u)$ of them need be chosen. This is done in the same way as we chose the edges which are likely to be in the matching. The $b(u)^{th}$ heaviest successful edge (u,v_{SH}) is selected by the P_u processors. The weight of edge (u,v_{SH}) is made known to processors in P_u by a broadcast. Each processor in P_u then goes through its successful edges and sets LIKELY[(u,v)] to $+1$, 0 or -1 if $\omega((u,v))$ is greater than, equal to or less than $\omega((u,v_{SH}))$ respectively. New sets G(u), E(u) and L(u) are obtained. The element (u,v_{SL}) in E(u) which has the $(b(u)-|G(u)|)^{th}$ least numbered vertex v_{SL} is selected by all processors in P_u. Each processor gets to know the number of v_{SL} by a broadcast as before. Each processor then checks for each of its successful edges if LIKELY[(u,v)] is equal to 0 and sets LIKELY[(u,v)] to $+1$ if the number of vertex v is less than or equal to that of vertex v_{SL} and -1 otherwise. The edges with LIKELY equal to $+1$ are the edges which are to be in the maximum weight $b-$matching.

Algorithm Parallel WTSELECTMATCH

Step 0:
Compute the postorder number of the vertices using the algorithm POSTORDERNUMBER.

Step 1:
Assign a set P_u of processors to vertex u, by executing the algorithm ASSIGN.

Step 2:
for each set of processors P_u **do in parallel**
  for all processors $P_{u,k} \in P_u$ **do in parallel**
    for each incident edges (u,v) **do**
      (a) **if** $(b(v) = 0)$ **then** $\omega(u,v) \leftarrow -\infty$ **end if**
        {This implicitly deletes v from the adjacency list.}
      (b) SUCCESSFUL[(u,v)] $\leftarrow 0$
    end for
  end for
end for

Step 3:
for each set of processors P_u **do in parallel**
  (a) select the $b(u)^{th}$ heaviest edge $(u,v_{b(u)})$ incident on vertex u.
  (b) broadcast $\omega((u,v_{b(u)}))$ to all processors in P_u
  (c) **for** all processors in the set P_u **do in parallel**
    for each assigned incident edge (u,v) **do**
      if $(\omega((u,v)) > \omega((u,v_{b(u)})))$ **then** LIKELY[(u,v)] $\leftarrow +1$ **end if**
      if $(\omega((u,v)) = \omega((u,v_{b(u)})))$ **then** LIKELY[(u,v)] $\leftarrow 0$ **end if**
      if $(\omega((u,v)) < \omega((u,v_{b(u)})))$ **then** LIKELY[(u,v)] $\leftarrow -1$ **end if**
    end for

```
            end for
        (d) set TALLY(u) to the number of vertices marked +1
    end for

Step 4:
    for each set of processors P_u do in parallel
        for all processors P_{u,k} ∈ P_u do in parallel
            for all incident edges (u,v) with LIKELY[(u,v)] equal to 0 do
                (a) select the edge (u,v_L) with v_L being (b(u)−TALLY(u))^th least numbered
                    vertex, v_L, adjacent to vertex u.
                (b) broadcast the number of v_L to all processors in P_u
                (c) for all processors in the set P_u do in parallel
                        for each assigned incident edge (u,v) do
                            if (the number of v is less than or equal to that of v_L)
                                then LIKELY[(u,v)] ← +1
                                else LIKELY[(u,v)] ← −1.
                            end if
                        end for
                    end for
            end for
        end for
    end for

Step 5:
    for each set of processors P_u do in parallel
        for all processors P_{u,k} ∈ P_u do in parallel
            for each assigned edge (u,v) of P_{u k}∈P_u do
                if (LIKELY[(v,u)] = +1) then SUCCESSFUL[(u,v)] ← 1 end if
            end for
        end for
    end for

Step 6:
    for each set of processors P_u do in parallel
        (a) select the b(u)^th heaviest edge (u,v_SH), such that SUCCESSFUL[(u,v_SH)]=1
        (b) broadcast ω((u,v_SH)) to all processors in P_u
        (c) for all processors in the set P_u do in parallel
                for each (u,v) such that SUCCESSFUL[(u,v)]=1 do
                    if (ω((u,v)) > ω((u,v_SH))) then LIKELY[(u,v)] ← +1 end if
                    if (ω((u,v)) = ω((u,v_SH))) then LIKELY[(u,v)] ←  0 end if
                    if (ω((u,v)) < ω((u,v_SH))) then LIKELY[(u,v)] ← −1 end if
                end for
            end for
        (d) set TALLY(u) to the number of vertices with LIKELY[(u,v)] = +1
    end for

Step 7:
    for each set of processors P_u do in parallel
        for all processors in P_u do in parallel
            for each incident edges (u,v) with LIKELY[(u,v)] equal to 0 do
                (a) select edge (u,v_SL) with the (b(u)−TALLY(u))^th least numbered vertex, v_SL,
                    such that SUCCESSFUL[(u,v_SL)]=1
                (b) broadcast the number of v_SL to all processors in P_u
                (c) for all processors in the set P_u do in parallel
                        for each incident edge (u,v) do
                            if (the number of v is less than or equal to that of v_SL)
```

```
                                    and (SUCCESSFUL[(u,v)] = 1)
                                 then LIKELY[(u,v)] ← +1
                                 else LIKELY[(u,v)] ← -1
                              end if
                           end for
                        end for
                  end for
               end for
            end for
```

Step 8.

```
   for each set of processors P_u do in parallel
       (a)  for all processors in the set P_u do in parallel
                for each assigned incident edge (u,v) do
                    if (LIKELY[(u,v)] = +1) then
                        if (the number of u is less than that of v)
                            then mark edge {u,v} as in the maximum weight b-matching
                        end if
                    end if
                end for
            end for
       (b)  for all processors in the set P_u do in parallel
                for each assigned incident edge (u,v) do
                    if (LIKELY[(u,v)] = +1) then
                        if  ((the number of u is greater than that of v) and (edge {u,v} is not
                            marked as in the maximum weight b-matching))
                                then mark edge {u,v} as in the maximum weight b-matching.
                        end if
                    end if
                end for
            end for
   end for
```

Theorem 4.2.
Given a weighted undirected tree T, with bounds $b(v)$, $0 \leq b(v) \leq deg(v)$, associated with the vertices of the tree, the maximum weight $b-$matching of the tree can be computed in $O(n/p + \log n)$ time in parallel using p processors.

Proof.
Step 0 is the execution of the $O(n/p + \log n)$ postorder numbering algorithm POSTORDERNUMBER and Step 1 is the execution of the algorithm ASSIGN which takes $O(n/p + \log p)$ time. Step 2 is also executable in $O(n/p)$ time. The execution time of Steps 3(a), 4(a), 6(a) and 7(a) can be done by the selection algorithm of Akl [1984] in time $O(n/p)$. The broadcast operations of Steps 3, 4, 6 and 7 each needs $O(n/p + \log p)$ time as explained in the proof of Theorem 3.3. Part (c) of Steps 3, 4, 6 and 7 can each be done in $O(n/p)$ while the (d) parts of Steps 3 and 6 are each executable in $O(n/p + \log p)$ time. Step 5 is executable in $O(n/p)$ time as is Step 8. Thus the running time of the algorithm is $O(n/p + \log n)$. □

We illustrate the computation of the maximum weight $b-$matching of our sample tree using the above algorithm in Figure 17. Figure 17(a) is the distribution of the adjacency lists among the four processors obtained at the end of Section 3.4. Executing Steps 1 through 4 of the algorithm, we obtain the values of the array LIKELY as shown in Figure 17(b). Going through Step 5 we obtain the values of SUCCESSFUL in Figure 17(c). Steps 6 and 7 produces the values of the array LIKELY as shown in Figure 17(d). Executing Step 8 gives the edges of the maximum weight $b-$matching in Figure 17(e).

EDGES

v_1v_4	v_4v_1	v_4v_7	v_4v_5	v_7v_9	v_7v_4	v_9v_7	v_2v_5	v_5v_8	v_5v_3	v_5v_2	v_5v_4	v_8v_{10}	v_8v_{11}	v_8v_5	$v_{10}v_8$	$v_{11}v_8$	v_3v_6	v_3v_5	v_6v_3
P_1				P_2				P_3							P_4				

(a)

LIKELY

+1	-1	+1	+1	-1	+1	+1	+1	+1	+1	+1	+1	-1	+1	+1	-1	+1	+1	-1	-1

(b)

SUCCESSFUL

0	0	1	1	0	1	0	1	1	0	1	1	1	1	1	0	1	0	1	1

(c)

LIKELY

-1	-1	+1	+1	-1	+1	-1	+1	+1	-1	+1	+1	+1	+1	+1	-1	+1	-1	+1	+1

(d)

MAXIMUM WEIGHT *b*-MATCHNG

0	0	1	1	0	0	0	1	1	0	0	0	0	1	0	0	0	0	1	0

(e)

Figure 17. (a) Adjacency Lists with assigned processors, (b) Edges likely to be in the *b*-matching, (c) Edges most likely to be in the *b*-matching, (d) Most likely edges satisfying the bounds, (e) Edges the Maximum Weight *b*-Matching.

5. Applications of $b-$Matchings in Trees

In this section we describe two applications of $b-$matchings in trees. We show how the *minimum cover set* problem and the *maximum independent set* problem in trees can be solved in parallel by an algorithm whose cost is optimal.

Definition 5.1.
A *covering set* of a tree $T = (V_T, E_T)$ is a subset $C \subseteq E_T$ such that for any edge $\{u, v\} \in E_T$, $\{u, v\} \cap C \neq \emptyset$. The *minimum covering set* problem is to find a covering set of T with minimum cardinality.

Theorem 5.1.
The algorithm TREEMATCH of Goodman, Hedetniemi and Tarjan [1976] can be used to solve the minimum covering set problem in trees.
Proof.
We construct the minimum covering set of T as follows:

- (i) Pick any vertex u with degree 1.
- (ii) Pick the other vertex v of the edge $\{u,v\}$ as in the minimum covering set.
- (iii) Remove the vertex v, all edges incident on v and all vertices adjacent to v whose degree becomes zero from the tree.
- (iv) Repeat these three steps until there are no more vertices to consider.

This technique is an extension of the method of Goodman, Hedetniemi and Tarjan [1976] for solving the maximum cardinality $b-$matching problem when the bounds of the vertices are set to 1. The extra work that needs to be done does not increase the time complexity of algorithm TREEMATCH. □

Theorem 5.2.
The minimum covering set problem on trees can be solved in $O(n/p + \log n)$ time using p processors on the EREW PRAM model.
Proof.
In this case, the minimum covering set is determined by first constructing the maximum cardinality $b-$matching of the tree T by setting the bound of each vertex to 1. The minimum covering set is obtained by picking, for every edge $\{u, v\}$ in the maximum cardinality $b-$matching, the vertex with larger number. To do this we can assign $(2n-2)/p$ edges to a processor as before. Each processor checks whether its edge is in the $b-$matching and includes the vertex with larger number in the minimum covering set. This requires an extra $O(n/p)$ time. Thus by Theorem 3.3, the proof is complete. □

Definition 5.2.
An *independent set* of a tree $T = (V_T, E_T)$ is a subset $I \subseteq V_T$ such that no two vertices in I are adjacent. A *maximum independent set* is an independent set with maximum cardinality.

Theorem 5.3.
The maximum independent set problem in trees can be solved in $O(n/p + \log n)$ time using p processors on the EREW PRAM model.
Proof.
Bondy and Murty [1976] have shown that $C \subseteq V_T$ is a minimum covering set of T if and only if $V_T - C$ is a maximum independent set of T. By Theorem 5.2 we can determine the set C in time $O(n/p + \log n)$. The set I is determined by assigning n/p vertices to a processor. Each processor checks each of its vertices v if it is not in the set C and marks it as in the set I. This extra work takes $O(n/p)$ time in parallel. □

6. Conclusion

Optimal parallel algorithms for computing matchings in trees have been presented. The special structure of trees was exploited to obtain the linear time sequential algorithm which led to the optimality of the parallel algorithm.

One problem left open is whether a weaker model can be used to obtain the same time and processor bounds, if at all possible. Also, can some special feature of other graphs be exploited to design efficient parallel matching algorithms as was done for trees?

7. References

Aho, A.V., J.E. Hopcroft and J.D. Ullman, [1974], *The Design and Analysis of Computer Algorithms,* Addison-Wesley, Reading, Massachusetts, 1974.

Akl, S.G., [1989], *The Design and Analysis of Parallel Algorithms,* Prentice Hall, Englewood Cliffs, New Jersey, 1989.

Akl, S.G., [1985], *Parallel Sorting Algorithms,* Academic Press, Orlando, Florida, 1985.

Akl, S.G., [1984], "An Optimal Algorithm for Parallel Selection", *Information Processing Letters,* Vol. 19, No. 1, 1984, pp. 47-50.

Berge, C. [1973], *Graphs and Hypergraphs,* (translated by E. Minieka) North-Holland, New York, New York, 1973.

Blum, M., R.W. Floyd, V. Pratt, R.L. Rivest and R.E. Tarjan, [1973], "Time Bounds for Selection", *Journal of Computer and System Sciences,* Vol. 7, 1973, pp. 448-461.

Bondy, J.A. and U.S.R. Murty, [1976], *Graph Theory with Applications,* North-Holland, New York, New York, 1976.

Brent, R.P., [1974], "The Parallel Evaluation of General Arithmetic Expressions", *Journal of the Association for Computing Machinery,* Vol. 21, No. 2, April 1974, pp. 201-206.

Cole, R. and U. Vishkin, [1988], "Approximate Parallel Scheduling. Part I: The Basic Technique with Applications to Optimal List Ranking in Logarithmic Time", *SIAM Journal on Computing,* Vol. 17, No. 1, February 1988, pp. 128-142.

Cole, R. and U. Vishkin, [1986], "Approximate Parallel and Exact Parallel Scheduling with Applications to Lists, Tree and Graph Problems", *Proceedings of the IEEE 27th Symposium on Foundations of Computer Science,* October 27-29, 1986, pp. 478-491.

Edmonds, J., [1965a], "Paths, Trees and Flowers", *Canadian Journal of Mathematics* Vol. 17, No. 3, 1965, pp. 449-467.

Edmonds, J., [1965b], "Matching and Polyhedrons with 0,1 Vertices", *Journal of Research of the National Bureau of Standards B.* Mathematics and Mathematical Physics Vol. 69B, Nos. 1 and 2, Jan.-June 1965, pp. 125-130.

Gabow, H.N., [1976], "An Efficient Implementation of Edmonds Algorithm for Maximal Matching on Graphs", *Journal of the Association for Computing Machinery,* Vol. 23, No. 2, April 1976, pp. 221-234.

Gabow, H.N., [1974], "Implementation of Algorithms for Maximum Matching on Nonbipartite Graphs", *Ph.D. Dissertation,* Department of Computer Science, Stanford University, Stanford, California, 1974.

Goodman, S., S. Hedetniemi and R.E. Tarjan, [1976], "b-Matchings in Trees", *SIAM Journal on Computing,* Vol. 5, No. 1, March 1976, pp. 104-108.

He, X. and Y. Yesha, [1988], "Binary Algebraic Computation and Parallel Algorithms for Simple Graphs", *Journal of Algorithms,* Vol. 9, 1988, pp. 92-113.

Hyafil, L., [1976], "Bounds for Selection", *SIAM Journal on Computing*, Vol. 5, No. 1, March 1976, pp. 109-114.

Knuth, D.E., [1976], *The Art of Computer Programming, Vol. I: Fundamental Algorithms.* Addison-Wesley, Reading, Massachusetts, 1968.

Lawler, E.L. [1976], *Combinatorial Optimization: Networks and Matroids,* Holt-Rinehart-Winston, New York 1976.

Papadimitriou, C.H. and K Steiglitz, [1982], *Combinatorial Optimization: Algorithms and Complexity,* Prentice-Hall, Englewood Cliffs, New Jersey, 1982.

Pawagi, S., [1987], "Parallel Algorithms for Maximum Weight Matching in Trees", *Proceedings of the 1987 International Conference on Parallel Processing,* Aug 12-21 1987, pp. 204-206.

Tarjan, R.E. and U. Vishkin, [1984], "Finding Biconnected Components and Computing Tree Functions in Logarithmic Parallel Time (Extended Summary)", *Proceedings of the 25th Annual IEEE Symposium on Foundation of Computer Science,* IEEE, New York, 1984, pp. 12-20.

Tarjan, R.E. [1972], "Depth-First Search and Linear Graph Algorithms", *SIAM Journal on Computing,* Vol. 1, No. 2, June 1972, pp. 146-160.